"A unique and interesting portrait of the development of the atomic bomb and the brilliant man who oversaw the process. . . . Conant gives the reader one story after another, revealing the humanity of these people within the framework of the project that ushered the world into the Atomic Age. . . . Highly readable." —*San Antonio Express-News*

"Bears the weight of inexorable drama. . . . Excels in showing how bedeviled the brilliant, oddly spineless but extraordinarily powerful Oppenheimer was." —*St. Petersburg Times*

"A spellbinding account of a venture that often teetered on the brink while the future of the world lay at stake. . . . Vividly told, the interplay of personalities that would ultimately transform the world." —*Kirkus Reviews* (starred)

"More than any other account of Los Alamos that I've read, Conant's narrative evokes the texture of life there. . . . A well-told narrative of daily life in a top-secret operation." —*Newsday*

"An engaging portrait of life on the remote mesa that served as backdrop for the world's most audacious scientific enterprise. . . . Conant packs her book with colorful, little-known details that bring the quotidian side of the bomb-building effort to life." —*Baltimore Sun*

Praise for *109 East Palace*

Winner of the Mountains & Plains Booksellers Association's 2006 Regional Book Award for Nonfiction

"For a broader picture of Los Alamos as a unique human settlement, part Western boom town, part scientific prison camp, turn to *109 East Palace*. . . . An entertaining picture of day-to-day life in a deadly serious wartime enclave."

—*Time*

"Conant has turned up what are perhaps the last few nuggets in the well-mined story of Los Alamos: Dorothy McKibbin's unpublished personal diary, as well as a handful of letters between her and her boss [Robert Oppenheimer]. . . . The story of the relationship between the two, told against the backdrop of the bomb project and the small town in which it took place, is a touching and sometimes amazing story. . . . [A] wonderful story, well told."

—*The Boston Globe*

"[A] haunting, beautifully realized, and highly entertaining story. . . . In revealing the human side of the Manhattan Project . . . Conant has performed an invaluable service. That she has done so with tremendous wit and style makes this book a stunning accomplishment."

—*Edmonton Journal*

"Conant carefully and effectively re-creates the daily anxieties, frustrations and exhilarations of the scientists and their wives who transformed a boys' school on a desert mountain mesa into a small city populated by the world's best scientific minds. . . . The books serves, in many ways, as a testament to their 'courage, service, and sacrifice.' "

—*Chicago Tribune*

"Lovingly details how a boys' school near Santa Fe was transformed into a lab said to look like a cross between 'an alpine resort and a mining camp.' "

—*Newsweek*

"Conant has done a lively job of describing the Los Alamos atmosphere."

—*The New York Times*

"*109 East Palace* reveals the human side of the Los Alamos story. . . . This book is a compassionate look at the hopes, loves, and anguish of people obliged to live closely together in the desolate New Mexican desert."

—*Houston Chronicle*

ALSO BY JENNET CONANT

Tuxedo Park:
A Wall Street Tycoon and the Secret Palace of Science
That Changed the Course of World War II

109 EAST

LOS ALAMOS
PROJECT
MAIN GATE
PASSES MUST BE
PRESENTED TO
GUARDS
POST
No. 1

PALACE

ROBERT OPPENHEIMER AND THE SECRET CITY OF LOS ALAMOS

Jennet Conant

SIMON & SCHUSTER PAPERBACKS
NEW YORK LONDON TORONTO SYDNEY

SIMON & SCHUSTER PAPERBACKS
Rockefeller Center
1230 Avenue of the Americas
New York, NY 10020

First Simon & Schuster paperback edition 2005

SIMON & SCHUSTER PAPERBACKS and colophon are registered trademarks
of Simon & Schuster, Inc.

For information regarding special discounts for bulk purchases,
please contact Simon & Schuster Special Sales at
1-800-456-6798 or business@simonandschuster.com.

Designed by Dana Sloan

Manufactured in the United States of America

21 23 25 27 29 30 28 26 24 22

Library of Congress Cataloging-in-Publication Data
Conant, Jennet.
109 East Palace : Robert Oppenheimer and the secret city of Los Alamos /
Jennet Conant.
p. cm.
Includes bibliographical references and index.
1. Los Alamos Scientific Laboratory—History. 2. Manhattan Project (U.S.)—
History. 3. Atomic bomb—United States—History. 4. McKibbin,
Dorothy Scarritt, 1897–1985. 5. Oppenheimer, J. Robert, 1904–1967.
6. Los Alamos Scientific Laboratory—Biography. 7. Manhattan Project (U.S.)—
Biography. 8. Physicists—Biography. I. Title.
QC773.A1C66 2005 623.4'5119'0973—dc22 2005042497

ISBN-13: 978-0-7432-5007-8
ISBN-10: 0-7432-5007-9
ISBN-13: 978-0-7432-5008-5 (Pbk)
ISBN-10: 0-7432-5008-7 (Pbk)

For Grandpa

CONTENTS

They won't believe you, when the time comes
that this can be told.
It is more fantastic than Jules Verne.

—JAMES B. CONANT
TO THE *NEW YORK TIMES'* WILLIAM L. LAURENCE
IN SPRING 1945

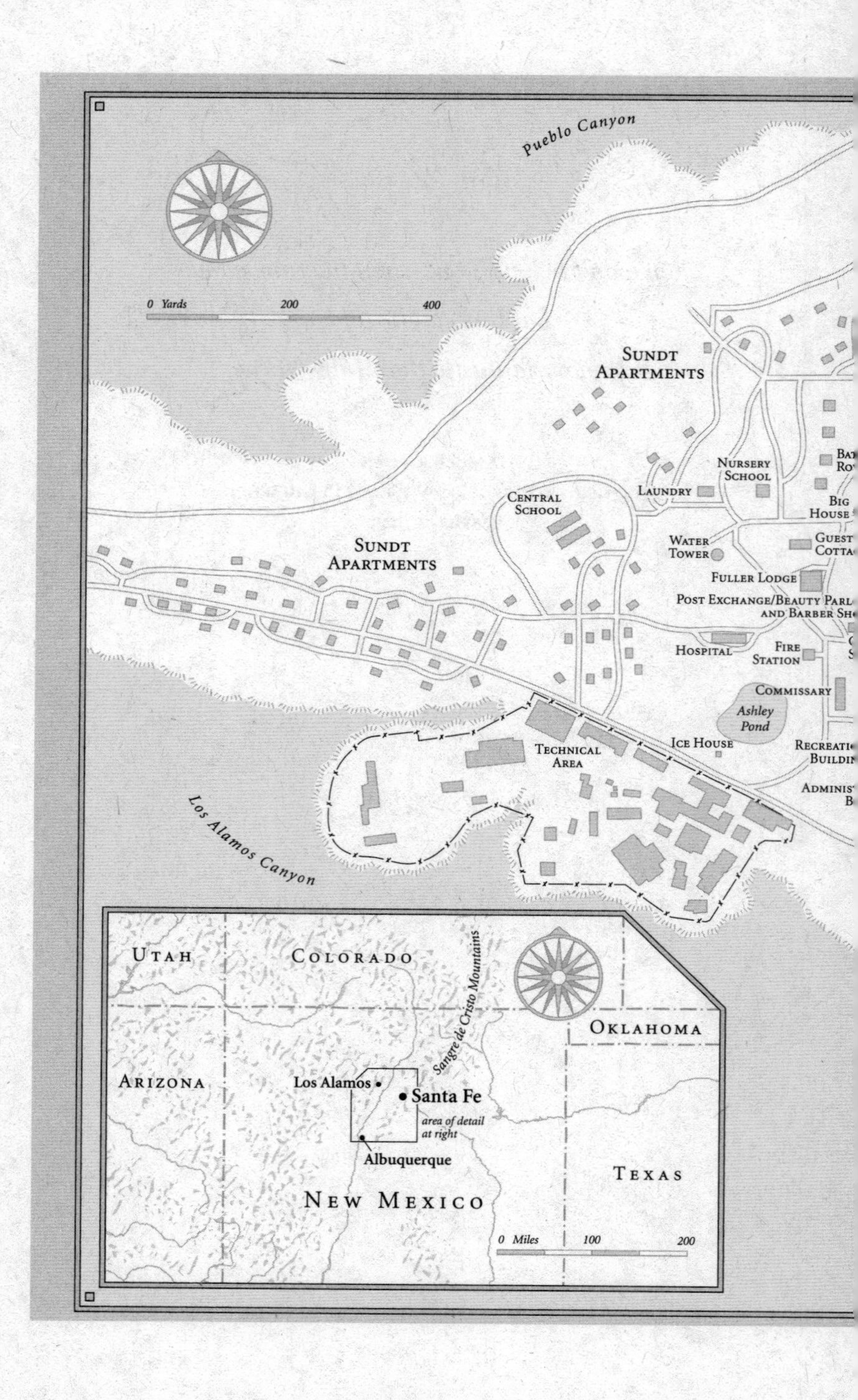

Pueblo Canyon
0 Yards 200 400
Sundt Apartments
Nursery School
Laundry
Central School
Sundt Apartments
Water Tower
Fuller Lodge
Hospital
Fire Station
Commissary
Ashley Pond
Technical Area
Ice House
Los Alamos Canyon
Utah
Colorado
Sangre de Cristo Mountains
Oklahoma
Arizona
Los Alamos
Santa Fe
area of detail at right
Albuquerque
Texas
New Mexico
0 Miles 100 200

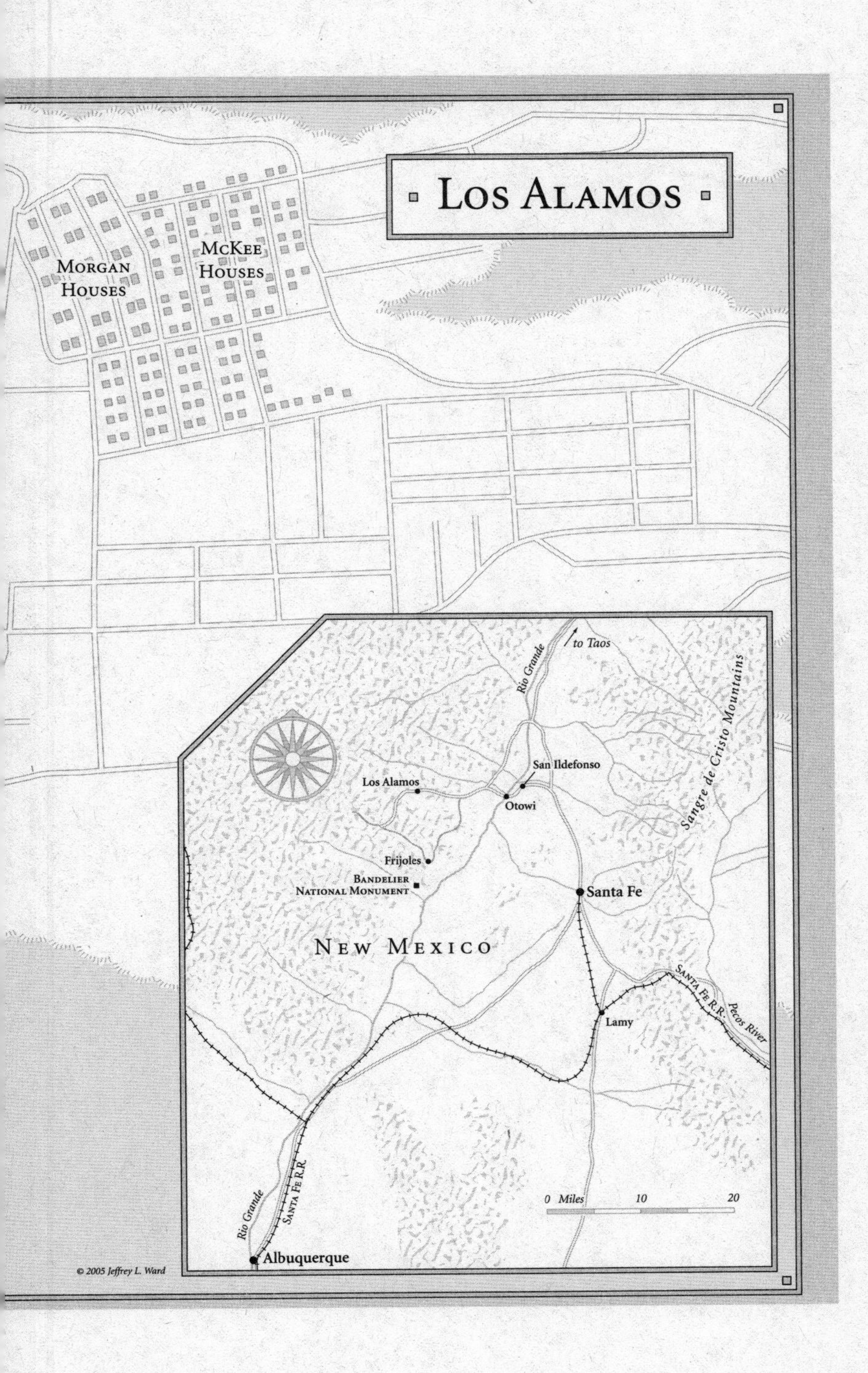
Los Alamos
Morgan Houses
McKee Houses
to Taos
Rio Grande
Sangre de Cristo Mountains
San Ildefonso
Los Alamos
Otowi
Frijoles
Bandelier National Monument
Santa Fe
New Mexico
Santa Fe R.R.
Pecos River
Lamy
0 Miles 10 20
Rio Grande
Santa Fe R.R.
Albuquerque
© 2005 Jeffrey L. Ward

PREFACE

For as long as I can remember, my grandfather James B. Conant kept a memento of Los Alamos on the desk in his study. It was a sample of trinitite, fused sand from the crater in the desert floor formed by the first explosion of an atomic bomb at the Trinity test site near Alamogordo, New Mexico, on July 16, 1945. It had been embedded in Lucite for safekeeping, but I was often warned not to play with it as a child because it was still "hot" and emitted enough low-level radiation to set a Geiger counter madly clicking. The rock was a potent talisman from my grandfather's past—a tumultuous time during which he took on the secret assignment of investigating the feasibility of designing and building a nuclear weapon for use in the war against Germany. He ultimately became the administrator of the Manhattan Project and the classified Los Alamos bomb laboratory, located on a remote mountaintop in New Mexico. After the successive bombings of Hiroshima and Nagasaki brought World War II to a swift and decisive end, he emerged as one of the country's great scientific leaders in the eyes of some, and as a mass murderer to others, responsible for helping to create the most diabolical weapon in the history of the world and for recommending its use against Japan.

Los Alamos was the chief morality tale of my childhood, as intrinsic, formative, and fraught as the most morbid of Mother Goose nursery rhymes are to other children. Like most good stories, it featured a cast of heroes and villains; only they changed depending on who was telling the tale. My white-haired, bespectacled grandfather, a lean, austere Yankee mellowed by age and made approachable by the twinkle in his eyes, never expressed regret over his role in World War II. I loved him, and I would listen raptly to the stories of those urgent, exciting days, when the army often dispatched a special military plane to retrieve him from the isolated cabin in Randolph, New Hampshire, at the foot of the Presidential Range, which was for more than three decades our family's summer home. Blue Cottage, as the house was known, was so far removed from the nearest town, and communication so uncertain, that the army was forced to run a special phone line down the dirt road to ensure they could reach him in an emergency. When the scientists who were his old friends would come to visit, they would reminisce about their pioneer days at Los Alamos, where the most brilliant, sophisticated men in the world attempted to do nuclear physics while cooking on Bunsen burners and camping out. In their company, my grandfather would relax his guard, and hearing their laughter and stories of absurd mix-ups, near misses, and desperate last-minute saves, I found it impossible not to come away with the impression that despite the pressure and grim purpose, theirs had been the adventure of a lifetime.

At the same time, even as a very young child, I was aware that my grandfather's war service was a loaded subject, publicly as well as privately. The mere mention of the time period could elicit a withering remark from my grandmother or a sarcastic rebuke from my father. "My own family did not escape the poison of deception," my grandfather wrote in a rare moment of introspection in his 1970 autobiography, *My Several Lives*. As one of the leaders of the Manhattan Project, he was bound by a very high degree of secrecy, or "compartmentalization," and conceded that the amount of "bald-faced lying" that he was forced to engage in permanently impaired his credibility with those closest to him. Security was extraordinarily tight, and he considered it a "fact of wartime" that he could tell his wife and two sons nothing about his work or frequent absences. Even after my grandmother discovered a Santa Fe railroad matchbook in one of his suit pockets after a trip that he had said took him no farther west than Chicago, he remained silent, leaving her to imagine in her barely suppressed rage what other betrayals he might be concealing. The estrangement that developed during those years left deep wounds on all sides and was still palpable years later when we gathered for tense family dinners.

When I was growing up in Cambridge, Massachusetts, in the 1960s, the heroics of World War II had faded from our national memory, patriotism was out of vogue, and anti-war rallies protesting America's involvement in Vietnam regularly blockaded Harvard Square. My liberal parents were full of anger and recriminations toward my grandfather, his complicity in the secret military effort to develop chemical weapons and the bomb, and the subsequent—and in their view, cruel and unnecessary—destruction of Hiroshima and Nagasaki. When I was twelve, they took my brother and me to the scene of the crime. At the Hiroshima Peace Memorial Museum, we sat and watched the horrifyingly graphic documentary made in the aftermath of the attack, showing the black and burning city and the unspeakable suffering of those who survived the blast, their scorched skin hanging down like torn rags from their bones, radiation eating away at their insides as they slowly and painfully died. My mother walked out of the museum theater in the middle, sick to her stomach. We were living in Japan then, expatriates in a foreign land, and during the long train ride back to our home in Tokyo, I looked at the Japanese faces staring back at me and wondered what they would think if they knew.

In the years after the war, every decision made by the leaders of the Manhattan Project was subject to review, second-guessing, censure, and the inevitable assignment of blame. There was an enormous sense of guilt and remorse among some of the scientists actively involved in the bomb project, coupled with a tremendous sense of responsibility that they must do everything in their power to convince the nations of the world to stop developing weapons of mass destruction. Others steadfastly maintained a sense of pride in having accomplished what had to be done in those dark days, when the shadow of totalitarianism had fallen across the map of Europe and England stood alone, on the verge of being engulfed by Nazi domination. Still others were determined to build more powerful bombs to maintain America's technological superiority.

In the protracted debate over the development of the hydrogen bomb, my grandfather, along with the director of Los Alamos, J. Robert Oppenheimer, strongly opposed the more militarist faction, represented by Ernest Lawrence and Edward Teller, in the bitterly divided scientific community. In the cruel betrayals and unimaginable reversals that exemplified the Cold War era and Joseph McCarthy's red-baiting campaign of terror, Oppenheimer, the celebrated "Father of the Atomic Bomb," was stripped of his clearance and barred from government work on the grounds that he represented a risk to national security. My grandfather, who testified in Oppenheimer's defense at his hearing before the Atomic Energy Commission's Personnel Security Board, considered the whole proceeding little more than a "kangaroo court" and left Washington in disgust.

The detonation of the first atomic bomb triggered a chain of events the scientists could not stop, and Oppenheimer's tragic downfall was, in a sense, history's verdict on their collective failure to properly safeguard such a devastating new weapon. In an essay entitled, "How Well We Meant," published after the fortieth reunion of Los Alamos scientists in 1983, the physicist I. I. Rabi reflected on how people have tended to forget the terrible state of affairs, both in Europe and the Pacific, that brought the United States to enter the atomic bomb race in the first place, and equally have failed to learn from the mistakes that resulted in the proliferation of nuclear weapons and competitive stockpiling of armaments. Rabi wanted to remind people that the story of Los Alamos was one of "human greatness and human folly." The scientists' motivating impulse was to protect people, not to destroy them. For all the cynicism and fatalism that exists about the arms race fifty years later, it was important, Rabi argued, not to lose sight of "science as the highest achievement of mankind, and as the process that takes us out of ourselves to view both the universe and our place in it." The scientists who built the bomb were of all different nationalities and religions. They banded together out of a sense of patriotism, not just for America, but for Western civilization and for the ideals of humanity that they all embraced. "It was the example of the United States that toppled most of the kingdoms and empires of the world," wrote Rabi. "We did this by asserting the greatness of the human spirit. Somehow, rather than this calculus of destruction, we must get back to our true nature as a nation."

My grandfather firmly believed that what he and his colleagues accomplished at Los Alamos saved the country and protected Western civilization. His misgivings were never about his part in facilitating the design and construction of the bomb, or its ultimate use against Japan, a decision that he always maintained was "correct." Instead, he was plagued by a sense of personal responsibility and failure. From the moment of the Trinity test he realized that something irreversible had happened—that they had given humanity a tremendous new form of power and they would now have to dedicate themselves to trying to control it. As the years passed, he was distressed by the many missed opportunities. If only they had not been quite so naïve in handing the bomb over to a government and military not wise enough, or responsible enough, to understand the full ramifications of what they possessed. If only the scientists had not abdicated their moral obligation to see the atomic bomb placed under a strong international authority. If only they had succeeded in preventing the further development of the hydrogen bomb. My grandfather was haunted by these "if onlys" until the day he died, worrying about them even as his heart gave out and his mind slipped away in that cold New Hampshire winter of 1978.

My purpose in writing this book was to try to gain insight into the greatness and the folly of Los Alamos, not by retracing the saga of scientific discovery, which was chronicled in Richard Rhodes' authoritative book *The Making of the Atomic Bomb,* but by reexamining the very personal stories of the project's key personnel. When she was swept up into the Manhattan Project, Dorothy McKibbin was plunged into a secret world of great men and ideas beyond anything she could have imagined. As the gatekeeper to Los Alamos, she presented herself as a peculiarly compelling witness to history, registering the full scope of the momentous change and moral upheaval the scientists' work unleashed. She was not objective in any real sense, but for that matter, neither am I. She was smitten with Robert Oppenheimer from the moment they met and unreservedly embraced both him and his brilliant crew of scientists, including my grandfather, whom she liked and admired. But as an intelligent, articulate, and knowing observer, she made the human element all the more vivid and understandable, offering a unique view of the atomic pioneers that led the way to the Trinity test and a dangerous new world. Dorothy chronicled the dramatic times, problems, and mounting conflicts in her unpublished memoir and many letters to Oppenheimer. She made no claim to heroism for her own small role in history, but her record of courage, service, and sacrifice, along with that of Oppenheimer and the hundreds of remarkable men and women who followed him to that high mesa in New Mexico, may give us a new understanding of what they achieved in those unparalleled twenty-seven months. And with America at war again over weapons of mass destruction, her story also gives us a new appreciation for the need to promote peace and defuse the nuclear predicament they unwittingly helped to create.

ONE

Charmed

THERE WAS SOMETHING about the man, that was all there was to it. He was six feet tall and very slender, and had on a trench coat and a porkpie hat, which he wore at rakish angle, so that people, women in particular, could not help taking notice. His face had a refined quality, with closely cropped black curls framing high cheekbones and startling blue eyes that radiated a strange intensity. He stuck out in Santa Fe like a sore thumb. But it was not his unusual looks, his city clothes, or even the pipe that he waved about in one hand while talking that caught Dorothy McKibbin's attention. It was something in his bearing, the way he walked on the balls of his feet, which "gave the impression he was hardly touching the ground."

Someone might have mentioned his name when they were introduced, not that it would have meant anything to her at the time. She had done little more than shake his hand, but she felt instinctively that their meeting was about to change everything about her quiet life. She had never intended to make a decision so quickly. She had only planned to come in for an interview, but she was so struck by the man's compelling personality that she blurted out the words "I'll take the job" before she had any idea what she was saying. In less than a minute, she had agreed to go to work for a complete stranger, for some kind of government project no one in Santa Fe seemed to know a thing about, doing God only knew what. She was forty-five, a widow with a twelve-year-old son, and flustered as a schoolgirl.

Dorothy McKibbin had come to the hotel La Fonda in the center of the old Santa Fe Plaza that March afternoon in 1943 in search of employment. It was no secret she was looking for work. Everyone in town knew the venerable old Spanish and Indian Trading Company, where she had been bookkeeper for ten years, had closed on account of the war, which had been raging on two fronts for more than a year with no end in sight. The store's owners had gone to Washington to do their part and were working for one of the many government agencies. A lot of people were leaving. Most of the native Spanish American population had already headed to California to work in the big factories and defense plants. Tourism had slowed to a trickle. The whole town was emptying out. She had worried about how she would earn a living until an old friend, George Bloom, who was president of the First National Bank, had found her something in the loan department. To qualify for the position, Bloom told her, she would first have to pass the Civil Service exam. The first time she took the required typing test, she flunked it. She was typing so fast she ran clean off the lines, and the "old machine did nothing to help. No bells rang; no horns blew." She was so infuriated by her humiliating performance, she went home and practiced for five days. When she took the test the second time, she passed it with flying colors. The job was hers. Her dignity restored, Dorothy informed Bloom she would have to "think it over." She hated the thought of working in a bank. Still, it paid $120 a month, and with all the men going off to war, she knew it would not be too difficult to work her way up to "a high position." Dorothy promised to let Bloom know in a day or two.

She was still mulling over his offer when she ran into Joe Stevenson, a local entrepreneur, who told her about another job that might be available and that would pay as much as $150 a month. Under any other circumstances, Dorothy would have jumped at the chance to get a higher-paying position, but there was something about the way Stevenson talked—he was vague, almost evasive—that left her feeling unsettled. She was crossing Palace Avenue when Stevenson approached her, and they carried on their awkward little exchange in the middle of the street with the traffic steering around them.

"How would you like a job as a secretary?" he proposed cheerfully.

"Secretary to what?" she asked.

"Secretary," he said, smiling. Dorothy took in the smile. She knew Stevenson had recently returned to town from California, where he had been enrolled in some kind of government training program. Rumor had it that the training was war related, and she could not help wondering if this might be, too.

"Well, what would I do?" she persisted.

"You would be a secretary," he repeated. "Don't you know what a secretary does?"

"Not always," she answered smartly, wondering how long they were going to go round and round like this.

"Well, think it over," was all Stevenson said before continuing across the street. "I'll give you twenty-four hours."

She spent most of that night on the telephone, calling virtually all the friends she had in Santa Fe to find out what, if anything, they knew about a big company or project coming to town. Nobody had heard a thing about it. She was leaning toward the bank job on the grounds that it was permanent, whereas the project Stevenson had mentioned sounded temporary and would last only as long as the war. Still undecided, and with the twenty-four hours almost up, she headed for her appointed meeting with Stevenson at La Fonda no better informed than she had been when she had last seen him.

La Fonda was the central meeting place for locals and tourists alike in the tiny state capital, and legend had it that there had been an inn of some sort on the corner of the Plaza since the early 1600s. Situated at the end of the Old Santa Fe Trail, where the wagon trains from Missouri used to come rolling to a stop, the building once served as a casino and lively brothel and had been host to generations of trappers, traders, merchants, gamblers, politicians, and thieves, from Kit Carson to Billy the Kid. Since 1925, La Fonda had been run by the Harvey Houses hotel chain, which operated fine establishments throughout the Southwest, and so these days it catered to a somewhat better clientele. There was always a lively crowd milling around the dark, elegant lobby, so she was not surprised to see Stevenson deep in conversation with an out-of-towner.

He was a businessman, probably a Californian, she judged, comparing Stevenson's typical Santa Fe attire of open-neck, blue work shirt and faded Levis with the other's brown silk gabardine suit, matching coat, white shirt, tie, and shiny dress shoes. Deciding to wait in the side hall where she was until they were finished talking, she sat in one of the big leather chairs favored by the bellboys when they were idle.

As soon as Stevenson saw her, he waved her over and introduced her to Duane Muncy, explaining that he was the business manager of the new government housing project that was interested in hiring her. Dorothy, regarding Muncy skeptically, asked exactly what the job would entail and what her duties would be. All he would say was that she would be "secretary to the assistant of the project manager." Finally, baffled by the wall of silence, Dorothy inquired somewhat tentatively if the project had "anything to do with the war." Lowering his voice, Muncy allowed as how it did. He began to tell her a little about the project, his carefully circumscribed answers implying that it was too secret to comfortably discuss, when the man in the porkpie hat sauntered over and stopped to join them. He lingered only long enough to exchange a few pleasantries, and to look her over with apparent curiosity, before abruptly excusing himself, murmuring, "All right, I'll leave you, then."

Dorothy watched him walk down the long foyer toward the heavy double doors of the hotel. He had not gone six feet when she turned to the two men and, before they could say another word, announced that they could count her in.

She was surprised that beyond her astonishment at her own behavior she had few misgivings, and none whatsoever about the man who was to be her boss. She could not explain why she was immediately taken with this tall stranger, a New Yorker in a funny hat who observed the social formalities of Park Avenue in the wilds of New Mexico. There were just moments in life when one found oneself open to opportunity, to the unexpected, whatever the risks. Dorothy was from pioneer stock. A decade earlier, after burying her husband, she had come out west to make a fresh start for herself and her baby. She had the strong sense that this man's offer promised a new adventure. "I never met a person with a magnetism that hit you so fast, and so completely," she told our interviewer years later. "I just wanted to be allied with, have something to do with, a personality such as his."

In the days that followed, she learned that the man she had met, who went by the name of Mr. Bradley, was actually J. Robert Oppenheimer, a famous American physicist from the University of California at Berkeley and the leader of a secret wartime project. Bradley was the name he would use in and around town, and the way she was to address him in public. She would learn never to mention his real name and, for that matter, never to mention him at all. Not to anyone. She was told never, under any circumstances, to use the word "physicist." "I was told never to ask questions, never to have a name repeated," she wrote, recalling her initiation into the most momentous scientific project of the twentieth century. "I soon knew that I was working for a project of great importance and urgency."

Though not a native Santa Fean, Dorothy knew the area as well as anyone, if not better, and she supposed that in her own way she belonged there. The old town, with its bright sunshine and crystalline air, had been lucky for her once years ago, and she had claimed it as her own. That was back in 1925, when thin and weakened by the tuberculosis infecting her lungs, she had come to the Sunmount Sanitarium seeking a cure for a disease that had killed more Americans, young and old, rich and poor, than any other affliction up to that time. Her mother had brought her by train, and they had taken a room at La Fonda, unaware they would be forced to bide their time for a full month because of the long waiting list for entrance to the famed sanitarium. Finally, after money and influence had secured her daughter a place at the popular hospice, Dorothy's mother prepared to leave. When they said their tearful good-byes on December 9, both of them knew the parting might be final. Dorothy was twenty-nine then and had already lost two beloved sisters: Frances Margaret, who had died of tuberculosis six years earlier, and Virginia, to a respiratory infection when she was still a child. Dorothy had been devastated by their deaths. Now that she, too, had fallen ill, she was determined not to spread the contagion to her family and friends. To avoid any chance of being the cause of that misfortune, Dorothy chose banishment and isolation. Their family doctor had confirmed the wisdom of seeking complete rest in a sanitarium, sealing her fate as an outcast. She had no choice but to try to rid herself of the debilitating disease once and for all, or risk never having a normal life. "The tuberculosis was holding her back," said Betty Lilienthal, an old friend. "She was going to beat it, no matter how long it took." In those days before antibiotics, the period of recuperation for "lungers" in a sanitarium was of indeterminate length—from six months to several years, even a lifetime.

It was a far cry from the wealth of opportunities that had awaited her on her graduation from Smith College in 1919, where Dorothy Ann Scarritt had been voted president of her freshman class and was one of the most popular in her year. "Not bad," she liked to boast, "for a little girl from Kansas City." Far from being intimidated by the East Coast sophisticates, she once confided to a friend that she had arrived at the elite women's college in Northampton, Massachusetts, with every intention of becoming a campus leader. She was outgoing and engaging, and from early on had demonstrated the great appetite for life that was her father's greatest gift to her. As the fourth of five children in the large, noisy, and devoted Scarritt family, Dorothy had grown up with a firm sense of her place in the world.

The Scarritts were an exceptionally upright, educated, and substantial midwestern clan. Her father, William Chick Scarritt, had earned a law degree from Boston University and was a prominent corporate lawyer and leading citizen in their town. He was active in civic affairs and politics, and he was a power at the Melrose Methodist Episcopal Church, which had been founded by the family patriarch, the Reverend Nathan Scarritt, a well-known local minister. The Reverend Scarritt had come to Missouri from Edwardsville, Illinois, in 1848 to "teach the classics" in the barren prairie outpost and had earned a reputation as a fine preacher, and an even better businessman. By the time he died in 1890, he had accumulated 260 acres of farmland in the Northeast District, which made him one of the largest landowners in the area and one of the first millionaires. He left each of his six children a substantial sum and enough acreage to build on, so that when Dorothy was growing up, the Scarritts were a powerful

Kansas City family and among them owned nearly a dozen mansions on the banks of the Missouri River. Dorothy's mother, Frances Davis, was from a southern family that had migrated to Missouri, and the lovely redhead retained the genteel manners and impassioned Confederate politics of her old Virginia roots. She was a leading light in Kansas City's social committees and garden clubs and a strong supporter of the arts.

Dorothy's father was the chief influence in her early life, and it was from him she got her buoyant personality and undeniable presence. He raised his children to be strong, self-reliant, and athletic, and Dorothy and her siblings pursued tennis, swimming, hiking, and mountain climbing. William Scarritt spared no expense to see that his children benefited from the finest academic training, taking the unusual step at the time of sending both his sons and his daughters to the only private preparatory school in the area, and then east to college. He loved to travel and had a taste for dramatic scenery, taking his children on expeditions to Alaska, across Canada to the Rockies, and to Yosemite National Park and the western United States. After college, he escorted Dorothy on a grand tour of Europe. He encouraged her to pursue a life of pleasure, interest, and adventure. But as forward-thinking as he was in many ways in regard to women, he was also conventional enough to expect his only surviving daughter would eventually come home and settle down. On her return from Europe, however, Dorothy had difficulty resigning herself to the slow, predictable rhythms of her parents' country club life and to the succession of tea dances and balls, all intended to lead her into a union with a boy from another good family.

In the fall of 1923, four years after graduating from Smith, she received an invitation to visit a college classmate, Alida Bigelow, who was vacationing at White Bear Lake in Minnesota. Dorothy seized the opportunity to escape Kansas City and renew her old acquaintance. While she was there, she met and immediately fell in love with her friend's cousin, Joseph Chambers McKibbin.* He had sailed his own boat on White Bear Lake since boyhood, and as he took her out on the water, he regaled her with tales of his racing exploits. He was thirty, a Princeton graduate, and a World War I vet, though, as he was quick to acknowledge, he had spent

* No relation to the physicist Joseph McKibben who worked at Los Alamos.

all his time as an army major training at Fort Sill in Oklahoma. He was tall and an excellent athlete: he made varsity crew in college, was the squash champion of the Twin Cities' team, and rode to the hounds. He told amusing tales of his years in New York City, where he and a group of fellow Princetonians had embarked on careers in investment banking. Dorothy thought he was infinitely more interesting than any of the boys back home and would later say she knew at once that "he was the only man" for her.

At the time he and Dorothy met, he had returned home to St. Paul to fulfill family obligations and was learning to manage his father's prosperous fur business with the expectation that he would one day succeed him. Both Joseph and Dorothy were chafing at the bounds of family ties, the constraints of small-town life, and the sudden narrowing of their horizons after college. Together, they saw a chance to realize their ambitions. They became engaged and talked of starting an exciting life together in New York. But they had to put their plans on hold when Dorothy, on her return from a trip to Latin America with her father in the winter of 1925, learned that she had contracted tuberculosis. After much soul searching, she decided that despite his repeated objections, it was not fair for her to hold her fiancé to his promise. She broke off her engagement, left everyone who was dear to her miles behind, and set off on the bleak and uncertain journey to New Mexico.

Once they agreed to send Dorothy away, the Scarritts had a great many facilities to choose from, as it was then the fashion for health seekers to go west. Since the 1840s, men and women with symptoms of consumption had been heading to the mountains and deserts, and by the 1870s, the completion of the transcontinental railroad brought throngs of invalids to the plateau between Mississippi and California. By 1900, they accounted for a quarter of all the newcomers to California and a third of Colorado's growing population. No place was deemed more restorative than New Mexico, where medical researchers had found tubercular consumption among the native Indian tribes to be almost nonexistent. Every western state to some degree owed its development to the burgeoning new health industry: railroads distributed brochures touting the advantages of the region, and promoters and developers competed to lure patients to their towns and subdivisions with tales of remarkable recoveries. As more and more travelers headed to the arid plains as an antidote to the feverish cities, those hospitals and institutions catering to itinerant health seekers cropped up all across the West, advertising their "pure air," "wholesome climate," and exaggerated claims of rejuvenation in line with the myths of El Dorado gold. Inevitably, some travelers, weakened by the ravages of illness, died in the course of the journey, but such discouraging stories rarely found their way into print. Boosterism was the order of the day, so much so that Mark Twain was moved to parody the lavish endorsements in *Roughing It,* noting that three months in Lake Tahoe would "restore an Egyptian mummy to his pristine vigor." The West's reputation as a natural paradise was given full expression by Daniel Drake, a leading Cincinnati physician, who published a series of journal articles and books widely publicizing the health benefits of frontier life and the renewing influence of "the voiceless solitudes of the desert." If Drake's fervent belief in the curative powers of the western climate and lifestyle went well beyond empirical science to wishful thinking, consumptives and their worried families could not be blamed for keeping the faith.

After investigating reputable sanitariums everywhere from Colorado Springs to the Sierra Madre, the Scarritts eventually settled on Sunmount as the ideal place for Dorothy to wile away those disappointing days. The high-desert plateau and crisp, clear air were regarded as therapeutic from a medical standpoint, and the stunning mountain vistas equally invigorating from a psychological one. Sunmount had first opened its doors to patients in 1903, but the lovely salmon-colored stucco compound on Camino del Monte Sol that greeted Dorothy on her arrival was built in 1914. It was designed so that the exterior took on the luscious contours of adobe and conformed to the Spanish Pueblo Revival style that was then coming into vogue. Under the guidance of its director, Dr. Frank Mera, Sunmount's patients were prescribed a European-style regimen featuring rest, a nutritious diet, and plenty of fresh air. They enjoyed large, comfortable rooms with big, open windows and slept year-round on sun porches constructed to face southwest so as to capture the winter light. Dorothy quickly became practiced in the art of wrapping herself tightly in blankets to ward off the cold of the desert nights.

Sunmount advertised itself as a modern health resort rather than a hospital and, in contrast with the grim prisonlike isolation enforced by many sanitariums at the time, placed no serious restrictions on its patients' activities or sociability. Mera was something of a Southwest aficionado, and together with his brother, Harry, a local archaeologist, made sure those in his care took full advantage of Santa Fe's cultural attractions and saw to it that the hospital became closely integrated into the local community. Sunmount sponsored tours of the old town and Indian pueblos, organized educational events that celebrated the ancient culture of the Indians and the Spanish settlers who came in the seventeenth and eighteenth centuries, and staged concerts, poetry readings, and dance recitals to help pass the time fruitfully. Whatever quaintness had originally characterized his efforts gave way to a busy and profitable service industry. There were classes available in everything from painting and pottery making to architecture, and the Pueblo Indians and area craftsmen were invited to sell their artwork to the rich and idle in residence.

To Santa Feans, TB was a respectable affliction, and they maintained a tolerant attitude toward the moneyed easterners who flocked there for the cure in the 1920s. Even then, Santa Fe was not the typical tourist town, and had a lively bohemian culture that qualified it as "the Paris of the Southwest." The recent boom in archaeology, and growing government interest in preserving the ancient Indian ruins in nearby Frijoles Canyon, had served to greatly raise the town's profile and importance, attracting academics, collectors, and affluent hobbyists. New Mexico was becoming part of the vast western vacationland, "the land of enchantment," and the Harvey Houses company was running "Indian Detours" out of La Fonda, packaging trips for wealthy tourists eager to "rough it in style" and view a formerly dangerous land, still rugged and untamed enough to thrill.

Santa Fe and more particularly Taos, its smaller rival seventy miles to the north, had long been a destination for artists, dating back to 1883, when the popular Western painter Joseph Henry Sharpe first passed through the area on a sketching trip. His idyllic paintings of Native American life brought attention to the region and inspired other artists to travel west in search of the enchanted landscape and magical quality of the light. Because of its stunning vistas, Taos soon captured the imagination of such painters as Joseph Henry Cottons, Ernest Blumenschein, Bert Phillips, and Irving Couse, and in 1913 they joined together to form the Taos Society of Artists. Not long afterward, the heiress and arts patron Mabel Dodge Luhan took up residence in Taos, and her fabled hospitality drew many of the era's great names in art, literature, philosophy, and psychology. On his first visit to Luhan's ranch in 1922, the writer D. H. Lawrence was so moved by the bold colors and exotic scenery of New Mexico that, he wrote, the place changed him forever and liberated him for the "greatness of beauty." He returned to recuperate from tuberculosis and remained until 1925, producing a series of paintings, in addition to turning out some of his finest poems, stories, and essays, and the novel *The Plumed Serpent*.

The Atchison, Topeka and Santa Fe and the Denver and Rio Grande railways opened the West to many more artists and collectors, and they came in droves in search of the unspoiled land and simplicity they could not find elsewhere. The new arrivals discovered Santa Fe, turning it into a thriving cultural center. The city's first resident artist was Carlos Vierra, a photographer who came from California in 1904 seeking relief for his lung problems in the arid climate, and liked it so much he made it his home, opening up a studio on the Plaza. As Santa Fe's reputation spread, Sunmount prospered, playing host to a stream of well-known figures: among them the New York painters Sheldon Parsons and Gerald Cassidy; the architect John Gaw Meem; and the poet Alice Corbin, who prior to her illness had been an associate editor of the highly influential magazine *Poetry*.

Alice Corbin Henderson was thirty-five when she traveled to Sunmount in 1916 accompanied by her husband and daughter, and she feared she would never leave the tuberculosis sanitarium, writing that she "had been thrown out in the desert to die, like a piece of old scrap iron." Unexpectedly, she survived, and she ended up building a home nearby on Camino del Monte Sol and inviting colleagues such as Ezra Pound, Carl Sandburg, and Witter Bynner to visit and give readings at the sanitarium. Bynner, who briefly stayed at Sunmount while recuperating from an illness, found life in Santa Fe so pleasant he remained for the rest of his life and, with Corbin, helped establish the writer's colony. In 1917, Corbin, Bynner, and Sandburg helped publish a special issue of

Poetry magazine inspired by the chants and spoken poems of the Pueblo Indians. Meanwhile, Corbin's husband, the painter William Penhallow Henderson, went on to found the Canyon Road art colony and contributed significantly to local architecture and design, becoming a leading figure in town. The couple's busy, productive life encouraged others. Many Sunmount patients stayed on for years, receiving aftercare treatment in town, ultimately choosing to settle permanently in the area, and, as in the case of John Gaw Meem, building large airy houses, carrying on extremely active careers, and using their well-heeled connections at the sanitarium to secure an abundance of commissions. "Sunmount was really a magical place," explained Meem's daughter, Nancy Wirth. "Its founder, Frank Mera, wanted much more for his patients than just a cure. He pushed them to do more, to get out and explore. Many patients, like my father, became imbued with passion for the place."

The avant garde art scene lured a string of early Modernist painters from New York, most notably Robert Henri and John Sloan, members of the Ashcan school, who became frequent visitors to the city. In the 1920s, the penniless young Philadelphia painter Will Shuster founded the Santa Fe Art Colony with four other aspiring artists, known locally as the Los Cinco Pintores ("the five painters"), and developed a communal artists' compound, where for five years they studied and showed their work together. Other young artists soon followed, including the watercolorist Cady Wells, as well as Charles Barrows and Jim Morris and the writers Mary Austin and her friend Willa Cather, who came and went, and whose novel *Death Comes for the Archbishop,* based on the life of Santa Fe archbishop Jean-Baptiste Lamy, was published in 1927. By the end of the decade, when Georgia O'Keeffe spent her first summer in the area, eventually purchasing a seven-acre spread at Ghost Ranch in the village of Abiquiu, Santa Fe's Roaring Twenties art scene had acquired international fame.

Like Corbin, Dorothy had come to Sunmount expecting to die and was greatly surprised to find herself enjoying her exile. She was captivated by the romantic aura of the place, the artists in their bright smocks and blue jeans and the Santa Feans themselves in their cowboy clothes, silver belts, silk scarves, and fringed deerskins. She attended poetry recitals, lectures, political discussions, and amateur theatricals. "Raymond Duncan, the brother of Isadora Duncan, appeared in a sack cloth and shepherd's crook and told of his experiences and hopes," Dorothy wrote in an unpublished memoir. "People from Russia enlivened us with stories of their theatre and ballet. I fell in love with the place because of its beauty and the cultural and intellectual atmosphere."

Dorothy put her time at Sunmount to good use, hobnobbing with the most interesting patients, as well as with the healthy clientele in town, former "TBs" who were busy making a name for themselves in Santa Fe. One such friend was George Bloom, who would later become head of the First National Bank. Another was Katherine Stinson, who was already something of a celebrity when Dorothy met her at the sanitarium. In 1912, at the age of twenty-one, Stinson had become only the fourth woman in the country ever to receive a pilot's license, and a year later she had founded her own company, Stinson Aviation. She proceeded to bust records of every kind in the ensuing decade: she was the first woman to fly at night; the first to fly to China; the first to perform skywriting, spelling "C-A-L" over Los Angeles. Toward the end of World War I, she contracted tuberculosis in France, and spent seven years on and off at Sunmount gradually regaining her health. Although Stinson's barnstorming days were over, Dorothy admired her independence and fearlessness, even in confinement, and they would become lifelong friends.

Dorothy spent eleven months at Sunmount and was pronounced "cured." Joe McKibbin renewed his attentions to her, after being given permission to visit her toward the end of 1926, by which time her health had substantially improved. In the spring of 1927, he proposed again, and she accepted. When she compared herself with Katherine Stinson, Dorothy knew she had much to be grateful for. But she was also keenly aware of the time she had lost, and that all her schoolmates and cousins had married and moved on with their lives. It was only natural that a young woman, after watching her sisters die and then falling under the same shadow herself, and after having seen so many friends languish and fail, would come away with a certain mental toughness. What may have begun as a defensive posture in Dorothy became a deeply ingrained characteristic, and her stoicism and stubborn optimism would be her chief assets in the years ahead, always within reach, no matter how trying the circumstances.

On October 5, 1927, Dorothy and Joe McKibbin were married in the garden of her parents' home. She often joked that she had a small wedding owing to its belatedness and their advanced ages—she was nearly thirty, and he was thirty-four. The Kansas City society pages diplomatically covered the restrained celebration, observing that the bride demonstrated her "distinct individuality in her wedding gown . . . [and] chose a sapphire blue velvet dress and wore a small hat to match." The couple honeymooned in Rio de Janeiro and then settled in St. Paul. Joe continued to manage the fur trade at McKibbin, Driscoll & Dorsey, while quietly reassuring his Princeton friends he would soon be joining their brokerage business. Those plans had to be postponed again when Dorothy discovered she was expecting a child. On December 6, 1930, they welcomed their son Kevin into the world.

For Dorothy, it was a time of pure and unexpected joy. "None of this valley of the shadow of death but peaks of excitement of life," she wrote days after his birth, putting her feelings to paper as she often did during fraught moments in her life:

> There is nothing personal about happiness. It is an aliveness. It is a oneness with nature and humanity. It is a deep knowledge of suffering and desolation. It is an intenseness of feeling, of understanding good and evil. Compassion. Vital awareness of being a part of the universe in all its manifestations. Harmony. Acute knowledge of kinship with all the peoples of the earth. Recognition of true values. Preoccupation with truth.

But even as she wrote those lines, Dorothy knew that she had not escaped the "shadow of death" for long. She had already had a harbinger of the dark days that lay ahead. A year and a half into their marriage, Joe McKibbin was diagnosed with Hodgkin's disease, a progressive cancer of the lymph system. Unlike tuberculosis, it was then considered incurable. His condition was hopeless from the start, though neither Dorothy nor his doctor ever told him of his terminal illness. His death, when it finally came, was hard and long. Dorothy endeavored to be as strong and steadfast in her husband's sickness as he had been in hers, and she nursed him over the last painful year to the end. He died on October 27, 1931.

Before her husband became bedridden, she had taken him on a trip across Colorado to New Mexico, hoping it would improve his health. He had taken to the luminous landscape as quickly as she had and, with his strength already faltering, had agreed that "someday they would settle there where they could have horses and dogs and raise their family under the incredible blue skies and golden sun." This had always been more her dream than his, and now it was hers alone. "I was ten months old, and she was suddenly on her own," Kevin reflected years later. "She folded up everything in St. Paul and went down to Kansas City to her parents' house. She stayed a couple of months, but she couldn't stand it. In April of 1932, she loaded me up in the little Model A coupe she had and we drove out to Santa Fe." The trip took almost a month. Although he was too little to understand, Dorothy told Kevin they were going away to a place "where they would sit under a piñon tree and spend all their days in peace and happiness."

Dorothy was drawn to Santa Fe as a refuge, and by the memories of her days at Sunmount and of the interesting people she had met there. She had fraternized with a very different sort than she had known in her Kansas City milieu, and those experiences and friendships had awakened a fascination with the region's unique culture and redemptive climate. Something in the atmosphere, in the strangely beautiful windswept landscape and extravagant purple twilights, encouraged a sense of well-being and possibility. In Santa Fe, there was a ready-made community to welcome her, people who were no strangers to pain and loss, who had gone there to recover and had remained because they had discovered a new way of life. She had felt more alive there, more aware of every breath she took and the feel of the sun on her back, than she ever had been before. "She saw staying in Kansas City as being trapped," said her nephew Jim Scarritt. "Getting away was liberating." She seemed to feel as O'Keeffe did when the artist observed of New Mexico, "The world is wide here." Eager to escape the pitying faces of family, and the suffocating loss of freedom she experienced at being back under her parents' roof again, Dorothy headed west.

"I think she may have missed the independence that she had when she was in Santa Fe," said Kevin. "It was a very creative community and kind of unusual in the way it operated. It had some real characters in it, and everyone went their own way. She felt right at home."

When Dorothy arrived in Santa Fe, she found little had changed in the bustling little city at the base of the mountains. Many of the streets lacked paving, the old Spanish women still wore their traditional long black skirts over wide boots, and the familiar, pungent smell of piñon from all the small wood-burning fireplaces in the shops filled the air. Piñon wood heated the city, and she had not realized how much she had missed its smoky perfume. The Indians continued to hawk their wares every day on the sidewalk in the shade of the old Governors' Palace as they had for centuries. The Woolworth's was new, and there were more curio shops and dry-goods stores, two drugstores, and innumerable small cafés. In the Plaza, the oldest residents of the region mingled with the new European immigrants, the poorest of people with the wealthiest, the illiterate with the highly educated. The supporting cast consisted of the same oddball mix of artists, writers, consumptives, neurotics, speculators, adventurers, escapists, and dreamers of every description. It was the colorful crossroads she remembered so well, with everyone in his or her own way making a last stand for rugged individualism. Presiding over it all was the dignified bronze statue of Archbishop Lamy in front of St. Francis Cathedral, a dramatic Romanesque structure that always struck her as somewhat incongruous amid all the humble Pueblo architecture.

Despite its historic position as the terminus of the Santa Fe Trail, the city had stubbornly resisted growth, and even after the admission of New Mexico to the union in 1912, the city had consisted of only 5,000 to 6,000 inhabitants. In the 1920s, the tourist trade had brought some development, and by the time she returned in 1932 the population hovered around 11,000. But since the Great Depression, business was down. The rich private collectors had all but disappeared, and many of the roughly one hundred artists living in town were in dire straits. The dry years had only compounded the problems. A sustained drought and particularly brutal winter had driven the local economy even further down, and many homesteaders in the middle Rio Grande area were fighting for survival. There were not a lot of good jobs to be had, but former patients had the advantage over the local residents because they were generally better educated, and Dorothy was no exception. She soon found work as a bookkeeper at the Spanish and Indian Trading Company. Kevin teased that sitting up at that high desk with her pencil and heavy ledger, she "looked like Scrooge."

Over time, it became Dorothy's job to balance the books while the two shop owners, Norman McGee and Jim McMillan, went on buying sprees around the state. They thought nothing of writing $10,000 checks for consignments of art treasures, but could never quite remember to pay the rent. They were far too much in demand to attend to such mundane matters and increasingly relied on Dorothy to keep the business end of the trading company functioning. It was the most noted store of its kind in New Mexico, or for that matter the entire Southwest, and Indians from the surrounding pueblos, Navajo reservations, and Hopi villages brought their finest blankets, pottery, and woven baskets to sell or barter. Spanish traders brought hand-carved furniture and the rarest of old *santos* (images of saints), *bultos* (carved wooden sculptures), and *retablos* (painted wood or tin altarpieces). The store was always busy and lively, and along the way Dorothy was receiving a first-class education in Spanish and Indian arts and crafts and developing an astute collector's eye. She earned fifty cents an hour keeping "a very complicated set of books with 14 columns," but the workday was flexible and allowed her time to look after her little boy. She had the small insurance pension her husband had left her, and in a small town like Santa Fe, the little money she earned went a long way. It was enough for them to live on.

Dorothy's father, who had never ceased to admire his daughter's spirit and courage, came out to visit several times. He had lost much of his fortune in the Depression, but he scraped together enough money to buy her a house. Instead of purchasing a home, however, Dorothy was intent on building one. She was inspired in part by her friend Katherine Stinson, who in the intervening years had married Miguel Otero, Jr., a New Mexican attorney and aviator who was the son of one of the state's territorial governors. During her many years in Santa Fe, Stinson had become fascinated with the local architecture. No longer strong enough to fly, she had taken up her new hobby while at Sunmount, where she met John Gaw Meem, who had become a reknowned local architect and had offered to act as her mentor in what would become her new career. Stinson had built her own large adobe house in Meem's Pueblo Revival style, and Dorothy asked her if she would design a modest home for her and Kevin on one and a half acres of piñon-studded land on the Old Santa Fe Trail road. The house was in a rural area on the outskirts of town, two miles down a dusty, red-dirt road, and many of her friends thought she was mad at the time. But the price was right—the land and construction came to $10,000—and Dorothy loved the view, which took in the blue Sangre de Cristo Mountains to the east and the Sandia and Manzano ranges curving off to the south. On clear days, even Mt. Taylor, the legendary "Sentinel of the Navajo Land," was visible, its snowcapped peak glittering in the winter sun.

Dorothy and Katherine Stinson designed the house together, working every morning at the Oteros' kitchen table. Dorothy had her heart set on an exact replica of an ancient Spanish-style farmhouse, laid out in a U shape with a traditional patio or interior courtyard, flat roof, thick walls, and a long sheltering portal. She insisted on incorporating into the architecture the old beams and other antique pieces she had collected, giving it the intentionally wobbly silhouette and rounded corners of time-worn adobes. She and Stinson went on shopping sprees while the construction was under way, scouring the countryside for distinctive elements. They salvaged the front portal from a crumbling rural farmhouse and found the rare, hand-carved corbels used to support it in the neighboring village of Agua Fria. A centuries-old door was bought for five dollars. To Dorothy's delight, Stinson covered the living room floor with dark red bricks made by prisoners at the old penitentiary. The cozy 1,800-square-foot structure was completed in 1936, and Dorothy furnished it with native woven hangings and tinwork lamps she had acquired at the trading company. Thanks to the Depression, it was possible to find beautiful hand-woven Navajo and Chimayo rugs for three to five dollars at roadside gas stations on the outskirts of town, where the Indians sold them or traded them for fuel. She had one obsessive goal as she put in her garden and planted yellow Castilian roses on the border of her patio—that it be "a happy house."

The adobe house was a rare luxury. On the last Saturday of every month, Dorothy would take Kevin to town and settle her accounts at all the shops where she had charged things. By the end of their rounds, there was usually just enough left over for them to buy Cokes at the Capital Pharmacy before heading home. "Money was tight," said Kevin. "My father left her pretty well fixed-up in that she had some stock, and a little something put away, so we were comfortable. But it wasn't much."

In those days, there was only a small group of prominent "Anglos" in town, and as Dorothy already knew a number of them from Sunmount, she was quickly ushered into their tight-knit social world. She was close to John Meem, as well as his brother, both of whom lived in adjacent houses on the same street, only a short distance from her home. "She was never part of the 'horsey' set of well-to-do women from back east who kept ranches just outside town," recalled Bill Hudgins, a boyhood friend of Kevin's. "You could always tell those women by their tailored wool suits and mannish hats. Dorothy wasn't like that at all. She was fun and informal and liked to wear bright Navajo skirts and fiesta blouses. She really knew the local culture and was part of it. I'd have sworn she was born in Santa Fe and lived here all her life, if I didn't know better."

Just when everything seemed to be falling into place for Dorothy, illness again threatened to destroy her world. In the summer of 1937, Kevin was diagnosed with rheumatic endocarditis, a potentially fatal cardiac disease. The doctor's best recommendation for the ailing six-year-old was a year of bed rest. Too much exertion could kill him. There would be no going to school for him that fall. All they could do was wait and see. Dorothy shut her mind against the terrible possibilities. She put everything aside to tend to her little boy, rushing home for lunch every day and returning straight after work each afternoon. Kevin was bored and lonely, and she hated seeing him lying there looking so miserable. She spent hours a day sitting on his bed reading to him, playing games, and devising ways to keep him amused. It was a long, trying winter, made more so by the death of her father from pneumonia. The following summer, Dorothy took Kevin to Los Angeles Children's Hospital to see if there had been any improvement. "We drove out to Los Angeles, accompanied by my grandmother and aunt, who followed in a LaSalle limousine with a liveried driver," recalled Kevin. "It was quite a convoy." Dorothy rented a tiny house near the shore in Long Beach, hoping the sun and sea air would do Kevin some good. The doctors' verdict was that Kevin had been misdiagnosed. He was suffering from nothing worse than a bad case of tonsillitis. They removed the swollen glands and sent him home.

Over the next few years, many of Dorothy's Santa Fe friends tried to play matchmaker. She kept company with several gentlemen, including the painter Cady Wells, whom she almost married. But after burying her husband and nearly losing her son, Dorothy may have been inclined to take her happiness where she found it, without asking for more. "She had suitors," said her friend Betty Lilienthal. "But I don't think there was anyone in town who really interested her that way, who was really appropriate. She had a great many friends, and kept herself busy, and took care of Kevin. I think she was content."

If Dorothy had once yearned for a more adventurous life, she had put those thoughts aside and reconciled herself to slow, sun-drenched days in the mesas and mountains of the high desert. Though she counted her blessings that both she and Kevin were doing well, she could not help feeling uneasy at times about the events taking place in Europe. She looked sadly back on those youthful sojourns abroad with her father and wondered if his generation's sacrifices in the Great War had all been a waste. She read the increasingly alarmist newspaper reports about the rise of fascism and Hitler's treaty-breaking demands; and when German troops invaded Poland on September 1, 1939, she realized that the ideological tensions had finally erupted into a military conflict. As much as she, like most Americans, was determined that Europe would fight it out alone, and that this country would remain neutral and uninvolved, she understood that even in Santa Fe there would be no respite from worry and strife.

In her memoir, she recorded the morning in May 1940 when the harsh reality of war suddenly intruded on her peaceful world and wrenched her heart and filled her with foreboding. She and Cady Wells had accompanied their friends Eliot Porter, a well-known landscape photographer, and his wife, Aline, on a trip to the small Hopi village of Supai, located thirteen miles down, at the bottom of the Grand Canyon, and reachable only by foot or horseback. In this unhurried, isolated corner of the world, they sat lazily listening to the radio in a trader's house, when a report crackled over the airwaves that the Nazis had marched into Holland:

> There had been a moment of stunned disbelief. The sky was just as bright and deeply blue. The wild celery was just as green on the bank of the small stream which wandered over the rocks on its way to Havasupi Falls. The green backs of the frogs leaping in and out of the stream glistened as always with crystals of brook water. Everything was the same. But nothing would be the same again. There would never again be a piñon tree under which was eternal peace and everlasting happiness.

When war was declared on December 8, 1941, the day after the Japanese attack on Pearl Harbor, Dorothy's first reaction was that at least she and Kevin "would be safe," tucked away in a sleepy backwater such as Santa Fe. She had lost family in the Great War and felt no good had come from that or any other war. She had seen enough misery and death to last her a lifetime and had come out west in part to escape all that. Despite a strong sense of duty that tugged at her conscience, she turned her back on the drumbeat of patriotism and the colorful propaganda banners that filled the Plaza. But the war soon came home to Santa Fe in ways she could not ignore. The small city fell into an economic slump as all the able-bodied men enlisted and one business after another was liquidated. Not far from town, in Casa Solana, a large internment camp was built, and surrounded by barbed wire. Rumor had it that the prison held Japanese Americans and Japanese from Latin America who were deemed "dangerous persons," and the U.S. government was holding them as bargaining chips for potential prisoner exchanges with Japan. By the beginning of 1943, Dorothy, whose travels had given her a lasting interest in foreign affairs, found it impossible not to feel "quite worked up about the world situation." She followed the reports of fierce fighting in the Pacific, where America had mobilized almost half a million troops to prepare for General Douglas MacArthur's offensive. In North Africa, Rommel's panzer army had routed Allied forces at Kasserine Pass in Tunisia, leaving nearly 1,000 dead. The war was not going well, and Dorothy kept her small kitchen radio tuned to the news.

By the time she met Robert Oppenheimer, Dorothy had been a widow for twelve years. At his request, she plunged herself into the clandestine wartime project. She did not have the slightest idea what he was doing in the high country, or what would be asked of her. She did not care, she wrote, "if he was digging trenches to put in a new road." He was the most compelling man she had ever met, and she would have done anything to be associated with him and his work. Perhaps the desert had worked its cure on her a second time, and she was strong again. Her heart, like her scarred lungs, had healed. Maybe after so many years the town had become a bit too small, and she felt the stirrings of an old restlessness. It may also have been that her father's spirit of adventure ran deeper in her than she knew, and she was ready to see what else life held in store for her. Oppenheimer asked her to start right away, and she agreed.

To people in town, she remained the same sweet widow, working at a nondescript office around the corner from her old job and spending all her free time with her boy. She was told to attract as little attention to herself and her new position as possible, and she did as she was asked, sticking carefully to the daily rhythms of her previous life. No one, not even her son, was aware of what she was really doing. To people inside the project, however, she became known as Oppenheimer's loyal recruit, his most inspired hire, and the indispensable head of the Santa Fe office. For the next twenty-seven months, she would lead a double life, serving as their confidante, conduit, and only reliable link to the outside. She would come to know everyone involved in the project and virtually everything about it, except exactly what they were making, and even that she would guess in time. One of the few civilians with security clearance, she was on call night and day. As she soon discovered, she would learn to live with that one word—"security"—uppermost in her thoughts at all times. Everything was ruled by "secrecy, the conditions of secrecy," she wrote. "One's life changed. One could not speak of what one was doing, not even in the smallest detail, to one's family or friends. Every scrap of paper used in the office was burned every evening before closing." This was a well-known wartime practice, but part of a whole new world to her.

TWO

A Most Improbable Choice

STANDING A FEW FEET AWAY in the lobby of La Fonda, Oppenheimer's twenty-three-year-old secretary, Priscilla Greene, watched him work his magic on Dorothy McKibbin. The meeting could not have lasted more than a few minutes, but she had no doubt of the outcome. Dorothy appeared to be bright, lively, and intelligent, with rosy cheeks and fine-boned features topped by a mass of curls. She had an engaging manner, a gentle, assured way about her that was very attractive. Oppenheimer would like her, and there was no question of her liking him. In the short time she had worked for him, Greene had observed that it was the rare individual who was not beguiled by his Byronic looks, quick mind, and grave, courteous manner. "I don't think he really interviewed her. He just offered her the job," she recalled, "and she didn't hesitate for a minute to accept."

Priscilla Greene understood this all too well. She had fallen for Oppenheimer almost as quickly as Dorothy McKibbin had. Scarcely a year earlier, in February 1942, Greene had landed a job working for Ernest Lawrence, a Nobel Prize–winning physicist at the University of California at Berkeley. Not long after she had started, Lawrence had doubled her workload by loaning her out on a part-time basis to his good friend

"Oppie," yet another tall, handsome, flirtatious physicist. Oppenheimer (who had picked up the nickname "Opje" during a postdoctoral stint in Europe and would sign personal letters that way for the rest of his life, though the nickname eventually became Americanized as "Oppie") was head of Berkeley's theoretical physics department and had an office in Le Conte Hall, the same administrative building where Lawrence worked. Oppenheimer had been asked to hold a special wartime science conference that summer and needed a hand getting it organized. As it turned out, he had needed a lot of help, and Greene was delighted to find herself in the employ of such a dynamic figure.

At the time, Oppenheimer was thirty-seven, and had a reputation on Berkeley's campus as an inspiring lecturer. He was also known to be impatient, arrogant, and possessed of a razor-sharp tongue—and as a young teacher had been infamous for terrorizing anyone in his classroom he found plodding, dull-witted, or in any way crass. He was considered one of the very best interpreters of mathematical theory, and study with him guaranteed the ambitious a fast-track career in theoretical physics. Many people were intimidated by him, though those who knew him better claimed that he had mellowed in the decade since he had come to Berkeley in 1929 after a sojourn in Europe, where he had studied with a small colony of world-class physicists, including James Franck and Einstein's friend Paul Ehrenfast, and been a recognized participant in the quantum theory revolution. But there was always the sense with Oppenheimer that the mediocre offended him and that he did not regard the denizens of a West Coast university as quite his equals. John Manley, a refreshingly low-key experimental physicist at the University of Illinois whom Compton assigned to assist Oppie on the wartime physics project, recalled that when he met Oppenheimer for the first time, he was "somewhat frightened of his evident erudition" and "air of detachment from the affairs of ordinary mortals."

Oppenheimer could also be dismissive to the point of rudeness. He had a habit of interrupting people mid-sentence by nodding and saying quickly, in a slightly affected Germanic accent, *"Ja, ja, ja,"* as though he understood exactly what they were thinking and where their argument was headed—an argument that he would then proceed to rip apart in brutal fashion. After witnessing one such performance, Enrico Fermi, who was every bit as agile if not more so, observed that Oppie's cleverness sometimes allowed him to sound far more knowledgeable about a subject than he might be in practice. But with his magnetic presence, astonishing quickness of mind, and wide range of intellectual interests, Oppenheimer was an exciting figure to be around, and students and colleagues were drawn to him as much by his great capacity as a physicist as by his immense charm. "We were all completely under his spell," said Philip Morrison, one of the brightest of the young physicists who studied with him. "He was enormously impressive. There was no one like him."

His allure extended well beyond the lecture hall. Oppenheimer had the powerful charisma of those who know from birth that they are especially gifted. He expected to dazzle—the implacable blue eyes said as much in a glance. It was his mind that burned so brightly, with an intensity that he brought into every room, every relationship, every conversation, so that he somehow managed to invest even an offhand gesture or remark with some extra meaning or significance. Everyone wanted to be initiated into his inner circle. Even his younger brother, an astute observer of the Oppie effect, was not immune. "He wanted everything and everyone to be special and his enthusiasms communicated themselves and made these people feel special," said Frank, who was eight years his junior and idolized his talented brother, following him into physics even though he knew he would never be in the same league. "He couldn't be humdrum. He would even work up those enthusiasms for a brand of cigarettes, even elevating them to something special. His sunsets were always the best."

What drew people to Oppenheimer was that he was so very serious and he took those he collected around him so seriously, endowing them with rare qualities and facets they did not know they possessed. He would focus on them suddenly and relentlessly, showering them with phone calls, letters, favors, and unexpected, generous gifts. His attention could be unnerving, but at the same time exhilarating and gratifying. He was far from perfect, but his flaws, like his dark moods and savage sarcasm, were part of his fascination. He liked to show off, but the performance disguised a deep well of melancholy and self-loathing he carried with him from his cosseted New York childhood. It was the loneliness of a prodigy. He was named for his father, Julius Oppenheimer, a wealthy textile importer, but was always known simply as Robert or Bob until his early twenties, when he felt compelled to embellish his name, perhaps in the belief that "J. Robert Oppenheimer" sounded more distinguished. He suffered from serious bouts of depression as a student first at Harvard, and then later at Cavendish Laboratories in Cambridge, England, and even flirted with the idea of suicide. After failing to find satisfaction in psychiatry—one high-priced London doctor diagnosed his condition as "dementia praecox" and a "hopeless case"—he immersed himself in Eastern mysticism and became a fervent admirer of the Bhagavad Gita, the seven-hundred-stanza Hindu devotional poem, which he read in the original, after studying Sanskrit for that purpose. For a scientist, his search for wisdom in religion, philosophy, and politics was so unusual as to be considered "bohemian." While it got him into trouble at Caltech (the California Institute of Technology), where he also taught, and the Nobel Prize–winning physicist Robert Millikan refused him a promotion on the grounds he was too much of a dilettante, at Berkeley it only added to his appeal.

His style was to be the tormented genius, and his spare frame and angular face reflected his ascetic character, as if his desire to engage every moment fully and completely were consuming his inner resources. He had been a delicate child, and when he pushed himself too hard, he became almost skeletal, resembling a fifteenth-century portrait of a saint with eyes peering out of a hollowed face. There was something terribly vulnerable about him—a certain innocence, an idealized view of life that was only saved from being adolescent by the sheer force of his intellect—that touched both sexes. His students all adored him, and he inspired the kind of devoted following which led some jealous colleagues to sneer that it was more a cult of personality, that Oppenheimer was the high priest of his own posse. He was trailed everywhere by a tight, talented group of graduate students, the stars of their class, and Greene learned to easily identify them by their pompous attempts to imitate Oppenheimer's elegant speech, gestures, and highbrow allusions. She sometimes had the impression that Oppie was conscious of his ability to enthrall. It was no accident that people wanted to help him and would go to extraordinary lengths to earn his approbation.

Greene, who had graduated from Berkeley the previous year and still wore her long, blond hair loose on her shoulders like a schoolgirl, found him "unbelievably charming and gracious." His voice was one of the most mesmerizing things about him. When he singled her out for attention, he was "so warm and enveloping," he made her feel like the most pleasing guest at the party. "When he came into a room, my most characteristic memory of him is [his] coming across to shake your hand, with a slight tilt and a marvelous smile," she said. "And what secretary wasn't going to be absolutely overwhelmed by somebody who, in the middle of a letter—we all smoked in those days—whipped his lighter out of his pocket and lighted your cigarette while you were taking dictation and he was talking."

Compared to Ernest Lawrence, Oppenheimer was a person of enormous culture and education. Lawrence was celebrated for his invention of the cyclotron, the powerful atom smasher, but was proletarian in his pursuits outside of physics. Oppie was from a wealthy New York family, wore good suits, and tooled around campus in a Packard roadster he nicknamed "Garuda," in honor of the Sanskrit messenger to the gods. He spoke six languages, quoted poetry in the course of everyday conversation, and could be snobbish about music and art. "Bach, Mozart and Beethoven were acceptable," noted his protégé, Robert Serber. "Ditto the Impressionists." He had fierce opinions when it came to food and wine. "Martinis had to be strong. Coffee had to be black. . . . Steak had to be rare," listed the British physicist Rudolf Peierls. Once, Oppenheimer took Peierls and a group of graduate students out to a steak restaurant for dinner. He proceeded to order his entrée rare, and this was echoed by everyone in turn until the last student at the table requested his, "Well done." Oppie looked at him for a moment and said, "Why don't you have fish?"

He spent a great deal of time cultivating people and interests that had nothing to do with science, and even Greene could not help being struck by the wide variety of his correspondence. One of the first things he asked her to do was take down a letter to a San Francisco museum to which he was planning to give a painting by Van Gogh, which he had inherited from his father. He had pronounced the artist's name in the guttural German style with lots of breath—"Van Gaaaccchhh"—which was beyond her, and in the end he had had to spell it. "The people he thought about, wrote about, and talked to, he had such a wonderful *feeling* for, that you really wanted to be part of whatever he was doing," she said. "It was very hard to resist him."

His personal life was equally flamboyant, and subject of much comment. Two years earlier in 1940, he had shocked friends and colleagues by marrying Kitty Puening after a whirlwind romance, and their son Peter had been born so soon afterward that Oppie had attempted to jokingly defuse the scandal by dubbing him "Pronto." Kitty was dramatic, dark-haired, and petite; claimed to be a German princess; and was prone to putting on airs. She had also been married three times before the age of twenty-nine and had been with her previous husband for less than a year when Oppie met her at a Pasadena garden party. It was characteristic of Oppie that he would fall for someone so exotic, utterly unsuitable, and beyond reach as Kitty, who, among her many problems, was at the time another man's wife. Oppenheimer, who was besotted, called her "Golden." His close-knit circle was less charitable, considering the poetic young wunderkind—who was so bereft after his mother's death in 1930 that he described himself to a friend as "the loneliest man in the world"—easy prey for a calculating woman. The faculty wives who had doted on Oppie, who was known for bringing flowers to dinner, took an instant dislike to her. After his marriage, many of his peers felt he became more socially ambitious than ever, as though seeking to remove himself from the dreary confines of academic life, and came to regard him with a mixture of envy and resentment. To Greene, however, he seemed even more of a romantic figure. While she would never have admitted it at the time, she was, she said, "more than a little in love" with her boss.

Back in the spring of 1942, Oppenheimer had been summoned to the office of Arthur Compton, the director of the new Metallurgical Laboratory

(Met Lab) at the University of Chicago, and briefed on what was unofficially becoming known among physicists as the country's "bomb headquarters." Immediately after Pearl Harbor, Compton decided that America was moving far too slowly in its atomic bomb research and that the country needed to drastically step up its efforts if a weapon were to be developed in time to be used in the current war. Ever since December 1938, when the German and Austrian scientists Otto Hahn and Lise Meitner first reported their startling discovery that uranium atoms fissioned upon impact by neutrons, physicists in laboratories around the world had been working on how the process of fission, in which a large quantity of energy was released, along with neutrons, could possibly make a chain reaction—and a nuclear explosion. By the summer of 1939, with the drums of war beating in Europe, the Hungarian refugee scientist Leo Szilard had become so alarmed about reports that the Germans were working on a powerful new weapon that he prevailed upon Albert Einstein to write to President Franklin Delano Roosevelt warning him about the military application of nuclear fission. The president had approved the formation of a uranium-research committee, and $6,000 had been appropriated, but there was so much confusion about the new science of fission that little had been accomplished.

Impatient with the committee's progress, Compton decided to centralize all the different teams at the Met Lab—the misleading name was meant to disguise its main purpose—to make a concentrated push to develop a method for making a fission explosive. He told Oppenheimer that even as they spoke, the Nobel Prize–winning Italian physicist Enrico Fermi was secretly at work in a squash court, tucked under the west stands of the university's football field, to achieve the first chain reaction in a graphite pile. With all signs indicating that Fermi's pile would work, Compton needed Oppenheimer, whom he regarded as having a brilliant mind, to take charge of a division of the Met Lab and organize a group to study the physics of an explosive chain reaction—or bomb. It was going to be an extraordinarily difficult task, but Compton thought Oppenheimer, despite his rather flamboyant personality, might be just the man to do it.

As Oppenheimer commuted back and forth between Chicago and

Berkeley organizing the conference on bomb development, Greene noticed a change in his attitude, a tenseness and excitement that alerted her to its importance. It was May 1942, six months after Pearl Harbor, and everyone involved in war work had a heightened sense of purpose. Many of his close colleagues had already been called on to do defense research. Lawrence was doing double duty at Berkeley and a radar lab at the Massachusetts Institute of Technology (MIT), and Oppie had felt left behind and stuck with a heavier teaching load than usual because of all the absences. He had only been asked to step in and supervise the special summer conference because his predecessor, Gregory Breit, had succeeded in alienating almost everyone with his unreasonable demands and obsession with secrecy. Oppenheimer had been chafing on the sidelines, and he seized the chance to lead a brainy group exploring the possibilities of designing a nuclear bomb. At the same time, he knew that he had inherited a difficult position and that managing so many egotistical scientists, from such widely varied countries of origin, was not going to be an easy task.

In June, Oppenheimer gathered the top theoretical physicists in the country, and they met in two attic rooms at Le Conte Hall. He had assembled a high-powered group, among them the Hungarian physicist Edward Teller, the German physicist Hans Bethe, the leading Swiss physicist Felix Bloch, Richard Tolman, and Oppie's former student Robert Serber. Their meetings were veiled in secrecy, and the university had taken unusual measures to safeguard their privacy, including securing the windows and small balcony with heavy wire fencing and fitting the door with a special lock with a solitary key that was kept by Oppenheimer himself. Among the refugee scientists, there was a palpable anxiety to get under way immediately, amid worried conjecture about how far ahead the Germans might be and how much the strain of fighting a war might have slowed the Nazi weapons program.

Greene had discerned this was defense-related work from the official letters she had typed, but when she walked into the room across the hall from the office one afternoon and saw what was on the blackboard, she realized for the first time what all the security precautions were for. "Somebody had drawn a spherical shape and, from the various scribbles, well, it was obviously a bomb," she said. "So I knew then. I was glad to know what we were doing. Almost immediately after that, everyone started calling it 'the gadget.' "

The goal of the special summer session was for Oppenheimer and his team to calculate to the best of their ability the exact specifications for the design of an atomic bomb. What was known to date was fairly rudimentary—the British had done some preliminary work—and the most they could do was make an educated guess at the force an atomic fission bomb could exert. So they had to begin with the basics, such as the size and structure of the bomb, work up an estimated critical mass, figure out how it might be made to explode, and then begin to try to address the practical problems entailed in assembling such a device, and ultimately detonating it. Oppie delegated teams to research different problems, and they met regularly to exchange information and ideas. Teller, who hardly knew Oppenheimer prior to the Berkeley conference and had only a moderate opinion of his abilities as a theoretical physicist, was impressed by his "sure, informal touch," by his ability to motivate and guide the various participants, and by how much work was getting accomplished. "I don't know how he had acquired this facility for handling people," he said later. "Those who knew him well were really surprised."

By the end of July, after they had been meeting regularly for almost two months, Oppie called everyone to a meeting at Le Conte Hall to review their progress. After a discussion of the critical mass calculations presented by Serber's group, and some consideration of the potential damage from the blast, neutrons, and radioactivity, it was decided that the goal looked feasible. Oppie became almost chipper, and the theorists continued to settle important questions about the plant design, to offer predictions and suggestions about what work would have to be done to make the job a success, and, as far as Greene could tell, to have what to a bunch of physicists was clearly the time of their life.

So the days passed, with the physicists arguing back and forth, and the plans for the fission bomb progressing, until Teller introduced the idea of the Super—a hydrogen bomb based on the possibility of nuclear fusion—and everyone got sidetracked for weeks on end arguing about whether it would work or not and, as Serber put it, "forgot about the A-bomb, as if it were old hat." Oppie was already exasperated that so much valuable time was being wasted on an entirely new and difficult proposal for a hydrogen bomb, with the problems of the atomic bomb still far from settled, when Teller brought the proceedings to a grinding halt by asking if the enormously high temperature of an A-bomb could ignite the earth's atmosphere.

The apocalyptic scenario Teller outlined forced Oppenheimer to abruptly adjourn the conference. They had no choice but to look over his figures and determine what the effects of the fission reaction would be. As Hans Bethe was by far the quickest at calculations—he and Oppie often whipped out their slide rules and raced to see who could run the numbers the fastest—he was assigned to check Teller's work. In the meantime, Oppie had to alert Arthur Compton at the Met Lab.

Oppenheimer immediately phoned Compton and, after numerous frantic calls, finally tracked him down at his summer home in northern Michigan. Talking somewhat awkwardly, as Compton was calling from the tiny Otsego general store, Oppenheimer grimly reported that his group had "found something very disturbing—dangerously disturbing." Oppie explained that he had to see him in person "immediately, without an hour's delay." Oppenheimer caught the first train out. Compton picked him up at the Otsego train station the following morning, and they drove down to the lakefront. Staring out at the empty stretch of beach, Oppie laid out the dangers raised by Teller and his calculations. As Compton recalled in his memoir, these were questions that "could not be passed over lightly":

> Was there really any chance that an atomic bomb would trigger the explosion of the nitrogen in the atmosphere or of the hydrogen in the ocean? This would be the ultimate catastrophe. Better to accept the slavery of the Nazis than to run a chance of drawing the final curtain on mankind!
>
> We agreed there could only be one answer. Oppenheimer's team must go ahead with their calculations. Unless they came up with a firm and reliable conclusion that our atomic bombs could not explode the air or the sea, these bombs must never be made.

During the last sultry weeks of July, the Berkeley conference limped to a close, and the group went their separate ways. Bethe had reached the "reliable conclusion" that there was a flaw in Teller's theory, and while nitrogen and hydrogen were unstable, it was highly improbable that an atomic explosion would create the conditions to set them off. It was safe, at least as far as the atmosphere was concerned, to proceed with the atomic bomb. That fall, Oppenheimer continued to supervise bomb theory studies at Berkeley. As for Teller's Super bomb, Compton decided that the idea be kept a closely guarded secret, and it was shelved for the time being. But the lingering effects of that summer's tension with Teller would surface again and again. "Oppie had trouble with Teller in the summer of '42," said Greene. "After that, he [Oppie] always tried to keep him at arm's length."

By the time General Leslie R. Groves paid his first visit to Berkeley on October 8, Oppenheimer's plans for building an atomic weapon were in good order. Groves, however, wanted to satisfy himself that the many hurdles he foresaw had been addressed and that he and Oppenheimer were thinking along the same lines. A professional soldier and trained engineer, who had attended MIT and West Point and had just completed the construction of the Pentagon, a massive project that had included the building of everything from airfields and ports to factories, Groves was appointed by the Army Corps of Engineers to take charge of the entire atomic bomb project, known as the "Manhattan Engineer District" (and later as the "Manhattan Project") because Groves' predecessor had been based in New York City.

Oppenheimer was aware, as were many of the top nuclear physicists and chemists at the time, that the Manhattan Project had already been under way for more than a year. In June 1941, President Franklin D. Roosevelt had formed the Office of Scientific Research and Development (OSRD) to organize all of the various uranium and fission research projects in the country that could help lead to the achievement of a sustained chain reaction. The OSRD was run by Vannevar Bush, the former head of MIT and the Carnegie Institution, and his deputy, Harvard president James B. Conant. They were asked to form a new committee, code-named "S-1," which would be responsible for organizing and accelerating the atomic weapons research and for making sure authorized objectives were accomplished. Once S-1 delegated the bomb project to Groves, he would direct the development of the bomb and all the related projects, backed by a Military Policy Committee, which included Bush and Conant among others. This sent a clear signal that the scientists and army officers would have to put aside their natural suspicions and work together. Under Groves, the scientists would have to accept life under a military regime, with army representatives becoming a constant presence in their lives, laboratories, and meetings.

While some scientists had initially greeted this proposal with open hostility, the heat of the battle against the Nazis had an ameliorating effect. The idea that the Germans might have nuclear weapons before the Allies was a constant threat. There was also no arguing the fact that the army was better at procurement and would undoubtedly run such a large-scale building and engineering operation more efficiently than the scientists themselves or, for that matter, a building full of bureaucrats in Washington. As Groves made clear when he was introduced to some of the research leaders at an October 5 meeting at the Met Lab, time was of the essence. He demanded results, and fast. Although a number of the scientists objected to Groves' bullying tone, which Serber recalled as "You're working for me now so you'd better toe the line," there was not much they could do about it. Both groups recognized that it was vital to the success of the bomb project that they work smoothly together, and this hinged on a relationship of trust and understanding between the key players.

An unlikelier pair than Oppenheimer and Groves could not have been found, and it did not seem to bode well for their partnership. Where Oppenheimer was tall, trim, and elegant, the general was bulky and square, his unwieldy frame threatening to burst from the constraints of his tightly belted pants, his rumpled appearance comically at odds with his rank. But in character, Groves was all spit and polish, a perfectionist who expected discipline and absolute devotion to the task at hand. He obsessed over the smallest details and labored exhaustively to achieve his objectives. His military aide, Colonel Kenneth Nichols, who was appointed to work with Groves on the bomb project, later described him as "the biggest S.O.B." he had ever worked for: "He is most demanding. He is most critical. He is always a driver, never a praiser. He is abrasive and sarcastic." Nichols also conceded that he would opt to have Groves as his boss again, because he was "one of the most capable individuals."

Even Bush, who was a hard-nosed Yankee with a flinty manner when pressed, was left flustered by their first encounter. He found Groves "abrupt and lacking in tact" and was so pessimistic about how the general would get on with the physicists on the project that, he worried to Conant in a letter, "I fear we are in the soup." Compared with the subtle, soft-spoken Oppenheimer, Groves was a blunt and ruthless taskmaster, and few would have predicted that their first meeting would go well. Groves wasted no time in enlisting Oppenheimer in a detailed discussion of exactly what kind of bomb laboratory he envisioned. He not only wanted to take his measure of the man, he needed to see if the erudite theorist had a practical bone in his body and could possibly make a go of the weapons project. The general, raising his concerns about the assembly of the bomb, security, and other precautions he considered vital to the planning of the project, fired question after question at Oppenheimer. Oppie, who had always excelled as a teacher, patiently addressed each technical and organizational obstacle, so that the tough-minded Groves later reported to Compton that he was "strongly impressed by Oppenheimer's intelligence and quick grasp of the problem."

Groves reportedly made an equally strong impression on Oppenheimer, when on their subsequent meeting the general marched directly into his Berkeley office trailed by Colonel Nichols, immediately unbuttoned his jacket, handed it to his subordinate, and barked, "Find a tailor or dry cleaner and get this pressed!" According to Serber, who was working with Oppie in his office at the time, the colonel took the coat and walked out without a word, leaving the civilians in the room with a lasting impression of what Groves expected when he issued an order, petty or otherwise. "Treating a colonel like an errand boy," said Serber. "That was Groves's way." Greene formed an impression of Groves as harsh, ill-tempered, and more than a little terrifying. But she marveled that Oppie was not in the least cowed by Groves' brusque manner and immense command of power, and she watched as the two men quickly established a good rapport.

Groves had met dozens of physicists since taking charge of the bomb project on September 23, 1942, and much to his dismay, he had found any number of them thoroughly objectionable—he loathed the eccentric Hungarian Leo Szilard almost on sight and thought him completely unsuitable for such a monumental task. He worried that "none of them were go-getters; they preferred to move at a pipe-smoking, academic pace." So relieved was he at having found Oppenheimer, a man who he felt had a big brain but was capable of the kind of vigilant thinking he admired, that he immediately asked to meet with him again in a week's time. Groves requested that Oppie accompany him, Colonel Nichols, and another aide on their train trip back to New York from Chicago so they could talk at greater length. Given his marching orders, Oppie flew from Berkeley to Chicago and caught the Twentieth Century Limited, the high-speed passenger express, where he joined the three military men for dinner in the first-class dining car.

As the train sped into the night, Oppie and Groves discussed the plans for the new bomb laboratory and the high-level security measures the military considered necessary to safeguard their research. Sitting almost knee to knee in the tiny compartment, Oppenheimer voiced his reservations about Groves' plan to have scientists at various sites around the country working on separate projects, pointing out that it would be extremely difficult to coordinate the different groups.

In a bold move, or a moment of unguarded honesty, Oppenheimer had confided to Groves in an earlier meeting that he was less than happy with the progress of the scattered small groups assigned to work on the bomb project. Forced to work under the constraints of "compartmentalization," which mandated that every group work on a need-to-know basis with only a handful of senior people being fully informed, and which he knew to be Groves' particular innovation, "the little laboratories suffered from their isolation." Oppie expanded on his earlier complaint, arguing that work was being duplicated, the physicists were confused about what was being done elsewhere and had no sense of direction or hope. Looking to the future, communication between the projects would be extremely difficult. They could not relay classified material over the telephone, and teletype was tricky. As the bomb work accelerated, it would necessitate constant travel for all the project leaders that would be inevitably delaying. Oppenheimer was also afraid that morale would suffer over the long haul if the scientists were walled off from one another and not permitted to know the magnitude of what they were working on.

Few people knew how to argue their case as winningly as Oppie, and he now employed those skills to his full advantage. Aware by now that Groves was not sympathetic to amorphous pleas for free and open scientific discourse, he carefully anchored his argument to more practical concerns. Appealing directly to Groves' obsession with security, Oppie proposed that all the research be consolidated in a single laboratory, located in an isolated region, where the scientists could converse freely among themselves and secrecy could still be maintained. Oppenheimer argued that if they picked a site that was smaller than Lawrence's operation at Oak Ridge, in Tennessee, where large numbers of workers came and went every day, and was far removed from any population center, the lab could be completely protected from outsiders and enemy spies.

Groves was quick to recognize the merits of Oppie's plan. The idea of rounding up the troublesome scientists and keeping them in one guarded camp had a logic he understood. It was a compromise, but one Groves could live with and make work. If he could not keep the scientists from talking to one another, then at least he could make sure they did not talk to anyone else. He dispatched Lieutenant Colonel John H. Dudley to conduct a preliminary investigation of possible sites for this central facility. Now, with the question of where the lab would be located well on the way to being settled, all that was left to do was pick the right man to run it. Compton had the right background, but could not be spared from the Met Lab in Chicago and its important work on chain reactions and fast neutrons. And although Compton had appointed Oppenheimer to head the design and fabrication phase of the project, Groves did not feel committed to him, especially since "no one with whom [he] talked showed any great enthusiasm about Oppenheimer as a possible director."

Despite Oppenheimer's technical competence and delicate handling of the prickly egos at the Berkeley seminar, it is doubtful that he would have been chosen to head the bomb project had Ernest Lawrence been available. Not only was Lawrence a Nobel laureate who was highly regarded by both his peers and the Washington establishment, he was the country's foremost experimental physicist, with a long record of success at building large machines. Groves had immediately hit it off with the enthusiastic, down-to-earth Lawrence, and he certainly would have been Groves' first choice had it not been for the fact that Lawrence was desperately needed to oversee the magnetic separation process at Oak Ridge that was to produce the uranium 235 for the atomic bomb. They could not risk distracting Lawrence or upsetting the Oak Ridge operation in any way. It is also unlikely that Lawrence would have been easily persuaded to leave his own laboratory to go off and collaborate on such a questionable venture under quasi-military auspices. Instead, Lawrence recommended one of his protégés, Edwin McMillan, who was an outstanding theorist but a somewhat reticent person and not the sort of commanding figure Groves had in mind. Teller favored his close friend Hans Bethe, who, he argued, was "unquestionably a superior theoretician," not to mention considerably more popular than Oppenheimer. But as Bethe was a foreign national, Groves, of course, considered him out of the question.

The selection process continued, with Groves meeting with small groups to talk over a short list of names. As the end of October drew near, Oppenheimer clearly emerged as his candidate to head the new laboratory. It promised to be an extraordinarily difficult job, and as Compton and Groves concurred, one that required unique qualifications. There were not only the massive scientific and technical challenges to consider, but as Compton looked ahead, extremely difficult experiments would have to be conducted under conditions of "almost unprecedented seclusion." What's more, the leader of this enterprise would have to be "a person of such human understanding that he could keep a group of high-strung specialists working smoothly together while largely separated from the outside world." Not only was Oppenheimer a specialist in the problems of nuclear physics, but in Compton's view, "he was one of the very best interpreters of the mathematical theories to those of us who were working more directly with the experiments."

Both Compton and Groves were well aware of Oppenheimer's left-wing sympathies, including his attendance at a number of Communist Party meetings and anti-fascist rallies, and his Teachers' Union work.

They had also reviewed the thick FBI file enumerating his many "pink" associations, including a former fiancée, Jean Tatlock, who had encouraged his nascent political activism and remained a close friend and confidante; his new wife, Kitty, who had joined the Communist Party in solidarity with her second husband, a party organizer; and his brother Frank and Frank's Canadian wife, Jacquenette (Jackie) Quann, both of whom had joined the party in the late 1930s. Many of Oppenheimer's Berkeley protégés had similar ties to leftist politics, including Robert Serber, whose wife, Charlotte, was the daughter of Morris Leof, a Russian Jew and prominent Philadelphia physician who ran a sort of liberal political-literary salon whose habitués included the playwright Clifford Odets and the journalist I. F. Stone. Compton knew firsthand of Oppie's participation in "certain Communist activities" in the late 1930s but, after talking those over with him, was satisfied by Oppie's explanation that they had only been attempts to educate himself. "He felt that a responsible citizen ought to have reliable knowledge of this growing new movement," Compton recalled in his memoir, and he gave Compton reason to believe he had come to regard it as "dangerous."

Despite people's expressed concerns about Oppenheimer's past radicalism, Lawrence had vouched for him in the embryonic days of the atomic program and had insisted on his being included in top-level meetings. Oppenheimer was, in a sense, *his* man, and in the end, Lawrence backed him, with the condition that the new laboratory be an administrative extension of the University of California, which also presided over Lawrence's lab. By 1942, Compton had also come to the conclusion that Oppie was far too knowledgeable to allow himself to be entangled in the "Communist net." "The important matter now," Oppenheimer had assured him, "is the nation's defense. I'm cutting off all my Communist connections. For if I don't the government will find it difficult to use me. I don't want to let anything interfere with my usefulness to the government." Oppie did not succeed in silencing the FBI's doubts about him, however, and the close surveillance that was kept up throughout the war years would later come back to haunt him.

From the very outset, Oppenheimer was regarded as a complex, problematic, and controversial character. Even as a scientist, he was an unlikely choice. As Manley had earlier pointed out to Compton, one of his principal reservations about Oppenheimer was that he "had essentially zero laboratory experience" and although "he did understand laboratory techniques, almost *anybody* would have had more experimental experience." He lacked the stature, the administrative skills, and, given his previous lack of interest in mundane affairs, possibly even the temperament for the job. There was nothing in his background or résumé to indicate that he was suited for a position of such awesome responsibility. He was second-best by any measure, but according to Groves, by then it had become apparent that they "were not going to find a better man." With few alternatives left to him by the end of that October, Groves named Oppenheimer director of the bomb project.

Given the prevailing sentiment at the time about Oppie's weaknesses, the general's decision did not inspire confidence. "To most physicists it came as a great surprise," said Isidor I. Rabi, the formidable Columbia University physicist who was, at the ripe old age of forty-four, regarded as an elder statesman by the young researchers he worked with on the radar project, adding that he considered Oppenheimer to be "a most improbable appointment."

Years later, when asked why he was chosen as director of the Manhattan Project, Oppenheimer would say simply, "By default." He would add that it was not a job anyone wanted. "The truth is, that the obvious people were already taken and that the project had a bad name."

In that moment, pressed to make a decision, Groves must have glimpsed some potential in Oppie—the same innate authority he had observed time and again in men who had taken an unruly group of soldiers and molded them into a disciplined unit—that would ultimately prove invaluable in bringing the atomic task to successful completion. He may also have thought he saw in Oppie something of himself. There was, after all, the same intense ambition, the same presumption of excellence—from themselves above all, and by extension anyone associated with them—that made them both demanding and compelling leaders. Oppenheimer, like Groves, did not suffer fools gladly, and both men had made their share of enemies in the course of their careers. Perhaps as a consequence, in spite of strong records and impressive credentials, neither had advanced to the highest level of his profession. Oppenheimer had majored in chemistry in college and had specialized in mathematical theory late in his training. He had made important contributions in the 1930s in the field of cosmic rays and stellar objects—later known as neutron stars and black holes—but was too diverse in his interests to mine one narrow band of research. He was already past his prime as a theoretician and was generally regarded as a better critic than original thinker. Groves had lingered too long at the rank of lieutenant and, at forty-six, was considered long in the tooth for an army colonel. Now that his overdue promotion to brigadier general had finally come through, it was pegged to what he regarded as a minor administrative post, another desk job, and not to the overseas combat assignment, and glory, he desired.

Here was an opportunity for both men to prove themselves and, in doing so, make a significant contribution to the war. The atomic weapons project was a highly risky proposition, and failure would certainly end their careers. But in their own ways, both Groves and Oppenheimer realized that in working together to build the bomb, they might achieve the greatness that had eluded them thus far. "They recognized in each other the absolute commitment to making the thing work, and a broad commitment to winning the war," said Morrison, who in the early days of the war served as a technical advisor to Groves and knew both men well. "So they had to give in on matters of style. You have to understand that nobody believed in the project, and besides, they couldn't tell anybody about it. Groves was a hard guy to work for, and he was mean, but he made things happen. He had to be tough, and he was very tough. But Robert handled him very well. He called him 'His Nibs' because he always had a colonel carrying his briefcase."

THREE

The Bluest Eyes I've Ever Seen

OPPENHEIMER WAS ALREADY knee-deep in the bomb project when Groves handed it to him. With the matter of his security clearance still pending—though the general would see that it was rammed through—Groves promptly arranged for Oppie to accompany him on a tour of possible locations for Site Y, as the classified bomb laboratory was called within the project. As secrecy was their first priority, isolation was the single most important criterion in selecting the site, but there was a long list of mitigating factors: a mild climate that would permit work year-round, access by road and rail, a large enough area for an adequate testing ground, and sparse neighboring population within a hundred-mile radius to enforce safety and security.

As Colonel Dudley recalled, Oppenheimer also advised him that in evaluating possible sites, he should bear in mind that some existing facilities were needed: "The idea was that the six scientists would move in and start a think-tank operation immediately." Dudley surveyed parts of California, Nevada, Utah, Arizona, and New Mexico, but as the air force and navy had already been through these areas and picked off the best spots, and as coastal areas were ruled out because of the possibility of attack by sea, Dudley quickly narrowed the list to five sites: Gallup, Las Vegas, La

Ventana, Otowi, and Jemez Springs. After being evaluated, the first three failed to meet the requirements, and Dudley concluded that based on what he had seen of the remaining two, their best bet was Jemez Springs, New Mexico.

It was decided that they should all meet there and take a look for themselves. On November 16, 1942, Dudley took Oppenheimer and Ed McMillan out to Jemez Springs, which turned out to be a narrow, deep canyon cut out of the Jemez Mountains. While it technically met all their requirements, Oppenheimer worried that being hemmed in by cliffs on three sides might have a depressing effect on laboratory personnel. There was also the politically sensitive problem of relocating the many Indian families who farmed there. They were in the midst of discussing the difficulties of inspecting the steep perimeter when Groves arrived. After a quick look around, Groves vetoed the site, declaring tersely, "This will never do." It was then that Oppie, who had had the Otowi site in mind all along and had been maneuvering toward it in his subtle way, suggested that while they were there, they might as well inspect another location nearby. He told Groves, "If you go on up the canyon, you come out on top of the mesa, and there's a boy's school there which might be a usable site."

They all piled into cars and drove up a perilously winding dirt road to the Los Alamos Ranch School. It was late afternoon when they reached the top of the mesa, and a light snow was falling. They did not enter the grounds, as McMillan recalled, but stood at the gate looking in:

> It was cold and there were the boys and their masters out on the playing fields in shorts. I remarked that they really believed in hardening up the youth. As soon as Groves saw it, he said, in effect, "This is the place."

Four days later, on November 20, after a hasty meeting in Washington, Oppenheimer and McMillan went back to carefully inspect the site, this time accompanied by Ernest Lawrence. Oppie, as it turned out, had been there before and had privately wanted Los Alamos from the beginning. He and his brother owned a ranch nearby in the Pecos Valley, just across the Sangre de Cristo Mountains, whose snowcapped peaks glowing dusky red at sunset had inspired the early Spanish settlers to name them "Blood of Christ." It seemed typical somehow of Oppie to choose such a dramatic setting, perched on the cone of an extinct volcano. The broad tabletop of the mesa, two miles long and almost 7,300 feet above sea level, afforded majestic views of the whole mountain range. It was bordered on the north and south by steep rock canyons, which offered natural protection. A beautiful grove of cottonwood trees, for which Los Alamos was named, grew beside a shallow stream that ran along the edge of the canyon to the south. The school grounds provided a scattering of buildings and crude log cabins that could be used until additional laboratory space and housing were constructed. The surrounding area was unpopulated, with empty canyons, flat green valleys, and acres of open fields for future testing grounds.

Although only a thirty-five-mile drive from Santa Fe, the Los Alamos site was isolated in the extreme and virtually inaccessible. The last, treacherous, ten-mile track leading up to the school was unpaved and riddled with deep ruts left by the dry creek beds called "arroyos," that flooded each spring and turned the road into an almost impassable bog. The road was precipitously steep as it approached the narrow Otowi Bridge leading up to the mesa, and parts of it were cut out of the solid wall of the canyon and featured half a dozen hair-raising switchbacks leading up the face of almost perpendicular cliffs. The nearest railroad was more than sixty miles away, which meant all the truckloads of goods, building materials, and heavy laboratory equipment would have to be hauled up over hill and vale. There were numerous objections to the site besides the road's obvious shortcomings, including the lack of water and power. But Oppie was happy. He had gotten what he wanted, and the problems were not insurmountable. Groves immediately began formal steps to acquire the school and surrounding land.

Oppenheimer returned to Berkeley ecstatic about his trip. Greene knew he had been investigating possible sites and asked him how that had gone. He told her he had found the ideal spot. "He said he had had this marvelous weekend on a beautiful mesa in New Mexico and was going out there to do a project." She had been intrigued by the rhapsodic descriptions of the countryside in letters she had typed and without hesitating asked, "Can't I go too?" Oppie said he would be delighted to have her along, which annoyed Lawrence no end, as he now had to replace another secretary. Lawrence reluctantly agreed to let her go, on the condition she find her own replacement. Greene went to work for Project Y right away, using the office of a physics professor who was on leave. "I was twenty-three," she recalled. "It was a very exciting time. There was no formality—potential staff visited and I got to go to all the parties."

There was an almost constant round of parties as Oppenheimer threw himself into courting the very best men for the Los Alamos laboratory. It was not an easy task, and Oppie chased all around the country wooing potential staff members. By late 1942, unemployed physicists were hard to find. Ernest Lawrence had already raided the universities and recruited the cream of the crop for the secret Radiation Laboratory (Rad Lab) at MIT. Oppenheimer knew if he were too aggressive in luring away the Rad Lab's supply of talent, he would raise the hackles of his old friend and ally. On the other hand, now that he was focusing on the technical difficulties of actually engineering an atomic weapon, it was becoming increasingly clear that he would need a "very large number of men of the first rank."

On November 30, 1942, Oppenheimer sent a detailed report to Conant, who was already familiar with Los Alamos, having once considered sending his youngest son to the elite preparatory school, which was known for taking sickly city boys and toughening them up with a strenuous regimen of academics and outdoor activity. Oppenheimer caught Conant up on the latest developments at the Met Lab and then launched into the main purpose of his letter, namely, the recruitment of his key staff. It was a tricky area, and he confessed he felt himself "on less secure ground." At the same time, he warned Conant that he could not afford to proceed too cautiously: "The job we have to do will not be possible without personnel substantially greater than that which we now have available."

Conant promised his support and said he would look into getting some top people released from the MIT Rad Lab. Observing Conant's skillful behind-the-scenes politicking, Oppenheimer realized he had much to learn about maneuvering within the treacherous Washington political and military establishments, and that the shrewd, businesslike Harvard president would be an ideal mentor. "Oppenheimer saw this faculty of Conant and wanted to learn from it," said John Manley, who watched their relationship develop into one of mutual fondness and trust. "[Oppie] relied on him a great deal."

Bolstered by Conant's assurances and advice, Oppie flew to Cambridge and succeeded in convincing Hans Bethe, and his wife, Rose, to come out west with him. Securing the participation of Bethe was a coup. Not only was the thirty-six-year-old Bethe an exceptionally gifted theorist, but since moving to the United States in 1935 and joining the faculty of Cornell University, he had earned a reputation as a singularly energetic, confident, and productive scientist. A bull of a man, with an affable manner and a booming laugh, he was married to the lovely daughter of a famous German physicist, Peter Paul Ewald, who like Bethe had left Nazi Germany when the climate became threatening to those with Jewish ancestry. Bethe was as close to being a natural leader as anyone in the field and was well liked and respected by both his colleagues and competitors—all of which made him invaluable to Oppenheimer's cause.

While in Cambridge, Oppenheimer had also hoped to recruit I. I. Rabi and even offered him associate directorship of the laboratory. In those early months, however, Rabi believed the project's overall odds for success were fifty-fifty at best. "I thought we could lose the war because of the lack of radar," he wrote later, explaining why he felt he could not leave the MIT Rad Lab. "As far as the fission bomb was concerned, it was very iffy." His wife, Helen, had also made it clear she did not want to go, telling him, "That's no place to raise children." The Rabis were not alone in their reluctance to become involved in building a bomb. According to Robert Bacher, whom Rabi eventually sent in his place, "People at the Radiation Laboratory thought that this was just absolutely crazy to take people off radar and put them on this fool's project out there."

Oppenheimer was painfully aware of the skepticism with which a great many physicists regarded the bomb project. "There was a great fear that this was a boondoggle, which would in fact have nothing to do with the war," he admitted later.

As a consequence, Oppenheimer felt he badly needed Rabi on his side and attempted to persuade him again. Failing that, he called on Rabi often for his help and counsel during the formative stages of the project. Of all the various matters that had to be settled that winter, perhaps none was more important, or more controversial, than the proposed "militarization" of the laboratory. Oppenheimer had been so eager for the top job that he had readily agreed to Groves' demand that the laboratory be run as a military installation, with the scientists all outfitted in army green and assigned rank. From the laboratory's opening day on, Oppenheimer would be a lieutenant colonel. He had already ordered his uniforms. "I would have been glad to be an officer," Oppie said later. "I thought maybe the others would." But Rabi was appalled at the prospect of scientists becoming commissioned officers in the army, and after talking it over with three other Rad Lab recruits—Bob Bacher, Ed McMillan, and Luis Alvarez—he was convinced the plan would not work. He was adamantly opposed to the whole idea. If Oppie went along with Groves' plan, Rabi told him in no uncertain terms, "none of [them] would come."

Caught between Groves' orders and Rabi's refusal, Oppenheimer wrote Conant on February 1, 1943, that he feared he was facing insurrection from the best scientists if he did not meet their "indispensable conditions for the success of the project." After summarizing their demands in detail, Oppenheimer worriedly told Conant that he was uncertain how Groves would respond to the scientists' ultimatum that the project be demilitarized. "I believe that he realizes the seriousness of these requests, but I am not sure that he feels that they can be met." But by the end of the letter, Oppenheimer mustered his confidence and argued firmly and eloquently that their services were needed too urgently to risk losing them, and ignoring their concerns would certainly result in further problems and delays:

> At the present time I believe the solidarity of physicists is such that if these conditions are not met, we shall not only fail to have the men from MIT with us, but that many men who have already planned to join the new Laboratory will reconsider their commitments or come with such misgivings as to reduce their useful-

> ness. I therefore regard the fulfillment of these conditions as necessary if we are to carry on the work with anything like the speed that is required.

Conant was not entirely persuaded, as he had served as a chemical officer in World War I and regarded enlisting in the army as a mere formality. But at Oppenheimer's insistence, he helped negotiate a compromise with Groves to leave the laboratory under civilian administration during the early experimental stages of the work. The military was to assume control at a later, more dangerous stage, at which point the scientists would become commissioned officers (a transition that ultimately was deemed unwise and never came to pass). This setup was formally communicated to Oppenheimer in a letter of February 25, 1943, from Conant and Groves. To help Oppenheimer with his recruiting, the letter was written in such a way as to assure scientists that there would be no military censorship of information. It stipulated that not only would Los Alamos researchers be exempt from the usual wartime restrictions that prevented them from learning what was going on at other laboratories, but Conant would serve as Groves' technical advisor and liaison: "Through Dr. Conant complete access to the scientific world is guaranteed."

With the crisis thus defused, Oppenheimer wrote to Rabi the next day, this time asking him to compromise and come out to Los Alamos only for the opening session in April. Clearly cognizant of the extent of Rabi's doubts, about both his leadership and the project, Oppenheimer makes, in his letter of February 26, an abject appeal for help. Admitting that he did not know if the arrangements outlined in the Groves-Conant letter would work, Oppenheimer asserted that he was going to make a "faithful effort" to go forward with the project, as he did not feel the Nazis allowed him any other option:

> I think if I believed with you that this project was "the culmination of three centuries of physics," I should take a different stand. To me it is primarily the development in time of war of a military weapon of some consequence. I know that you have good per-

sonal reasons for not wanting to join the project, and I am not asking you to do so. Like Toscanini's violinist, you do not like music.

Since he could not win him over, Oppenheimer continued, he was asking for only two things, "within the limits set by [Rabi's] own conscience"—first, that Rabi agree to give the project "the benefit of [his] advice at a critical time" and, second, that as he exercised a great deal of influence over Bethe and Bacher, he "use that influence to persuade them to come rather than to stay away." The last line of the letter reveals just how unsure he was of Rabi's support: "I am sending a copy of this letter to Dr. Conant and General Groves to keep the record straight." In the end, Rabi acquiesced to Oppenheimer's requests. He agreed to participate in the opening physics conference at Los Alamos and saw to it that the Rad Lab sent a half dozen of its most productive physicists out west to work on the project, including Bethe, Bacher, Alvarez, Kenneth Bainbridge, and Norman Ramsey.

Throughout those rushed months of planning, Rabi and Bethe, among the wiser, more experienced hands, worked on Oppenheimer, urging him to convert his abstract thoughts about the project into a workable plan on paper. Without Rabi's practical advice, Bethe said, "It would have been a mess":

Oppie did not want to have an organization. Rabi and [Lee] DuBridge [head of the physics department at the University of Rochester] came to Oppie and said, "You have to have an organization. The laboratory has to be organized in divisions and the divisions into groups. Otherwise nothing will ever come of it." And, Oppie, well, that was all new to him.

Rose Bethe, whom Oppie asked to run the housing office at the new installation, probably had a better idea than most about what they were getting themselves into, having already toured that part of New Mexico and having received in late December a long letter from Oppie addressing some of her questions about the living conditions. His warmth and concern were touching, but his rough sketch of their hypothetical mountain hideaway, which read in places like something out of *Boy's Life,* was hardly reassuring, nor was his airy promise to keep her list of questions "as a reminder of what we shall have to do":

> There will be a sort of city manager. . . . There will also be a city engineer and together they will take care of the problems outlined by you. We hope to persuade one of the teachers at the school to stay on to be our professional teacher. It is true that both Kay Manley and Elsie McMillan are professional school teachers and there will no doubt be others, but it seems to me very unlikely that anyone with a very young child will be able to devote very much time to the community. There will be two hospitals, one in town and one in the M.P. camp. . . .
>
> Room is being provided for a laundry; each house will have its washtub; and we shall be able to send laundry to Santa Fe regularly. It may be necessary for us to provide the equipment for the group laundry since this is now frozen, but this is a point that is not yet settled.
>
> We plan to have two eating places. There will be a regular mess for unmarried people which will be, when we are running at full capacity, just large enough to take care of these. The Army will take care of the help for this and I do not know whether the personnel will be Army or civilian. We will also arrange to have a café where married people can eat out. This will probably be able to handle about twenty people at a time and will be a little fancy, and may be by appointment only. We are trying to persuade one of the natives to do this and we have a good building for it.
>
> There will be a recreation officer who will make it his business to see that such things as libraries, pack trips, movies, and so on are taken care of. . . . The store will be a so-called Post-Exchange which is a combination of country store and mail order house. That is, there will be stocks on hand and the Exchange will be able to order for us what they do not carry. There will be a vet to inspect the meat and barbers and such like. There will

also be a cantina where we can have beer and Cokes and light lunches. . . .

Oppenheimer concluded by saying that he was "a little reluctant to do too much writing about the details of our life there until people are actually on the job." He added that he had done his best to provide some rudimentary information in the enclosed sheet and that she should let him know "if any of the arrangements that we have made so far are definitely unsatisfactory."

QUESTION	ANSWER
I. Housing	
1. Type of house	2 and 4-family units; bachelor quarters
Number of rooms	3, 4, 6 rooms (1, 2, 3 bedrooms)
Size of rooms	Kitchen, 10 x 14; livingroom 14 x 18
2. Type of heat	Hot air—one furnace for unit; wood fuel
Hot water heater	One heater for unit; wood fuel
Type of stove	Insulated wood & coal range, electric plate
Electricity?	Yes
Phone?	No
3. What furniture will be furnished?	Shelves, kitchen table, beds, chairs . . .
Ice box	Electric refrigerator
Ironing boards	Probably
Fireplace	Yes
4. Rent?	Free for Army;? For civilians*
What price utilities?	" " " " "
5. Will there be servants available for occasional help, especially for heavy work?	Yes
6. Small garden	Yes; water limited in Spring and Fall
II. Geography and clothing	
1. Will it be permissible to describe location in letter?	Altitude 7300; in Southwest

*For Army, most services free. For civilians, may be included in rent to make $150/mo. subsistence. Not yet decided.

QUESTION	ANSWER
2. Mean temperatures	Monthly means from 27.1 in January to 66.7 in July
Rain and snow fall	Heavy rain in summer; light snow in winter; temperate summers.
3. Type of clothing needed	Warm clothing; informal
4. Shoes	Boots for occasional use; skis
III. Community Services	
1. Public laundry charge	Probably a small charge
2. Garbage collection	Yes
3. Carpenters and other crafts	Yes
4. Store	Army Post Exchange
5. Hospital	Run by doctor
6. Mess Hall or cafeteria	Mess hall and café
7. City tax	Probably not

All that winter, Oppenheimer's home at One Eagle Hill Road, just north of Berkeley, became the project's informal headquarters, where he and Kitty played host to a steady stream of visiting scientists. Their house was a handsome, Spanish-style ranch, perched on a steep incline high above the city, with lush gardens and a sweeping view of San Francisco Bay below. The home was expensive and tasteful, decorated with fine oriental rugs and art, and bespoke a lifestyle few academics could afford. Oppenheimer might have calculated that this would play to his advantage, making him appear to be more of a strong, established leader as he surged ahead in his new role as Los Alamos's director. Serber, who had moved back to Berkeley with his wife, Charlotte, to help with the bomb project, had taken up temporary residence in Oppenheimer's garage apartment, and so after office hours, Eagle Hill became the place to gather. Much of the early planning for the new laboratory was done there over Oppie's superb "Vodkatinis," which were expertly prepared and generously distributed.

Standing in front of the fireplace, jabbing at the air with his large pipe to emphasize a point, Oppie would expound on his plan to have about thirty physicists go off together to the desert to build the bomb. According to his utopian vision, most of the support jobs, such as the secretarial and administrative positions, would be filled by the scientists' wives, to keep outsiders to a minimum and assure security. "We shall all be one large family doing vital work within the wire," he assured them, sounding uncomfortably like the army propaganda films running in the local movie houses. Greene recalled late nights and long, rambling conversations between Oppie and Serber about whom he should invite to join the team, which physicists were talented and resourceful enough to tackle the obstacles ahead, and whose intellectual powers might have a catalytic effect on the less qualified. Oppie might as well have been Noah lining up exotic creatures for his ark. When Charlotte Serber overheard them making plans one night, she told them, "You aren't really serious? You fellows don't think you're going to run a project like this, you must be out of your minds!"

Back in Washington, Groves' bright young assistant, Anne Wilson, asked him to tell her what the newly appointed director of Los Alamos was really like as a person. Groves' office was run with steely efficiency by his administrative assistant, Mrs. Jean O'Leary, but Wilson's desk was just outside the door to Groves' office, and as Anne was very good-looking, Oppenheimer had often stopped to chat her up. She found him fascinating. His reputation as a dashing and urbane ladies' man was already legend within the War Department, but she wanted to hear Groves' blunt appraisal. They were traveling to work together that morning as usual, and as was his custom, the general drove while she read the papers aloud to him. He always insisted on beginning with "Mary Haworth's Mail," the *Washington Post*'s high-class advice-for-the-lovelorn column, followed by the sports pages, so that they did not get to the front page until they were practically pulling up to the office. Wilson knew her question was a little impertinent, and had Groves decided to ignore it completely, she would not have been surprised. At the same time, theirs was not the usual secretary and boss relationship: her father was an admiral, and she had grown up around the corner from Groves in Cleveland Park, in D.C., and she often played tennis with him at the Army-Navy Country Club. Groves was used to her boldness, and he teased her about it incessantly. He drove in silence for a few minutes, and when he finally answered, Wilson was so taken aback by what he said she never forgot it. "He has the bluest eyes I've ever seen," Groves told her. "He looks right through you. I feel like he can read my mind."

No matter what one person saw in Oppenheimer, another would assert the opposite was true. But that he possessed a certain brilliance, and a signal capacity for leadership, few would deny. Almost no one was neutral about him. "Oppenheimer was a very clever politician," said Teller grudgingly. "He understood people. He essentially knew how to influence them." He may have been the consummate actor, as his critics contend, questing after power, and calculatingly turning to Groves a face he knew the general would favor. Or he may have had so many masks that he lost track of his true self somewhere along the way. But for all of his intellect and ambition, he could be maddeningly obtuse at times, even careless, as if somehow unaware that the same rules applied to him as to everyone else. It may simply have been that the Manhattan Project was too big an adventure for Oppenheimer not to take part in, no matter his qualms or private misgivings. He had always hankered after a certain kind of authenticity, a defining experience. He had to be "near the center" of things; that very impulse, he once admitted to a friend, which had originally moved him to leave chemistry and Harvard for Cambridge's Cavendish Laboratory, mecca for bright young physicists.

Pressed into service, Oppenheimer rose to the challenge. Eloquent, inspiring, and elusive, perhaps deliberately so, he became the pied piper of Los Alamos. By the end of 1942, he had passionately embraced the bomb project as a means of ending the war, and was using all his wiles and powers of persuasion to entice the most important physicists in the world to leave their jobs, uproot their families, and join him on that lofty mesa in New Mexico. "Almost everyone realized that this was a great undertaking," he later wrote of the hundreds who followed him into the desert:

> Almost everyone knew if it were completed successfully and rapidly enough, it might determine the outcome of the war. Almost everyone knew that it was an unparalleled opportunity to bring

to bear the basic knowledge and art of science for the benefit of his country. Almost everyone knew that this job, if it were achieved, would be a part of history. The sense of excitement, of devotion and of patriotism in the end prevailed. Most of those with whom I talked came to Los Alamos.

FOUR

Cowboy Boots and All

WHEN DOROTHY MCKIBBIN reported to work at 109 East Palace Avenue on March 27, 1943, the morning after the meeting in the lobby of La Fonda, she found Robert Oppenheimer waiting for her on the other side of a shabby screen door. He was as polite and disarmingly solicitous as he had been that first day, but his face was pale beneath the sunburn, and on close inspection, he appeared tense and drawn. She had sensed from the beginning that this was going to be no ordinary job, but that impression was powerfully reinforced by the presence of two Spanish Americans standing guard by the door armed with rifles. From his few comments, it was also clear that Oppenheimer already knew all about her, which was why he had been able to hire her so casually on the spot, without any question or hesitation. Much later, she realized that he had probably already reviewed a full security dossier on her and had strolled across the lobby for the express purpose of looking her over.

The ground-floor offices, which Joe Stevenson had leased the previous week, were housed in a venerable adobe building one block north of La Fonda, in one of the most ancient sections of the old city. The faded blue portal and three-foot-thick walls dated back to at least the late 1600s, when the property was deeded to a Spanish conquistador, Captain Don Diego Arias de Quiros, as a reward for his part in the conquest, and reconquest, of New Mexico. The residence was built to be a fortress, with the stables behind it and a walled garden running nearly two blocks long. There was once a secret subterranean passage that had led to the Governor's Palace. Legend had it that the scalps of Indians had once hung on the weathered vigas outside, proudly displayed by the Spanish settlers who paid a bounty for every head. Nevertheless, Dorothy suspected the office was selected less for its historic value than for its location. While it was just steps away from the main Plaza, the building was somewhat inconspicuous, set back behind a long, narrow courtyard and a shaded grass patio. A heavy, wrought-iron gate at the entrance of the cobblestone passageway discouraged the curious from wandering in. For security reasons, the project's presence was unobtrusively marked by a small blue sign with red lettering:

U.S. ENG-

RS

The wooden placard was so short that the abbreviated form of "Engineers" had apparently not fit on the first line, so the remaining "rs" rather incongruously occupied the line below. If obfuscation was the army's main purpose, Dorothy thought the sign succeeded admirably—she had almost walked right by it. In the coming months, she was often reminded of her first impression of the secret headquarters, as dozens of the world's top scientists failed to notice the tiny sign and passed by the centuries-old courtyard without stopping, blundering around the Plaza for hours in search of the designated contact point where they were to receive their marching orders.

Oppenheimer and his wife had arrived by train on March 15 and were staying temporarily at La Fonda. Priscilla Greene had followed a few days later, accompanied by the Oppenheimers' two-year-old son, Peter, and his nurse. A few other members of his team were also in town, staying at La Fonda or spread among several small area inns. They had been working around the clock to get the project off the ground, but were shorthanded and needed all the help they could get. A quick glance at the chaotic state of their surroundings confirmed as much. The old adobe walls had been freshly whitewashed, the calcimine so new it came off on people's coats.

The rooms were sparsely furnished with an odd assortment of tables and kitchen chairs and piled high with oversized crates and partially unpacked boxes. The whole place resembled the luggage storeroom of a large train station more than an official U.S. government office.

There were five rooms in all, with Dorothy soon installed east of the gate, next to Joe Stevenson, in a cramped space designated as the "housing office." Duane Muncy, the director of the business office, was set up on the west side of the patio. Also on the west side were the lawyer's office and that of the procurement manager, Dana Mitchell, whose services, Dorothy would soon learn, were of the utmost importance to the project. Mitchell was revered for his stores and stocks of equipment, without which the scientists could not work, and his almost magical ability to find and obtain obscure and scarce supplies. The two rooms on the north were occupied by Priscilla Greene, in the small outer office, and Oppenheimer, in the larger back room. Oppenheimer's office had double French doors that led back to a walled garden. When the personnel manager, Edward U. Condon, and his assistant, Mrs. Isabel Bemis, arrived from Berkeley, they were squeezed into the last few feet of space on the rectangle.

All the offices were overcrowded. This was an old adobe, after all, and Dorothy knew the diminutive scale of the rooms was by design because the only source of heat came from the small corner fireplaces. Oppenheimer and company had wedged the maximum number of desks into each room, so they sat elbow to elbow, literally. When the office was really going, it was impossible to hear oneself think, but Dorothy was assured that it would only be a few weeks before they moved into their permanent quarters. They had only two typewriters between them, a small portable Corona that Greene had brought from Berkeley and a somewhat sturdier machine that Stevenson had purchased at Santa Fe Book, the last one the store had in stock. More had been requisitioned, but in the meantime they would have to make do. For the first few days, Dorothy could not make out exactly what she was supposed to be doing. "There was no agenda," she recalled. "No one knew what was to be done or how to go about doing it." She asked Joe Stevenson for directions, and his answer was quick and direct: "Your instructions are never to ask for a name to be repeated, and never to ask a question." The scientists alone knew the exact purpose of the project. The less said about it, the better. At the end of the first week, Stevenson paid Dorothy out of petty cash. As soon as all the paperwork was put through, he told her, she would become an employee of the University of California, which was administering the project. That arrangement enabled the project to issue contracts and pay salaries without ever revealing the exact nature of the undertaking.

Dorothy's job, as it evolved, was to be Oppenheimer's assistant in Santa Fe and operate the small, benign-looking "housing office," which was actually a front for a classified laboratory under construction on a sparsely populated mountaintop thirty-five miles outside of town. She was to meet the arriving scientists recruited to work on the project and brief them on the details of their mysterious final destination, which would not have been disclosed to them when they were issued their travel orders to Santa Fe. Most importantly, using one of two pass machines that would be kept under lock and key in her office, she was to supply them with security passes, without which they would be turned back at the first guard station. Until the pass machines arrived, however, she would have to supply everyone with typewritten letters issued on heavy-bond University of California stationery, complete with three onionskin carbons, each personally signed by Oppenheimer. Last but not least, she was to give them directions and arrange for their safe passage up to "the Hill," the office shorthand for the top-secret installation at Los Alamos.

It was not the first time Dorothy had heard the name Los Alamos. Like Oppenheimer, she was no stranger to that high mesa. Over the years, she had been up the mountain a number of times, though she had difficulty imagining the boarding school with its scattering of rustic buildings as the site of an advanced scientific laboratory. She was great friends with Peggy Pond Church, a fellow Smith alum and well-known Santa Fe writer, who was the daughter of the school's founder and was married to Fermor S. Church, a Harvard graduate who was a former headmaster. The Ranch School had been the dream of her friend's father, Ashley Pond, an idealistic businessman who had established the school in 1917 as a place where affluent city boys could "learn by doing" and receive a rugged, outdoor-oriented education that would teach them to ride horses, hunt, and fend for themselves in the wilderness in the tradition of Theodore Roosevelt's ideal of the vigorous life.

Pond had grown up in Detroit, Michigan, and had been afflicted throughout his school years by chronic bronchial infections that had disrupted his education and exacted a toll in poor marks and failure. As the only surviving son of an accomplished man, Church wrote that Pond "carried the burden of his father's disappointment through his boyhood." After contracting an almost fatal case of typhoid during the Spanish-American War, he was sent out west to regain his health. While in New Mexico, he became an active outdoorsman and was inspired by the wild, beautiful countryside to use his inheritance to create a school dedicated to the mental and physical well-being of boys. After a failed first attempt, he finally managed to raise the capital to acquire a distant settlement situated on the Pajarito ("little bird") Plateau, at the edge of a pine forest. At the time, it consisted of little more than a homesteader's farmhouse, some four hundred acres of pasture, and a large watering hole used for cattle, which was later named "Ashley Pond" in honor of the school's founder. In the "Twenty-Fifth Anniversary Report of the Harvard College Class of 1921," Fermor Church wrote: "The unique features of Los Alamos School were those which were indigenous—its program and its spirit stemmed from its surroundings. No circumscribed artificial campus life could develop here before the beckoning attraction of forest, stream, canyon, mesa, and mountain peak—with horse at hand. And amid the ruined cities of early pueblo and cliff-dweller we felt we knew that the thing men live for never dies."

By the summer of 1932, when Dorothy and her son first went up to Los Alamos for a two-week holiday with Peggy Pond Church and her family, the school was a thriving community, with an enrollment of forty students. Staff members, their families, and ranch employees accounted for another hundred or more people living on the plateau. It could not have been a more picturesque setting, framed by the hazy purple cardboardlike mountains in the distance and, in the foreground, the brimming pond and massive main school building, the Big House, rising impressively from the carefully tended green lawns and tennis courts. Across the way, the handsome, three-story Fuller Lodge, designed by Dorothy's old friend John Gaw Meem in 1928, had been constructed from eight hundred hand-picked ponderosa pines, with massive vertical logs forming the columns for the stately balcony that flanked two sides of the building. The lodge's large dining hall boasted a nineteen-foot-high beamed ceiling and wrought-iron chandeliers, and gazing down from above the enormous stone fireplace was the mounted head of a New Mexico elk. The only other landmarks were the tall, wooden water tower and the Trading Post, a Western-style building, made of rough, stained lumber, that served as the local store and carried canned goods and other staples, in addition to school supplies, lanterns, blankets, hardware, sacks of feed—almost everything the self-sufficient little community required. The school grounds also included a stable for horses, corrals, grazing fields, and, stretching eastward across the mesa as far as the eye could see, hundreds of acres of growing crops, including alfalfa, barley, corn, oats, and beans. At one end, just north of an old Indian ruin and near the edge of the canyon, was a row of faculty residences with the forest at their back door.

Dorothy had been saddened to hear that the school had been closed by order of the War Department and that its buildings and the surrounding public land, 54,000 acres in all, had been cordoned off by miles of steel fence and barbed wire as a military reservation. She remembered her visits there fondly, and the cheerful sight of the young boys in their khaki uniforms, jaunty bandannas, and Dakota Stetsons. They looked brawny and browned by the sun, and she knew from firsthand experience the heartening effects of sleeping outside year-round on screened porches, as they did in the Big House, with only those striped awnings to protect them from the rain that poured in through the open windows. She realized that all this would end now and that Peggy Pond Church, along with all the other ranch families, would have to surrender her home and abandon the mesa where she had lived for the better part of her life.

She later learned that on December 7, 1942, exactly one year after Pearl Harbor, the school had received a letter from Secretary of War Henry L. Stimson that was in effect a blunt notice of eviction. A. J. Connell, the school's director, had read it aloud at a special meeting at the Big House: "You are advised that it has been determined necessary to the interests of the United States in the prosecution of the War that the property of Los Alamos Ranch School be acquired for military purposes." They were further asked "to refrain from making the reasons for the closing of the school known to the public at large." Even so, most of the heartbroken students promptly told their parents the school would be used as a military reservation and demolition range. The Christmas holidays that year had been canceled, and the boys had to cram the full year's course of study into the remaining weeks, so that the last four graduates could be awarded their diplomas before they had to vacate the premises.

The school officially closed on February 8, 1943, barely a month before the first scientists arrived. But as Peggy Pond Church wrote in a poignant memoir of prewar Los Alamos, which she dedicated to Dorothy, in the days before they left, the sounds of planes buzzing overhead disrupted classes and men in military uniforms had begun to defile the unspoiled plateau:

> Bulldozers moved in, and other weird machines roared up and down digging ditches for the foundations of future buildings. Everything was conducted in an element of extreme haste and mystery. Civilian visitors were conducted on tours of inspection everywhere, even through our homes. One day I recognized Dr. Ernest Lawrence, whom my husband and I had met one summer in California. He seemed strangely different when I questioned him about mutual friends, and broke away as quickly as possible from my attempts at conversation.
>
> Another afternoon I was introduced to a young looking man by the name of Oppenheimer. Cowboy boots and all, he hurried in the front door and out the back, peering quickly into the kitchen and bedrooms. I was impressed, even in that brief meeting, by his nervous energy and by the intensity of the blue eyes that seemed to take in everything at a glance, like a bird flying from branch to branch in a deep forest.

The present was rudely imposing on the past, but there was no time for regret. Three thousand construction engineers had been working since mid-December to build the new laboratory buildings and living quarters, hastily slapping together the structures using rough lumber and building paper. The buildings were contracted as piecework, so no one builder knew the size of the project or the entire construction plan. The working conditions were confusing and difficult, and hampered by the freezing weather, the project was running behind schedule. In Priscilla Greene's blunt assessment, the site was "a mess." She had accompanied Oppenheimer on an inspection tour just after they arrived in Santa Fe and was stunned by the dismal spectacle that greeted them. Plumes of smoke marked the town site, and a layer of the soft New Mexico coal the workers burned for heat blackened the snow. The building crews' temporary huts stood to one side, like a row of dreary tenements. "It was a pretty appalling place," said Greene. "It was windy, dusty, cold, snowy a little bit at that point, and nothing was finished. It looked kind of amazing, though not as if one was ever going to be able to move into it."

The Bethes, who had come early by request—Rose arriving a week before her husband because Oppie had asked her to help assign living quarters to the incoming personnel—were similarly dismayed. "It was a shambles," recalled Hans Bethe. "It was a construction site. You stumbled over kegs of nails, over posts, over ladders."

Their immediate crisis was housing. Dozens of scientists were due to arrive imminently, and there was no place for them to stay. Many of them had wives who were pregnant or had babies and small children in tow. A handful of bachelors, along with the Serbers, had moved straight up to the Hill and were making do in the Big House, formerly the school dormitory. It was "a little rough," Serber noted, adding in his usual droll understatement, "There was only one big bathroom in the entire house, and two or three fellows were embarrassed by walking in on Charlotte while she was taking a shower." One of Rose Bethe's first jobs was trying to find locks for all the doors, so they would not all have to live quite as communally as the schoolboys had. The makeshift arrangements on the site would clearly not suffice for the more senior physicists Oppie had cajoled into joining his secret venture, and it was imperative the staff find lodging for them soon, preferably places large enough for several families, so the laboratory personnel would not be scattered far and wide and the whole county alerted to their presence. As a March 24 memorandum from Oppenheimer instructed, "considerations of security require that the housing be carried out according to a plan and not left to the individual." There were not enough hotels in town, and a visitor who ventured into the outlying valleys would soon find Mexican villages where the basic scheme of existence had not changed much in two centuries. Greene, who had never been to Santa Fe before, was overwhelmed and turned to Dorothy for help. "We were desperate," she said. "I don't know what we would have done without her."

At first, Dorothy found it somewhat unbelievable that all these top-notch scientists would need her assistance, when at Smith she had managed to earn not one but two zeroes in physics. Yet there was an amateurism about the whole enterprise that seemed distinctly at odds with an important government project. Pandemonium was the order of the day. "This quality may have been dangerous, and frequently was," she reflected in her memoir, "but there was something endearing about its lack of sophistication. Most of those who came to the Hill were boys, brilliant boys, but boys nonetheless. The average age was 24 years old. At 38, Oppenheimer was one of the old ones."

Fortunately for the scientists, Dorothy knew every road, pueblo, village, rancher, shopkeeper, carpenter, and craftsman for miles around. In her ten years with the Spanish and Indian Trading Company, she had dealt with merchants from all the neighboring pueblos and villages. There was no nook or cranny of the region where she did not have a contact or friend to call. She was, in her own modest way, a well-known and respected local personage. With her Girl Guide forthrightness, and a calm demeanor that allowed none of the surrounding bedlam to fluster her, she was exactly what they needed. She pitched in immediately and got on the phone to a small dude ranch in the Nambe Valley. In her fine, educated Smith voice, underscored with an irrepressible Missouri twang, she politely informed the owner, "I'm sorry, you cannot accept any reservations. *We* are taking this over." She offered no explanation, brooked no argument, and in short order had rented out three more ranches only ten to twenty miles outside town.

"No one quite knew what was going on or what to do," recalled Ben

Diven, a graduate student from Berkeley, who was among the handful of project members who were already on the scene when Dorothy opened the office at 109 East Palace. "The contractors were late, so Dorothy sent Hugh Bradner all over hunting for housing. She was just wonderful and wanted to do whatever she could to help. We all had to figure it out as we went along, but she learned very quickly." Bradner, a young Berkeley physicist who served as Oppenheimer's gofer and jack-of-all-trades, remembered that people soon took to calling Dorothy their guardian angel. "She took care of all of us," he said. "She would always say how lucky she felt being that she was just this little girl from Kansas City, but Oppie was lucky to have found her."

As the bewildered scientists arrived at the small adobe train station in Lamy, blinking in amazement and dismay at the deserted landscape and the tumbleweeds that blew back and forth in the high wind, Dorothy greeted them with a bright smile and did her best to soften their disappointment. After consulting with Oppie, she farmed each of them out to the different locations. Scientists without families or where both were employed were put up in relatively comfortable quarters such as the Ancon and Del Monte Ranches. Those with children, and wives who could help prepare meals and contribute to the maintenance of the facility, were assigned to the less-expensive Cable and Schuyler spreads and got a quicker immersion in the hardships of ranch life than Dorothy reckoned they had anticipated. She hired local managers to help operate each location and ran around making sure the newcomers had everything they needed. When Greene stumbled into the office one day later looking pale and exhausted, Dorothy packed her off to the Del Monte Ranch for a rest. "I felt terrible and everyone, including me, thought I was having a nervous breakdown," recalled Greene. It was the measles. "I was very relieved when I was diagnosed, even after I started to itch."

Despite the outbreak of German measles, Ed McMillan's wife, Elsie, arrived a few days later with their baby daughter, Ann. In a book about her wartime experience at Los Alamos, Elsie described the kind reception she received after the long, uncomfortable journey on a packed train: "At the Santa Fe office, Dorothy McKibbin welcomed me and answered the few questions I had at that hour before the drive to our temporary, luxurious quarters." Dorothy had secured them a room and bath in the guesthouse of the Ancon Ranch, which was owned by Major John Post, and his wife, Marian. Two other families, Pat and John Wieneke and Dorothy and Ken Jensen, who were newlyweds, were also boarding there. The Posts were ex-service and so knew there was a military base being built at Los Alamos. Never asking what so many scientists were doing there, the Posts wined and dined their guests in style and lent the women a car so they could explore the Rio Grande Valley. Dorothy "smoothed the way for us all," wrote Elsie, "especially the soldiers sent to guard us, and often completely lost in Lamy, unbelieving that they had landed at the right spot for their already puzzling assignments. Only a Mrs. McKibbin could have assured and helped them on their way up the forty miles into a strange land inhabited by 'crazy' scientists and their 'crazy' families."

Some of the early arrivals could not believe that Oppenheimer had chosen such a godforsaken place, and it felt as if they were sneaking off to the ragged edge of the world to complete their sinister task. It took nearly two hours of driving on State Highway 4, a narrow, two-lane roller coaster, full of dips and hairpin turns, to reach the guarded military gate, beyond which the 54,000-acre Los Alamos compound sprawled. The final ascent up to the site was so steep and boulder-strewn as to give pause to all but the most stalwart. One look at the "sheer drop right beside us," recalled Elsie McMillan, and "I was scared." It was not reassuring to know that the Otowi Bridge, the old railroad trestle spanning the Rio Grande, was considered "too fragile" to carry the army's trucks and buses, which were forced to take State Highway 30, an even more circuitous route, by way of Espanola. Even Groves was unnerved and ordered the road paved, but the steady traffic of army trucks soon reduced it to rubble.

Others were left slack-jawed by their first glimpse of the barren mesa, only six miles by two, which the bulldozers had churned into a sea of mud. Joe Stevenson had unintentionally made things worse by sending out a fact sheet to all the senior laboratory people that, in language worthy of a chamber of commerce, claimed, among other things, "Los Alamos is situated on the shores of a small lake" and featured commissaries, beer halls, theaters, and other attractions. If Ashley Pond had once been a scenic attraction, after the long winter and months of construction, it was now little more than a brown puddle. The crude timber buildings and ramshackle Quonset huts looked as far removed from a modern research facility as anything they could have imagined.

Despite his lucidity in the classroom, Oppenheimer was hardly the most sensible of men, and he had a dreamy, scattered side that over the years had led to legions of absent-minded-professor stories of lost cars and forgotten dates. This was a man who by his own admission was so out of touch with everyday life that he never read a newspaper or listened to the radio, and only heard of the stock market crash of 1929 long after the event. He struck even his most ardent admirers as singularly ill-equipped to handle such a daunting administrative and organizational challenge, and the sheer folly of the Los Alamos setting confirmed their worst fears. Robert Wilson, a twenty-eight-year-old physicist from Wyoming who had achieved a reputation for brilliance in only a few short years at Princeton, remembered being troubled at the outset by Oppenheimer's naïveté. Wilson had just finished reading Thomas Mann's *The Magic Mountain,* and while he thought the whole idea of disappearing to a mountaintop laboratory sounded incredibly romantic, some of Oppenheimer's ideas struck him as impractical. When he would question the wisdom of a particular strategy, such as that of scientists becoming soldiers in uniform, Oppie would get "a faraway look in his eyes." Then he would wax philosophical about how "this war was different from any war ever fought before," that it was being fought for "the principles of freedom" and being fought by the "people's army." When Oppie "talked like that," Wilson recalled, "I thought he had a screw loose somewhere."

In the months leading up to the project, a number of the physicists who had signed on to the project had already begun to have serious doubts, not only about Oppie's choice of location, but also about the lack of planning and thought being given to basic concerns, everything from the number of people required to staff the laboratory and the division of labor to a reasonable operating schedule. John Manley worried that Oppenheimer failed to grasp just how big a job he had taken on: "What we were trying to do was build a new laboratory in the wilds of New Mexico with no initial equipment except the library of Horatio Alger books or whatever it was those boys in the Ranch School read, and the pack equipment that they used going horseback riding, none of which helped us very much in getting neutron-producing accelerators."

All during the early months of 1943, Manley tried to rustle up vital equipment from university laboratories around the country. He arranged for many pieces to be packed up and shipped to Los Alamos, including two Van de Graaff generators from Wisconsin and the cyclotron from Harvard, as well as his Cockcroft-Walton accelerator from Illinois. These were enormous, heavy machines, and getting them was not a simple task. He could not help wondering whether, "if Oppenheimer had been an experimental physicist and known that experimental physics is really 90 percent plumbing and you've *got* to have all that equipment and tools and so on, he would ever have agreed to try to start a laboratory in this isolated place." But there was another matter than bothered him as much, if not more: the organization of the new laboratory. "There were several reasons for concern," he reasoned. "People would tire in a very isolated place. They would be working under extreme time pressure and if there was not good organization from the point of view of the technical work, all of the services, the responsibilities, the whole enterprise could just really go flop."

Manley and Wilson kept needling Oppenheimer to focus on the organization of the laboratory so when they all assembled on the mesa, it would not be a madhouse. Each of them knew from separate visits to the site that the lack of planning was affecting the pace of construction, which was way behind schedule. Finally, after their repeated entreaties and pestering phone calls failed to illicit a response, they decided to team up and together present their concerns to Oppenheimer at Berkeley:

> We insisted that decisions had to be made, that people had to know what to do, when to come to Los Alamos, that priorities had to be established, that we had to come to a realistic understanding of where we stood with the Army people. We wanted a little organization, we wanted to know who was to be in charge of what, not just vague talk about the scientific problems nor the

> even vaguer ideas about democracy. There were immediate problems to be faced and, from our point of view, Oppy was not facing them.

After nagging him all day about his "indecisiveness," they continued to bug him for the better part of the night during a party at his home on Eagle Hill Road. "Typically," recalled Wilson, "the day's technical discussion drifted into the evening's socialities. The driest of dry martinis mixed by the hand of the master, sophisticated guests, gourmet food (but on the scant side), an amorphous buzz of conversation, smoke, alcohol, all these were the inevitable ingredients of an evening at the Oppenheimers. Manley and I never let up for a moment, and eventually we got to him. He exploded into a fit of cursing, acrimony and hysteria that left me aghast." Wilson departed, fearing that their critical assault had backfired, and that he had seen the end of his relationship with both Oppenheimer and the project, "that all was lost."

As was so often the case with Oppenheimer, the contrary proved to be true. One day in January, Manley, still frustrated at not having heard from Oppenheimer about the organization of the laboratory, decided to fly to Berkeley and confront him one last time. When he walked into his office in Le Conte Hall, Oppenheimer barely uttered a greeting and, without looking up from his desk, shoved a piece of paper at Manley and said, "Here's your damned organization chart."

Manley's eyes passed down the paper in some astonishment. "I checked through it and saw it not only covered big names in physics like Bethe and [Emilio] Segrè, but also practical things like organic chemistry and the stockroom," he recalled. "About the stockroom, he had got Dana Mitchell of Columbia for it, absolutely the country's best. I still don't know how he did it or what persuasion he used." For the first time, Manley believed it was just possible that Oppenheimer knew what he was doing.

The physicist Samuel Allison was not so easily persuaded. He was wary of Los Alamos's director from the start, recalling that when as a graduate student he had first encountered Oppenheimer at Berkeley, Oppie had been immersed in the Mahabharata, a story in the epic

Bhagavad Gita, which he was reading with Arthur Ryder, the chairman of the Sanskrit department. Among the first to be recruited for the project, Allison had accompanied Oppenheimer to Los Alamos in late December to plan the layout of the technical buildings that had to be erected and was taken aback by Oppie's quixotic view of the project:

> On the Mesa he and I sat down and planned the laboratory. He showed me what he called an organization chart for a hundred personnel. I looked at it and felt sure that something was wrong, but I didn't know what. The best I could do was poke at random.
>
> "Where are the shipping clerks," I asked.
>
> He gave me a thoughtful sympathetic look, "We're not going to ship anything," he answered.

Allison concluded that not only had Oppenheimer "completely underestimated the size of the installation," he also suspected that Oppie had allowed sentiment, rather than common sense, to dictate his selection of the site. "I thought the idea of a desert center was a mistake," he noted. "It would have looked more sensible to me to put it in a big industrial district. Certainly it would have been more sensible economically, but there was Oppenheimer's love for that country."

It was a topic of wry humor among Oppie's close friends and associates that he had ever managed to talk Groves and the Army Corps of Engineers into approving the Los Alamos site in the first place. The Serbers were better acquainted with Oppie's long love affair with the desert than most, having first visited his primitive ranch in the Pecos Valley in 1935 and having learned then of the health problems that had driven him out to the desert in his youth. During a holiday in the Harz Mountains in Germany after high school, Oppie had come down with a severe case of trench dysentery that over time became aggravated colitis. Too sick to enter Harvard in the fall of 1921, he passed a gloomy year at home, after which his parents hired Herbert Winslow Smith, a strapping young English teacher from his private New York day school, the Ethical Culture School, to chaperone the wan and melancholy teenager on a restorative sojourn out west. Smith suspected that some of the bright young boy's ailments might well be psychosomatic and dispensed with the prescribed treatment of rest in favor of camping and horseback riding.

While traveling through the mountains in New Mexico, Oppenheimer and Smith stayed at the fashionable Los Pinos Ranch in the Pecos Valley, north of Santa Fe, which during the heyday of dude ranches was known as a particularly elegant establishment with a high-class clientele. Los Pinos's appeal lay in its rustic lodge, glorious setting, and lovely proprietress, Katherine Chaves Page. The beautiful, aristocratic, twenty-eight-year-old daughter of an old Santa Fe family and prominent statesman, Page had studied in the East, held a master's degree from Northwestern University, and was newly married to a New Yorker named Bernard Winthrop Page. An accomplished horsewoman, Katherine Page knew the high country like the back of her hand, preferring to get off the beaten trails and explore out-of-the-way places. She utterly captivated the impressionable eighteen-year-old Oppenheimer, encouraging his interest in riding, as he showed an unusual feel for horses. Together, they went on long excursions into the Jemez and Sangre de Cristo Mountains. On one leisurely trip, the two came across a beautiful blue-green lake hidden high in the upper Pecos, and at the urging of the smitten teenager, they called their discovery "Lake Katherine," as it has been known to hikers ever since.

In the intimacy of quiet trails and campfires, Oppenheimer and Page struck up a close friendship, and for the rest of her life they would keep up a warm correspondence and he would return to the mountains, and to her, as often as he could. It was on one of their most memorable rides that Katherine Page took Oppie up to the Los Alamos Ranch School. "I first knew the Pajarito Plateau in the summer of 1922," he recalled later, "when we took a pack trip up from Frijoles and into the Valle Grande." It was a trip he would make again and again in the years to come, perhaps feeling some connection to the frail boys he saw playing on the fields.

During that trip to the Southwest, the solitary Oppenheimer permitted himself his first really close friends. He solidified his budding friendship with a high school classmate, Francis Fergusson, who was also bound for Harvard, and whose family was from Albuquerque. During his visit with Fergusson, he met another gifted student, Paul Horgan, and they instantly became "this great troika," as Horgan remembered, all "polymaths," precocious and driven to excel in their chosen fields—Oppie as a physicist, Fergusson as a literary scholar, and Horgan as a novelist and Pulitzer Prize–winning historian. "[Oppenheimer] was the most intelligent man I've ever known, the most brilliantly endowed intellectually," said Horgan. "He had this lovely social quality that permitted him to enter into the moment very strongly, wherever it was and whenever it was. So one didn't see him as eventually the incredibly great scientist or the celebrity at all. He had a great superiority but great charm with it, and great simplicity at that time." Even in the early days his young companions were aware that Oppenheimer battled "deep, deep depressions," during which he would withdraw and become incommunicado for a day or two. To Fergusson, it seemed that out west Oppie found some release from the pressures back home, "his Jewishness and his wealth, and his eastern connections, and [that] his going to New Mexico was partly to escape from that."

The change in Oppenheimer's outlook wrought by his time in the desert was evident in the witty, confident tone of a letter to Smith written during his freshman year at Harvard. Noting that he was amused to hear that his old teacher had decided to go west with "two new neurotics," he added wistfully, "I am of course insanely jealous":

> I see you riding down from the mountains to the desert at that hour when thunderstorms and sunsets caparison the sky; I see you in the Pecos "in September, when I'll want my friends to comfort me, you know," spending the moonlight on Grass mountain; I see you vending the marvels of the upper Loch, of the upper amphitheater at Ouray, of the waterfall at Telluride, the Punch Bowl at San Ysidro—even the prairies of Antonito—to philistine eyes.

In 1928, during his postdoctoral studies in Europe, a bout of tuberculosis brought Oppie back to the Pecos Valley for an extended stay. At Katherine Page's suggestion, Oppie and his brother leased a homestead in Cowles, a mile or so from her guest ranch, eventually purchasing it outright and turning it into a shared retreat. Katherine Page christened it Perro Caliente, Spanish for "hot dog," and a more mellifluous translation of the idiom Oppie reportedly uttered upon laying eyes on it for the first time. Once again, the climate proved therapeutic, and after a summer of "miscellaneous debauch," his condition had improved enough to allow him to return to Europe as a Rockefeller Foundation fellow. Oppenheimer divided his fellowship year between three European centers of physics—Leiden, Utrecht, and Zurich—but when a stubborn cough worsened even in the invigorating altitudes of Switzerland, it was recommended that he return a month early to the United States to prepare for his unusual joint appointment at Berkeley and Caltech.

Throughout the 1930s, Oppenheimer spent part of every summer in the mountains of New Mexico, and the Serbers, who were frequent visitors, took part in the spartan and vigorous life he led there. They discovered that for someone so slight, Oppie was deceptively strong, apparently immune to cold and hunger, and a fearless rider. He had a beautiful quarter horse, aptly named Crisis, that was every bit as high-strung as its owner. The "ranch," as he called it, was a bare-bones operation, consisting of little more than a rough-hewn log cabin and corral with a half dozen horses, situated on a meadow at the base of Grass Mountain. There was no heat, except for the large wood stove in the kitchen, and no plumbing, save for the makeshift outhouse at the far end of the porch. In his autobiography, Serber noted that Oppie had formed the habit of sleeping outdoors during his spot of TB and seemed not to notice that his guests froze at night on their cots on the porch and that, at two miles above sea level, the smallest exertion left them "gasping for air."

Although the Serbers were inexperienced riders, Oppie insisted on taking them on weeklong expeditions into the Sangre de Cristo Mountains, rising up 13,500 feet to the Truchas Peaks. "Always very solicitous about the horses and concerned they didn't bear too much weight," Serber wrote, Oppenheimer would pack plenty of oats for the animals, but barely enough provisions to keep his two-footed companions from starving to death. Oppie, who often neglected to eat, thought nothing of heading into the mountains alone for days on end with only a few chocolate bars in his pocket. In his autobiography, Serber repeats a telling, if apocryphal, description of a typical Oppie excursion as told to him by Ruth Valentine, a Pasadena psychologist and mutual friend: "It is midnight, and we are riding along a mountain ridge in a cold downpour, with lightning striking all around us. We come to a fork in the trail, and Oppie says, 'That way it's seven miles home, but this way it's only a little longer, and it's much more beautiful!'"

It was in the spring of 1940 during a quick getaway to New Mexico that Oppie asked the Serbers to bring the married Kitty Harrison along, saying in passing, "I'll leave it up to you. But if you do it might have serious consequences." Kitty came, and Oppie rode off with her for an overnight visit to the Los Pinos ranch. The following day, a bemused Katherine Page trotted over to the ranch and hand-delivered Kitty's nightgown, which had been discovered that morning under Oppie's pillow. That fall, Kitty went to Nevada to obtain a divorce. On November 1, the same day her divorce was finalized, she and Oppenheimer were quietly married in Virginia City, Nevada. It all happened so quickly that when Oppie took Serber aside and disclosed that he had "some news," Serber automatically assumed he had tied the knot with his longtime girlfriend, Jean Tatlock, with whom he had been having a tempestuous, on-again, off-again affair. Tatlock suffered from severe mood swings and was unstable. She had broken off with him once and for all a year earlier, and Oppie had been mourning her loss ever since. "I just gaped at him, trying to figure out whether he'd said Kitty or Jean," recalled Serber. "Charlotte had to kick me to remind me to make the appropriate salutary noises."

The Serbers were not the only ones Oppenheimer invited out to spend time with him in his favorite habitat. Over the years, many of his colleagues ventured to the ranch, including Ernest Lawrence, Ed McMillan, George Gamow, George Placzek, Victor Weisskopf, and Hans and Rose Bethe, whom he happened to run into one day hiking in the area and brought home. Oppie would treat his guests to one of his incendiary chili dinners, known to some as "nasty gory," and his brother, Frank, who played the flute almost as well as a professional, provided the entertainment. As a rule, talk of physics was forbidden, with exceptions made for special guests.

Sustained by the grandeur and serenity of the mountains, Oppenheimer had conquered first illness and then loneliness. "My two great loves are physics and desert country," he once wrote a friend. "It's a pity they can't be combined." As in most things up to then, Oppenheimer eventually got his way.

FIVE

The Gatekeeper

FOR THE FIRST FEW WEEKS in Santa Fe, Oppenheimer and his key staff worked out of the office at 109 East Palace Avenue in the early mornings and made daily trips up to Los Alamos to inspect the progress of the construction. "The laboratories at the site were in a sketchy state, but that did not deter the workers," Dorothy wrote of those hectic early days. "In the morning buses, consisting of station wagons, sedans or trucks, would leave 109 and pick up the men at the ranches and take them up the Hill. Occasionally, a driver would forget to stop at one or another of the ranches and the stranded and frustrated scientists would call in a white heat."

The "mañana spirit" indigenous to the area did not help matters. The local Spanish American drivers never understood why they had to make so many round-trips up the steep-sided, winding road, or why everyone was in such a hurry to get to such a barren settlement. Dorothy could never think of a good reason and instead tended to fall back on the project's overused catchphrase, "Don't you know there's a war on?" The drivers maintained such a slow, meandering pace that some scientists swore they would go mad. To hurry things along, Hugh Bradner commandeered anything with wheels and helped ferry men and equipment up to the site. Watching the motley-looking caravan come rolling back at the end of the day, Dorothy could only shake her head. It reminded her of what she had read about the Battle of the Marne in World War I, when the Parisians jumped into anything that could move and "hordes of taxis drove the French soldiers to the front."

Since none of the eating facilities on the mesa were operational yet, one of Dorothy's first tasks was to procure the two dozen picnic lunches every day that were taken up to the scientists for their noontime meal. Because so many of the battered cars consigned to carry the lunches were in poor condition and frequently waylaid by flat tires, just feeding the physicists working on the site during those first few weeks proved a challenge. In order to avoid any unnecessary questions about their activities, Dorothy was careful not to buy all the sandwiches in one place; she went all over Santa Fe in search of different restaurants and cafés. It was a snowy spring, and given the wet weather, she knew any local shopkeeper "would have thought you were crazy to have a lot of boxed lunches." She politely but firmly rebuffed the curious and, when required, told some tall tales to cover her tracks. Dorothy did not like lying, but became adept at it.

Oppenheimer's presence at 109 East Palace, and his importance to the project, not to mention the many classified documents he kept in his office, meant they were under surveillance at all times by army intelligence known as G-2. It was ironic that in this most secret of projects, Groves' field safe was the only secure file in the director's office and therefore by necessity served as a combination bank vault and lockbox for all the project's valuables, including reserves of cash, confidential reports, and registered mail, as well as precious laboratory materials, including platinum and gold foils. Despite the gravity of the situation, Dorothy could not help laughing at the sight of the G-2 agents, whom she learned to spot "a mile away." Dressed in their matching three-piece suits and wingtips and positioned at the street corners, casually leaning against the drugstore, and loitering on the sidewalk in front of the hotel, they stood out in sharp contrast to the other pedestrians on the street. They wore snap-brimmed felt hats in winter, and when the weather turned warmer, snap-brimmed straw hats. They cruised the Plaza in identical black Chevy sedans, and Kevin told her that at night they parked in neat rows behind the Chevrolet garage at the edge of town. She was none too amused, however, when two burly security guards manhandled her son at the entranceway to 109 not long after she started working there. When Kevin sauntered into the office after school one day in hopes of catching a ride home with his mother, they pulled him up short with a stern, "And where do you think you're going, sonny?" After pleading with the guards to check with Mrs. McKibbin inside, he had the satisfaction of seeing them told off by his mother, who rushed out onto the patio flushed with anger. "She got them set straight pretty quick," he said. After that, Kevin, who was fascinated by guns, took to dropping by after school every afternoon and became an expert on the different caliber guns the MPs carried.

As the weeks went by, the town was crawling with clean-cut young FBI agents. No matter how well Dorothy got to know some of the men, the rule held that if they passed on the street, there could be no sign of recognition. If a suspicious person showed up at the office, or someone pestered her with too many prying questions, she had only to make a quick call to G-2 and one of the agents would be on the trail "by the time he hit the street." For someone new to the spy game, Dorothy thought she showed great composure and believed she scared off her share of snoops. It seemed like a game at first, and she could not help being imaginative about German spies. But after a few unnerving encounters, she learned to be grateful for the added protection and the watchful gaze of her big gray standard poodle, Cloudy, who accompanied her to work every day.

G-2's secret office was located in the old post office across the street. The army had a separate office nearby in the Bishop Building, but there were plenty of mix-ups with people reporting to the wrong address, and they could not be too careful. On one memorable occasion, security called and alerted her that a suspected German operative was in the Plaza. "There is a spy coming your way and she's dressed in an American tweed suit and speaks very good English," the G-2 agent informed her, adding that she was in the company of an army sergeant. Dorothy decided she had better take a look for herself. She made a "constructive little trip" around the Plaza and peeked into one or two shops on Palace Avenue before spotting the woman in tweed, who was "very nice looking." Dorothy hurried back to her station and was sitting at her desk when they came in. The woman demanded a pass for the site, explaining that she "had a ride with the sergeant." Dorothy said pleasantly, "Well, that's nice, and who are you [here] to see?" When the woman could not name the person who had requested she visit the site, Dorothy told her, "Well, I'm awfully sorry. I cannot possibly issue a pass to anyone unless I have instructions from the Hill to do so." They went back and forth this way for five or ten minutes, the woman's voice rising and the sergeant just standing there, while the security guards at the door looked on impassively. Dorothy stood her ground, and finally they stalked out. "She didn't get up to the Hill," Dorothy noted with satisfaction.

Dorothy knew that most people were unaware of how many agents, or "creeps," as they were sometimes called, were in town. Operatives rode the trains and buses in and out of Santa Fe, and a soldier from the Hill, going home on furlough, might find himself engaged in casual conversation with another GI in a smoker. If one of the soldiers was indiscreet, he might find himself in the clutches of detectives before he reached home. In such cases, he would be packed off to some rear-echelon post without so much as a court-martial. Even a closed trial, it was pointed out to her, would involve records and stenographers. Instead, the soldier would be sent to a commanding officer with secret orders assigning him to a remote Pacific island where there was no chance of being captured. Suspected agents were not immediately arrested, but put under twenty-four hour surveillance, so security could follow their trail and see who their friends were. As an example of G-2's vigilance, Dorothy recalled the time a visitor made a random remark about machinery that caught G-2's attention. She later learned he was "tailed" for 1,500 miles before finally being cleared. She heard stories about tipplers sounding off in bars as to what was going on up at the Hill while an operative stood at their elbow taking notes. If the loudmouth turned out to be a young laboratory staffer, he was hauled in front of G-2 for violating security and given harsh punishment. If it turned out he was just a rancher who had imbibed one too many, he was saved from a good grilling by the fact that he was plainly talking through his hat.

Despite all the precautions and checkpoints, security agents worried constantly about espionage and the chance that an enemy agent would slip through their fingers. "We were haunted by G-2," she told a reporter as her thoughts turned back to when the threat of spies was on everyone's mind. "We needed to be. We were in a very dangerous spot as the frontier to Los Alamos."

The hours were long, and the pace fast, relentlessly so, as though "a spark was lighted day and night." But the scientists' sense of urgency and anticipation was contagious. Everything had to go through the office on East Palace, with many of the rules and ways of doing things improvised along the way. In the early days of the pass system, before the proper equipment arrived, Dorothy had to supply typewritten letters of identification to everyone going to the site, whether the person was a truck driver with a single load or a regular member of the staff. To top it off, because the letters were usually folded, read, and refolded repeatedly throughout the day, and carried in sweaty back pockets, a collection of torn illegible passes were turned in each week with a request for replacements. As Charlotte Serber, who was helping out in Oppie's office, recalled, "They were a nuisance to type since no erasures were allowed, and the system resulted in writer's cramp for the director, and bad tempers for the typists."

The phones rang incessantly. None of the lines were connected interoffice, so whenever Oppie got an important call, Dorothy would have to jump up and search the premises, drag him out of whatever office he happened to be in and then out across the courtyard, and force him to attend to the business at hand. When Oppie was on the Hill, he called down several times a day, usually inquiring about an overdue scientist and sending her on a frantic search for the missing individual. Usually, it turned out that the person was stuck out at a ranch in the valley and had been waiting for hours for one of the army's antediluvian buses to rattle by. Dorothy would promise to do her best to rustle up some sort of transport, and she cheerfully advised physicists who had never in their lives been on a horse, "If not, I'm afraid you'll just have to go on practicing your riding for a while."

Communication with the site was not easy, and the arduous process was enough to make an impatient scientist "tear his hair" while the operator repeated endlessly, "The line is busy." The only existing telephone at the school was a primitive ranger's phone that had a dozen parties on it and was operated by vigorously turning the small handle. The telephone line itself was made of iron wire, and there was no telling how many years ago the Forest Service had laid it down over the thirty-five miles between Los Alamos and Santa Fe. Charlotte Serber, who was working in the Tech Area one afternoon, went to answer the phone during a violent spring thunderstorm. Just as she reached for it, a bolt must have hit, and she saw a spark jump from the line to a lamp cord just inches away. Following that demonstration of the laws of physics, no one went near the phone in inclement weather. But even on a good day, it could take more than an hour to get a call through. Priscilla Greene, who was temporarily running the mail room, had to take over manning the switchboard at night because the scientists were getting so frustrated that they could not reach anyone. Moreover, the line, eaten away by chipmunks over the years, carried so much static, it was necessary to shout over and over again to make oneself understood. "We all yelled so loud you could probably hear us all the way from Santa Fe to Los Alamos," said Dorothy, who screamed herself hoarse every morning just trying to find out how many mouths she had to feed. Charlotte remembered one call from the Hill asking them to send up eight extra lunches: "The request as we heard it above the noise, but lucidly could not fill, was for eight extra-large trucks."

One way or another, the phone was their nemesis. The scientists were fond of sending telegrams, but the only means they had of transmitting them was through the telephone operators. Most of the telegrams sounded a little silly because of the code the Tech Area used when referring to classified material: "top" for "atom"; "boat" for "bomb"; "spinning" for "smashing"; and "igloo of urchin" for "isotope of uranium." The messages were never signed with proper names, and the Los Alamos physicists devised their own system of disguising their names, while Oppie and Groves employed a quadratic letter code that each man carried in his wallet throughout the war. "Western Union really must have hated us," observed Charlotte. "There were wires in makeshift codes, wires that said only 'Butane' or 'Yes.' There were wires in foreign languages. Wires and more wires." After a few weeks, they could spell words like PHYSIKALISCHE ZEITSHRIFT in their sleep.

The chaos in the office often spilled outside onto the street on East Palace Avenue. There the army drivers stopped hourly for personnel going up to the Hill, pulling up so close to the curb that the huge green buses would crash into the sagging corner of the old portal and its ancient posts. Young WACs (Women's Army Corps personnel) driving official cars, or "taxis," as they were euphemistically referred to by project members, parked out front until they were dispatched to Lamy to pick up important visitors. Nervous young scientists, driving trucks for the Procurement Office, paced outside, waiting for the delivery of equipment so sensitive it could only be shepherded up the bumpy, torturous mountain road by one of them. The narrow old street was never intended to accommodate such a circus, and the traffic would become backed up, and the police summoned. Invariably, one of the bus drivers would be given a summons for blocking the road and creating a hazard if the fire trucks needed to pass in a hurry. Worse yet, the physicists, many of whom came from abroad and were always driving on the wrong side of the road, would be ticketed and would come into her office waving their hands and protesting loudly. Dorothy would then have to trot round to the local magistrate, who was aware of the project's existence if not its purpose, and using all her feminine charms, plead with him to tear up all the tickets and set things right with the town authorities. In turn, she would promise that the army buses would stand at her door for only brief intervals. Then everything would be fine for a few days until the next crisis.

"There was never a dull moment," she recalled. "The office was a madhouse. It was bedlam. We worked six days a week but even so I couldn't wait to get back to work in the morning. There were always people who needed attention—they were hungry, exhausted, in a hurry. If there was anybody [left] at the end of the day I sometimes took them home with me." When she noticed that Bob Bacher had a cold, she insisted he stay the night with her in town, refusing to allow him to go up to the Hill because she could not be sure he would be properly looked after. From then on, her adobe farmhouse became known as a refuge for weary scientists, the only place they could steal away to for a good meal, warm bath, and peaceful night's sleep in a real home.

Because of her importance to the project, Dorothy held a Q badge, giving her laboratory security clearance and permitting her home on Old Pecos to be used as an officially sanctioned "safe house" by scientists who needed to overnight in town in order to catch an early train the next day. As the months went by, however, it turned into a popular getaway spot for couples who were desperate to escape the military post and enjoy a late night on the town. Kevin would return home some Friday nights to find cars stacked in the driveway, and bodies in sleeping bags strewn all over the lawn. "It was the only place Los Alamos people could fraternize off the Hill, so they would all come for the weekend," he said. "On weekends, the house was always full to overflowing." A hastily scribbled note from his mother on the kitchen table would inform him of the obvious: "All the beds are full. See if you can find a bedroll and a place to park it out back."

Dorothy was on call twenty-four hours a day and became accustomed to fielding frantic messages from Oppenheimer in the middle of the night. Once he phoned in a state explaining that Ed McMillan had returned late from a trip and had gone straight to La Fonda only to discover "there was no room at the inn." Dorothy promised to fetch him, hurried down to the hotel, and brought him home. McMillan had said an official car would be coming to take him to Los Alamos, so the following morning when Dorothy left for work at 7:00 A.M., she did not wake him. When Kevin got up, he found a stranger in his kitchen breaking eggs, and the two sat down for an amiable breakfast. They got to talking about Kevin's car, an old '27 Chevy parked in the driveway, which he had purchased with money he had been saving to buy a horse. He had gotten a good deal on the car, he told the Berkeley physicist, but it just would not start. McMillan offered to take a look under the hood, and happily employed, they lost track of time. "That's where my mother found us at noon after the lab had called, urgently searching for the scientist who'd missed his crucial appointment," said Kevin. Dorothy informed the Hill of McMillan's whereabouts, and shortly thereafter a car came and spirited him away.

All that month and the next, a steady stream of scientific luminaries stumbled into Dorothy McKibbin's office from the great universities across the country, from Harvard and MIT to the universities of Chicago, Wisconsin, and California. She took particular note of a fellow who came from Princeton, her husband's alma mater, a gangling youth named Richard Feynman, who hardly looked old enough to be an expert in anything. Many were foreigners, with oddly tailored clothes and battered briefcases, who looked poignantly out of place on the platform in Lamy. Displaced and disoriented, often showing up days after they were expected, the travelers collapsed into a chair as though they "could never move again," she wrote, so exhausted "one could almost see [their] fatigue dropping off and piling up against the old adobe walls":

> They arrived, breathless and sleepless and haggard, tired from riding on trains that were slow, trains that were held up for troop trains, trains that were too crowded to take on the hundreds of passengers waiting on the platforms, tired from missing connections, and having nothing to eat, and losing their luggage, or sitting the dawn hours in an airport waiting for a plane. . . . The new members were tense with expectancy and curiosity. They had left physics laboratories, chemistry, metallurgy, engineering projects, had sold their homes or rented them, had deceived their friends, had packed their *lares et penates* [personal belongings], and launched into the unknown and unheard of.

All they knew of their actual destination was contained in the "Arrival Procedure Memorandum" sent to all laboratory employees, instructing them to find their way to Santa Fe and report to the U.S. Army Corps of Engineers office at 109 East Palace Avenue. "The following procedure is suggested as the method by which your arrival in Santa Fe and the Site can best be simplified for you and for this office . . ." it began in language that was as bureaucratic as it was uninformative. East Palace Avenue was their last known address before the heavy door of secrecy shut behind them. It was the rabbit hole they fell into. In the jargon of the Manhattan District, Los Alamos was never referred to by name and was designated as "Site Y" or the "Zia Project." But as neither phrase really caught on, it was known generally as "the Site," "the Project," or simply "the Hill." These were all code words they learned to use interchangeably. Machinists dashed in and asked, "Where is the dance hall?" "Thirty-five miles to go" was all the reply they would get from Dorothy.

Joe Lehman, a civil engineer, drove out west in his own car, and expecting that the capital of New Mexico would be quite a large metropolis, blew right by it. "I drove through the square and kept on going looking for the main part of town and I found myself out in the desert again," he recalled. "So I turned around and came back and finally pulled into a service station right off the square and asked where downtown Santa Fe was. The guy said, 'This is it. You're in it.' " Unconvinced, Lehman asked for a telephone and called Dorothy McKibbin. "Where are you?" she asked. Lehman described his whereabouts as best he could, and she said, "Oh, yeah, I see you." She was right across the Plaza from where he had stopped and went outside to wave at him.

Most of the new arrivals could not get over the inconspicuous little office that was the entrance to the all-important Los Alamos laboratory. One afternoon, Dorothy got a call from the Hill asking her to go over to La Fonda and round up two lost scientists. They had been sent from Chicago with instructions to report to 109 East Palace and had walked up and down the sleepy town without seeing anything that looked like a government office. Apparently, a child had been playing on the gate outside Dorothy's office, and the iron-grill door had swung shut, further obscuring the miniature sign. Confused and tired, they had gone to the hotel, called the Met Lab in Chicago, and demanded, "Where the hell are we?" When Dorothy finally led them down the narrow passageway to her modest cubbyhole and invited them to fill out a few forms and turned the crank on the old-fashioned desk machine that produced their passes to the project, she saw the look of disbelief on their faces. "They were not impressed," she observed dryly.

A silent understanding of sorts developed between Dorothy and the townspeople. The phone would ring, and it would be the owner of the corner drugstore calling to report that some oddly dressed character had blundered in: "There is a party here who is lost." She would just laugh and reply, "Send him over right away." After wandering around the Plaza for hours, Leon Fisher, one of the young Berkeley physicists recruited to work on the project, and his wife, Phyllis, staggered into a nearby bakery by mistake and hesitantly told the girl behind the counter that they had been "told to come here." They stood staring in bewilderment at the freshly baked loaves, wondering if this was some elaborate ruse meant to mislead the general public, or if they were supposed to break open the bread to find a coded message bearing their final directions. Just then the bemused voice of the proprietor on the other side of the shop directed them to "the office down the way," adding in a bored drawl, "People are going in and out of there all the time."

For security reasons, even the word "physicist" was taboo. Some of the more jaded staffers took to calling physicists and chemists "fizzlers" and "stinkers," but Dorothy thought that was disrespectful. The most famous physicists traveled under assumed names: Enrico Fermi was known as Henry Farmer, Emilio Segrè was Eugene Samson, Ernest Lawrence was Ernest Lawson. When G-2 fretted that Lawrence's code name was becoming as well-known as his real one, they came up with an alternative—"Oscar Wilde"—reportedly because Wilde had written the play *The Importance of Being Ernest.* Dorothy was told to refrain from using their proper names at all times and to refer to them only by their "covers" when on the phone. Some came to stay permanently; others came only temporarily, as consultants. The latter included Fermi, who arrived in Dorothy's office accompanied by John Baudino, a "tremendous bodyguard, a big halfback of a football team." Baudino had been a lawyer before he was drafted into the army, and he now served as the famous Italian physicist's plainclothes protector, chauffeur, and messenger. Fermi was short and stockily built, and seemed dwarfed by the hulking Baudino, though the latter, Dorothy observed, "tried to efface himself as much as he could." Like most of the refugee scientists, Fermi was classified as an "enemy alien," and it was safe to assume that among Baudino's many duties was filing weekly reports to army intelligence.

Dorothy came to know Fermi quite well from his frequent trips, and he always stopped in, not only for his pass, but also to make a phone call. "He'd always call the Hill as soon as he got to Santa Fe and suggest they try such and such a computation. Then he'd call again from Lamy with another idea, and when he got back to Santa Fe, he'd call again with another suggestion." He would often look at her in an inquiring way as he talked on the phone, as though she might possess the answers to his questions, his eyes twinkling, as his fingers nervously played with a pencil on her desk. Fermi found his all-American code name quite funny, especially since as soon as he pronounced it in his heavily accented English, it immediately aroused the suspicions of the Los Alamos security guards. He was always stopped and questioned, and once the guards demanded he produce letters addressed to "Mr. Farmer" to prove there even was such a person. Dorothy, who more than once came to his rescue, overheard one guard swear under his breath that enforcing the security regulations was not half as hard as trying to understand all the strange accents.

The code names were an endless source of confusion, missed connections, and delays. The physicists often forgot their aliases, or got them mixed up. The problems were compounded by the fact that many of the scientists or their family members were never informed of their new identities in the first place, so they failed to respond to the WAC drivers waiting to pick them up at the train station to take them to Dorothy's office. Fermi, in particular, enjoyed making fun of all the cloak-and-dagger precautions, which struck him as fairly comical in the Wild West setting of New Mexico. He particularly loved the local Santa Fe policemen, who wore snappy Pancho Villa–type uniforms, all black and heavily studded with silver, and looked straight out of the movies. One afternoon, when Fermi and Sam Allison were in her office waiting for a car to take them up to the Hill, Dorothy, who had been instructed never to address any of the scientists using titles such as "Doctor" or "Professor," could not resist teasing them. "As I made out their passes, I tossed my head and informed them that we had to demote them down here and speak of them and to them as plain 'Mister,' " she wrote. But as usual, the two physicists had the last word on the matter:

> They strolled around the town while waiting. Near the Cathedral, they noticed a statue in the yard.
>
> "Who is that?" asked Allison of Fermi.
>
> "That is Archbishop Lamy," Fermi replied.

"Shhhhh," whispered Allison, grabbing Fermi's arm and glancing around cautiously. "Mrs. McKibbin would suggest we call him *Mister* Lamy."

Dorothy understood that all the secrecy was because the army feared too many visitors in the off-season would give rise to curiosity. It was also the case that too many assorted European accents might have the same effect. There was also the possibility that some of the more famous scientists might be recognized from their magazine covers or newspaper photos, and that could lead to talk. It had happened once, in the case of the much-photographed Albert Einstein. As Dorothy recounted in her memoir, the little old man with the distinctive shock of white hair was sitting alone on a bench in the Plaza one afternoon when he caught the attention of one of Santa Fe's more colorful gadflies:

> Brian Boru Dunne was by no means an everyday small town reporter. He wore a great ten gallon hat and rode a white horse to town which he would tether in front of the Post Office. He was from the East, had written books one of which had a forward by H. G. Wells. He not only wrote a column on visitors for the town newspaper, a Cutting property, but was business manager in Santa Fe for senator Bronson Cutting's interests. It was with undisguised excitement that he brought his column to the editor on a certain day. He had interviewed Einstein. The column was submitted to General Groves for approval, and brought the general storming into the office. The column was ordered killed and any mention of it forbidden. There was never an admission in the Los Alamos annals that Einstein ever visited Santa Fe.

While Dorothy tried to be vigilant about security, she comforted herself with the thought that Oppenheimer and company would probably be able to go about their business without people paying too much attention. "Santa Fe had always been a town of comings and goings," she wrote. "After all, its only industries were tourists and politics." For the most part, she was proved right. Will Harrison, who was editor of the daily paper, *The Santa Fe New Mexican,* instructed his staff not to bother her, cautioning them, "She won't tell you anything." In any case, the government had ordered the local papers to make no mention of the Los Alamos project until after the war. When the *Albuquerque Journal* slipped up and reported that Los Alamos personnel had helped put out a forest fire in the Jemez hills, the editor was raked over the coals by G-2. The official policy called for a complete media blackout—no newspaper and radio reports, no publicity of any kind, no mentions even indirectly related to the subject of atomic energy. The idea was to be able to separate the wheat from the chaff. If everybody kept quiet, it would be easier to track the talkers and to pinpoint suspects. Another reason for the policy was to lull the Nazis into underestimating the Allied effort and, as a consequence, not to accelerate their own atomic research program.

No matter how busy or bizarre the comings and goings at 109 East Palace, people in Santa Fe went about their business just as if it were life as usual. They all watched the coal trucks tear up the old road to the Pajarito Plateau and said nothing. They heard about the big moving vans that got stuck on the washboard road and had to be pulled out of knee-deep bogs, and shrugged. Neighboring shopkeepers were aware that Dorothy's office was in some way attached to a wartime project, but knew better than to ask about it. Dorothy did her part to keep it from becoming prematurely famous, using a combination of charm, soft soap, and double-talk to dismiss daily suspicions. When asked about people who had suddenly packed up and disappeared from town, as happened after she recommended her neighbor to fill the job as fire marshal on the site, or when a local baker went to work at Fuller Lodge, Dorothy made no reply. There were rumors that some folks had moved up to Los Alamos, but no one knew for certain. Over time, even her friends grew reluctant to stop by the office, where the most casual questions might be greeted by an awkward silence. They did not know what to say, so they stayed away.

At the end of March, Harold Agnew, a twenty-two-year-old graduate student, barged into Dorothy's office and demanded to know if all his boxes had arrived safely. Most of the young laboratory staffers arrived in a sweat, panic-stricken that their irreplaceable equipment had failed to arrive in one piece. Having witnessed some alarming displays of nervous agitation, Dorothy knew there was nothing she could say to relieve his distress. Seeing would be believing. She just smiled serenely, handed him his pass, and pointed the way to Wilson's Storage & Transfer. Agnew rushed straight to the shipping company, found a driver, jumped in the cab of the truck, and took off with his boxes without a clue where he was headed. All Dorothy had said was to "go on up" with the truck. She had supplied him with a yellow map covered with red pencil markings carefully indicating every mile and every turn of the last leg of the trip. He read the instructions: "Go around the Federal Building, turn north, Tesuque six miles, Pojoaque seventeen miles, turn west there and wind down the valley, cross the Rio Grande y Bravo, turn southwest ten miles and then ten miles directly west and up."

In every direction he looked there were dirt roads leading off into nowhere. They passed through sleepy villages and fields being worked by farmers still using horse-drawn plows, as though in a time warp. He and a handful of others had been recruited by John Manley to be part of his team at the new laboratory, and before coming they had spent a few days at the University of Illinois helping him painstakingly dismantle his Cockcroft-Walton accelerator. Agnew's bride of a year, Beverly, who had trained to be a schoolteacher, had stood there with a clipboard in hand, counting the number of boxes and making out the manifest. Some of the parts were made of hand-blown glass, and Agnew thought of them now, bouncing in the back of the truck. Manley had not been able to divulge their destination, but one night after dinner he had played them a record—"On the Santa Fe Trail." Looking out the window of the cab of the truck as they climbed past lava beds and black escarpments, and cut through strange volcanic rock formations and a bed of white pumice, bits of stone clattering down the canyon walls and bouncing off the side of the truck, Agnew thought he was beginning to understand why Manley had been so reticent: "It was a dusty, isolated, uncivilized place." A guard stopped him at the first gate and checked his pass and identification, and then it was another three miles of piñons and junipers to the settlement.

Taking Dorothy's last piece of advice, he stuck his head out the window and shouted for directions to the Tech Area.

Shortly after Agnew deposited his equipment, he set off for the Big House, where he was told he would be sleeping for the time being. On his way, he spotted Los Alamos's tall, thin director walking in the distance, wearing his distinctive porkpie hat. "Like a little puppy I bounded over to him and said, 'Hello, Dr. Oppenheimer!' He just looked at me and said, 'Hello, Harold. Where's Beverly?' " Oppie knew that Agnew's pretty blonde wife had worked for the head of the Met Lab back in Chicago and had handed out the security passes. Given how seriously shorthanded they were on the Hill, his only thought was that an experienced secretary was of far greater importance than the addition of yet another physics student. A much chagrined Agnew found himself stammering an apology for his wife's absence, explaining that Beverly had been given special dispensation to delay her arrival in order to visit her brother before he was shipped overseas and would be there shortly.

The project was in desperate need of day laborers, and the years of poverty in the region meant that volunteers were not in short supply. For the men and women from the neighboring pueblos of San Ildefonso, Santa Clara, and Tesuque, the meager wages the project offered seemed like a fortune. From her office at 109 East Palace, Dorothy helped screen hundreds of applicants, some of whom had packed their families into trailers and traveled across the state on the rumor of good wages. If they were at all qualified, they were given extensive applications to fill out. "We asked most of the questions," she wrote. "All kinds, ages, and conditions of people came in seeking employment." Some left an indelible impression. Under the heading "Occupation," one officer who had been overseas with several different combat units listed his profession as "murder." Under the heading "Equipment Used," a woman who had worked in mess halls described her personal arsenal as "blue uniform and hair nets." Waiting for security clearance to work on the site was hard on many of the prospective employees, particularly those who had traveled across the state for the interview, and Dorothy went through the long, painful process with them. "The two weeks required for that often consumed their funds and also their spirits," she wrote. "They were tossed out of their hotels every three days, and the scarcity of housing in the 'Land of Enchantment' made the waiting doubly difficult."

The worst part was that security had ruled that employees from certain defense projects could not work on the Hill. The decision, however arbitrary it seemed, was final. There would be no explanation provided, and no room for appeal. A highly skilled man might present himself at Dorothy's office and ask for work at Los Alamos, and she had no choice but to follow G-2's instructions and look him straight in the eye and deny the project's existence. "I can't understand wherever you got that idea," she would say through gritted teeth. "There's nothing of that sort in Santa Fe that we know of." Some of them had come so far, it broke her heart to send them away. "We were put in bad spots," she said. "It was tragic sometimes."

Eventually, the new telephone line was installed and enough of the main structures had been completed to allow nearly a hundred scientists and their families to move up to the Hill. Groves had ordered the construction of technical buildings, an administration building, a mess hall, a theater, an infirmary, officers' quarters, apartments, and barracks for the military personnel. On the south side of Ashley Pond was the Technical Area, consisting of five long, barnlike laboratory buildings, ringed by barbed wire and heavily guarded. West of the pond were row upon row of prefabricated apartment complexes all painted the regulation "o.d." (olive drab), ending in barracks and dormitories, which overlooked the horse pastures. Oppenheimer had directed that the apartment buildings be laid out along the natural contours of the land and at an angle that retained as much of the magnificent mountain views as possible, but the effect was rather odd, as if they had been scattered by a storm. The rest of the town conformed to the gridiron pattern common to most military posts, dirt roads outlining the minimal housing section from the business section.

To the east was what remained of the old school buildings, together with more ghastly green duplexes. The Big House served as bachelor quarters, recreation room, and library; the Arts and Crafts Building became schoolrooms and two residences; and the five-car garage was converted into a fire station. Fuller Lodge was turned into a hotel and restaurant, and the classrooms into the Post Exchange and assorted shops. The old Guest Cottage was reserved for General Groves and visiting dignitaries. There was a Commissary, where they could buy food, and the original Trading Post, which still sold an odd assortment of supplies, including animal feed, mousetraps, kerosene lamps, small boys' T-shirts, underwear, and moccasins. The army turned it into another PX, and added basic drugstore supplies, cigarettes, souvenirs, and newspapers and periodicals. Everyone knew that Oppenheimer had battled the army engineers, so fond of cheek-by-jowl construction, to preserve the few tall trees on the mesa and the remaining vestiges of the old school's charm. Groves had yielded to his wishes, but reportedly had grumbled, "All this nonsense because the families have to live here. If I could have my way and put all these scientists in uniform and in barracks, there would be no fuss and feathers."

The peaceful mesa Dorothy remembered was gone, replaced by the deafening roar of a rising war factory. Los Alamos looked like a frontier boomtown, "humming with rush and hurry," she recalled. "Military cars swarm all over its winding streets. Buildings are erected overnight and they look it and feel it." There was not much there, and what was there was raw and, for the most part, very ugly. The unpaved streets had no names, and the uniform military architecture made it next to impossible for Dorothy to get her bearings. The only way she could orient herself was by where the Serbers parked their red convertible. If they were not home, she had to rely on the tall wooden water tower, still prominent on the highest part of town.

Oppenheimer and his family were already ensconced in the old Ranch School headmaster's house at the end of a quiet road that was close to the heart of wartime Los Alamos. It was a sturdy cottage built of log and stone that was partly shielded by a small lawn, flowering shrubs, crab apple trees, and the wisteria that grew by the door. In the meritocracy of Los Alamos, individual importance to the laboratory dictated not only people's social prestige but also the quality of housing they were assigned. Without doubt, the eight attractive old masters' cottages were the most desirable, the Park Avenue of the plateau. The house opposite the Oppenheimers' would go to another top administrator, Captain William ("Deke") S. Parsons, an affable and experienced Annapolis graduate whom Groves brought in to run interference between the scientists and the military, and who would become assistant laboratory director. As one of the few men at Los Alamos with explosives training, he was also head of the Ordnance Division and would be in charge of dropping the bomb. His wife, Martha, was a "navy brat"—an admiral's daughter. She had lived on bases all her life and so adapted quickly to Los Alamos. Despite being military people, the Parsons immediately endeared themselves to the scientists with their big, informal house parties. The remaining masters' houses were divided into apartments, and the McMillans felt lucky to have snagged one next door to the Oppenheimers. Their residential street was a small oasis of serenity bequeathed by the previous owners, and was the first and most prominent class division in the pioneer town. Word quickly spread that Oppie and company were the privileged few to have private homes with all the amenities, and in honor of the luxury they alone enjoyed, the street was soon christened "Bathtub Row."

Over the next few weeks, the rest of the laboratory staff and their families migrated from the ranches in the valley to the new housing at Los Alamos as soon as the paint was dry. Nothing was ready, and almost no one's belongings arrived on schedule, so in the beginning they slept on army-issue cots, supplied with sheets and blankets stamped "USED" in bold black letters, which bothered people no end until they learned this was an acronym for "United States Engineer Detachment." Adjacent to Bathtub Row was the area that quickly became known as Snob Hollow, where a dozen identical, green, four-family apartment houses were built along a street that started on the summit of the town, by the water tower, and sloped down to the edge of the mesa, where it faded into the green forest. There, senior scientists shared a quadriplex, occupying mirror-image apartments that were small and spare, but sufficient. Each family was shoe-horned into a two- or three-bedroom unit, with front porches that actually faced the mountain views and were seldom used, and narrow back stoops just steps off the road that were the actual entrances, and where folks stood and sipped their morning coffee in the sun and traded gossip.

The close quarters were bound to make for bad neighbors. Some friction quickly developed between Alice and Cyril Smith and the Tellers, who lived downstairs from them, as Edward was a nocturnal creature who thought nothing of banging away on his grand piano at all hours of the night, disturbing those who worked more normal hours. But it could have been worse. The duplexes were even more claustrophobic; these were doled out to younger couples who did not have children, such as the Serbers and Agnews. Finally, all the Sundt apartments, named for the contractor, came with showers only and were much cursed for their shoddy construction and paper-thin walls, which permitted neighbors to hear every cough, quarrel, and baby's cry. But Dorothy told those with a roof over their heads to count themselves lucky. She had some young couples sleeping on the portal of the lodge while they waited for the construction on their unit to be completed. Still others had to be housed at the old guest lodge at Bandelier National Monument, the site of spectacular Indian ruins, where archaeological excavations had revealed the remnants of huge pueblos carved into the soft rock of the cliff sides. Dorothy ruefully reflected that the Anasazi had once managed to find housing for thousands on the mountain, creating a complex network of ancient cities that stretched over seventeen miles in the canyons below.

The whole administrative operation of Los Alamos was based at 109 East Palace from mid-March to the first of May, when Oppenheimer and company packed up their offices and disappeared "upstairs" for good. It was a big day, but also a moment for some sad farewells. It was not the distance so much as the heavy secrecy that made the separation seem so great. Dorothy promised to visit them all on the Hill as often as she could. Oppenheimer asked her to come with him and offered her a position on his personal staff. She also received similar invitations from Duane Muncy and his assistant, Isabel Bemis. But reluctant to leave her home in Santa Fe and uproot her son, Dorothy chose to stay behind. She would stick to her desk at the "housing office" and run the project's lone outpost in town. She was the gatekeeper, safeguarding their secret city and shepherding newcomers up the primitive road and through the labyrinthine canyons like so many stray goats or cattle. What she did not tell them was that she had come to enjoy the importance of her position and the unique responsibility of being the porthole to the unknown. The scientists who passed through her door, she wrote, left behind everything they knew and cared about and "walked into the thick-walled quietness of the old Spanish dwelling at 109 East Palace as if into another world."

As the dormitories and apartments were completed at Los Alamos, the scientists converged on the mountain, and the rudimentary beginnings of a miniature city, and a new culture, began to take shape. Another tribe was colonizing the Pajarito Plateau almost eight hundred years after the first known permanent inhabitants of the region, the Keres-speaking "Hopituh"—"the People of Peace," as they called themselves—had taken shelter there to escape the warring desert nomads below. Dorothy knew the rocky mesas had once been their sun temples, high and safe, and everywhere the invading scientists walked, they trampled on the remains of that ancient cliff-dwelling civilization, the fragments of pottery and hewn stone. She could not miss the irony that this settlement was for a very different reason than that earlier one, yet somehow she believed it followed from the same primal forces. Peace was again the objective, and to accomplish that, she wrote, the new inhabitants would build their "fantastic modern scientific laboratory" on the old pueblo ruins:

> One is scarcely aware that his foot descending from the jeep is crunching against ground full of potshards, and as the sun hits the steel and wire installations one can see a brighter shine than that of an arrowhead. But the arrowheads are there, and the awls, and the pots and feathers, and the fibre rings and the turquoise. And one can look across the angular buildings and survey the Sangre de Cristo Range in its myriad changes from Santa Fe to Taos, and know that at evening the mountains in snow or in green will be the color of amethyst. And the Tesuque, Nambe, Santa Cruz valleys contain villages, Rio en Medio, Chimayo, Sanctuario, Cordova, Truchas, where few speak English and burros carry wood on their backs as they wind their reluctant way down the mountains as other burros have done for centuries.

It was a strange twist of fate that Oppenheimer, who had originally ventured out to New Mexico for his health, should return twenty years later on such a different mission and cross paths with Dorothy McKibbin, a recovered consumptive and six years his senior, who had made the same journey, for the same reason. They had in common their love of the land—the red earth and the deep-rooted civilization—and the strength and freedom they found there. It was an unspoken bond that joined them from the outset and accounted for the warm and trusting relationship that was at once apparent, and the deep attachment that would form in the months to come. She was the most unlikely of people to help usher in the atomic age. But perhaps better than anyone except Oppenheimer himself, Dorothy understood the turbulent change they were bringing to the mountaintop and, in accordance with the ancestral rites and ceremonial songs of the great pueblos, the symbolic seven worlds through which man climbs on his evolutionary journey, the dangerous myth-making they were engaged in. The scientists would remake and reshape the world with their own powerful brand of magic, and nothing would ever be the same again. But believing absolutely in the rightness of Oppenheimer's vision, she never questioned him; she devoted herself completely to the laboratory and its leader, who, she wrote, "hold[s] the sun in new and streamlined machines on the old worn plateau."

SIX

The Professor and the General

ON APRIL 15, 1943, the Los Alamos laboratory formally opened for business. There was no fanfare or ribbon cutting that anyone can remember, but General Groves did put in an appearance, greeting Oppenheimer's scientific brain trust with his oddly limp handshake. "Like a dead fish," recalled Major Carlisle Smith, known to everyone as "Smitty," a chemist who served as the project's patent lawyer and all-purpose attorney. "I think that's one reason so many half-smart civilians here never appreciated him." Given his officious nature, Groves could not resist giving an orientation address, which amounted to a short lecture. He may have intended it as a pep talk, but to the gathering of young scientists, many of whom were still strongly opposed to being under the thumb of the military, he came across as boorish and doctrinaire.

Afterward the physicist David Inglis observed to Teller how "out of place" the general appeared, and while Teller shared his view, he thought Groves' speech was "about what would be expected from a person who knew nothing about the project he was supervising." The scientists could not conceal their disdain for Groves, and it showed in their mocking grins and flip asides. Their collective cockiness was fostered by the recent realization that they were no longer humble academics, but members of a profession of paramount importance to the defense of their country. They were immune to the draft, exempt from the usual "need to know" wartime restrictions on research, and they had been hand-picked for a classified project. In short, there was no telling them anything.

Groves more than returned their lack of regard, and he responded by reasserting his authority whenever possible. On a good day, he tolerated the scientists as "tempermental people" who "detested the uniform." On a bad day, he blamed them for making his life "hell on earth." His attitude of paternal condescension was made clear early on in the instructions he had given Oppenheimer, who duly passed them on to his troops: "Here at great expense the government has assembled the world's largest collection of crackpots. Take good care of them." It did not help morale that word had spread across the project that Groves had intended to bugle his ivory-tower recruits out of bed at dawn for an antiparachutist drill and was only talked out of this at the last minute. "You could see clear across the mesa and for miles and miles beyond," Sam Allison marveled later. "I thought it was the unlikeliest place in the world for parachutists."

By this time, many of the scientists had been at Los Alamos just long enough to be thoroughly exasperated by the military mind-set that ruled the mesa. They had been struggling for several weeks to unpack and assemble their equipment, and they felt they had been hampered at every turn by the army engineers, who not only were responsible for the jerry-built structures they now found themselves in, but refused to change so much as a screw, even when the invaluable accelerator parts being unloaded could not fit through the door. To make matters worse, many of the laboratories still had no power, and when the scientists pushed to get the work completed as soon as possible, they found the construction crews unresponsive and indifferent to their sense of urgency. Procurement was turning out to be an even bigger problem, with ordering even the smallest missing part necessitating directives, requisition forms, and a tediously involved procedure. For their part, the local crews had been laboring for months under very trying conditions and foresaw that the nervous scientists' constant requests for changes in the design would only result in chaos and more delays. They did not have the slightest idea what the project was about and regarded the whole thing as a giant boondoggle, probably the brainchild of some nut in Washington. Tempers erupted, and all the run-ins between the scientists and construction workers seemed to make the likelihood of their ever contributing anything to resolve the conflict half a world away that much unlikelier.

For many of the laboratory staff, it was extremely difficult to come to terms with the fact that Los Alamos was an army post and Groves was indisputably in charge. He had failed to get them all into uniform, but in every other respect, his word was the law of the land. Groves had overall executive responsibility for both the military and the technical administration, each of which was headed in turn by the commanding officer and director. The CO, who reported directly to Groves, ran Los Alamos on a day-to-day basis, overseeing everything from the maintenance of all the living conditions and the conduct of military personnel to security, including the large military police force that patrolled the fenced perimeter twenty-four hours a day, both on horseback and in jeeps. K9 Corps dogs guarded the base of the cliffs.

Groves was such a stickler for security that the first CO lasted only four months, leaving right around opening day. The rumor was that he had a weakness for alcohol that life on the isolated post had exacerbated. His successor, Lieutenant Colonel Whitney Ashbridge, was an engineer who had worked for Groves before and, fortunately for all involved, had attended the Los Alamos Ranch School and was already familiar with the lay of the land. His primary responsibility was to keep the makeshift community up and running and see to it that the scientists "stick to their knitting," as Groves put it.

Oppenheimer, as the laboratory director, also reported directly to Groves. From the very beginning, Groves had been apprehensive about the relations between the scientists and the military and had urged Oppie to appoint a special administrative assistant to manage the practical, housekeeping side of the project so that he could remain focused on the technical problems. Edward Condon was a first-rate physicist and had served as associate director of Westinghouse's experimental laboratory, which led Groves to believe he would be able to speak the same language as the construction engineers and ensure good communication with the commanding officer. Instead, trouble arose as soon as the first scientists set foot on the post.

As the new arrivals complained about everything from the roads and housing to the lack of facilities and a proper school for their children, the friction worsened. Groves felt that instead of proving himself an ally, Condon sided with the scientists and fought him on issue after issue. Their growing hostility culminated in a flare-up over compartmentalization, Groves' fetish for secrecy between the separate units of the project. "The basic security problem was to establish controls over the various members of the project that would minimize the likelihood of vital secrets falling into enemy hands," wrote Groves in his memoir. "Compartmentalization of knowledge, to me, was the very heart of security. My rule was simple and not capable of misinterpretation—each man should know everything he needed to know to do his job and nothing else."

Condon likened the policy to being asked to do "an extremely difficult job with three hands tied behind your back." While Groves had bowed somewhat to the scientists' demands and had eliminated the usual compartmentalization protocol within Los Alamos, he more than made up for this by drawing a tight ring of secrecy around the entire bomb laboratory. For Groves, the high steel fence surrounding the post was intended to contain every aspect of what went on there—to wall everyone and everything in, both literally and metaphorically. While the other classified Manhattan Project sites, including the uranium factories in Oak Ridge, Tennessee, and the graphite production piles in Hanford, Washington, sought to keep people from getting in, Los Alamos was expressly designed by Groves to keep any information or personnel from getting *out.* They were to be invisible to the outside world, cut off both physically and socially, and he would enforce this blanket security with the brass-knuckle vengeance he had been known for throughout his career. When he learned that Oppenheimer had violated security in late April by personally journeying to Chicago to ask Compton for a small amount of plutonium, and furthermore had disclosed exactly what he intended to do with it, Groves hit the roof. He called both Oppenheimer and Condon in on the carpet for a thorough dressing down.

Condon did not take the reprimand in stride and barely six weeks into the job tendered his resignation. Furious, Groves told Oppenheimer that for his own protection he should demand that his second in command put his reasons for leaving in writing. In his letter to Oppenheimer, written at the end of April, Condon cited the "considerable personal sacrifice" to himself and his family as the main reason for his departure, but then in a lengthy tirade, placed the blame squarely on Groves' policy of compartmentalization:

> The thing which upsets me the most is the extraordinarily close security policy. I do not feel qualified to question the wisdom of this since I am totally unaware of the extent of enemy espionage and sabotage activities. I only want to say that in my case I found that the extreme concern with security was morbidly depressing—especially the discussion about censoring mail and telephone calls, militarization and complete isolation of the personnel from the outside world. . . .
>
> While I had heard there were to be some restrictions, I can say that I was so shocked that I could hardly believe my ears when General Groves undertook to reprove us, though he did with exquisite tact and courtesy, for a discussion which you had concerning an important technical question with A. H. Compton. To me the absence of such men as A. H. Compton, E. O. Lawrence, and H. C. Urey was an unfortunate thing but up to that time in your office on Monday I had put it down simply to their being too busy with other matters.

Groves dismissed Condon's letter as less than honest, and cowardly. Condon had interpreted the decision by Compton, Lawrence, and Harold Urey to stay away from the mesa as a sign that they did not think much would come of it. The bottom line, Groves concluded, was that Condon feared "the work in which [they] were engaged would not be successful, that the Manhattan Project was going to fail, and that he did not want to be connected with it."

In fact, Groves was willing to be far more flexible on the subject of security than army intelligence would have preferred. At Los Alamos,

Groves confronted the problem that the army was dragging its feet when it came to providing clearances for the many foreign-born physicists because of their "enemy alien" status, even though a number of them had been working in highly sensitive areas for months already. Part of the army's reluctance was that because they were refugees from countries that were now at war, obtaining reliable information about their backgrounds and politics was next to impossible, which made the screening process extremely slow and incomplete at best. In addition, almost all of the physicists had been at foreign universities where Communist doctrines were popular and had friends and colleagues who were avowed members of the party. Given the limited number of first-class atomic scientists available, however, Groves was not prepared to part with someone as irreplaceable as Fermi on the basis of only vague suspicions. With so much at stake, Groves concluded it was worth taking the calculated risk that "among those whose employment would be an advantage to the United States, a reasonable distinction could be made between individuals whose use might be dangerous and individuals whose use would probably not be." All decisions on security, including the clearance of personnel, "had to be based on what was believed to be the overriding consideration—completion of the bomb. Speed of accomplishment was paramount."

Despite the tendency of a few outspoken foreign scientists to criticize the army's "gestapo" methods, the Italian physicist Emilio Segrè, a former student of Fermi's who had escaped Mussolini's Fascist regime and was awaiting clearance to work at Los Alamos, admired the American general's pragmatism when it came to issuing clearances. As Segrè noted in his wartime memoir, "Groves wiggled out of this impasse with good sense":

> When Groves saw that the usual security rules would preclude recruiting those he wanted, he invented new rules. Each of us was to guarantee some colleague he knew well. "Guarantee" sounded good, but how? Somebody proposed an oath on the Bible, but Groves objected: "Most of them are unbelievers." An Intelligence officer then proposed an oath of personal honor, but

> Groves replied, "They do not have any sense of honor." "Rather," he concluded, "let them swear on their scientific reputation. It seems to me that is the only thing they care for."

In the end, Segrè swore an oath on his reputation as a physicist that Fermi was loyal, while Bethe and Bacher guaranteed Segrè's loyalty, and the process continued on that way in a circle until every alien was cleared.

For his part, Eugene Wigner refused to allow himself to be fingerprinted by Groves' security people for fear of what might happen if the war did not go their way and the records fell into the hands of the Nazis. "I had no doubt that if the Germans won the war they would swiftly begin rounding up everyone in the Manhattan Project for execution," he explained. "And the roundup would go easier with fingerprints."

If the opening conference had not gone so well, there is no telling how many of Oppenheimer's recruits who endured this tense and trying time would have chosen not to stay on "for the duration," as the unforeseeable length of the project was always described. But in those first few weeks, what impressed them more than all the annoyances and minor discomforts was the extraordinary dynamic between the professor and the general, and the almost spontaneous organization and productivity that it seemed to generate. Rabi, who was a consultant and had come from Cambridge only for the meetings, later marveled at the ease with which things seemed to come together:

> First, Robert Oppenheimer—of all people to select as director. It was astonishing! He could drive a car with only occasional accidents but never fix it. But he was a man of brilliant insight with a command of language that was very elevating. He set a high-level tone. Then there was General Groves, his boss, whom most of the scientists who worked at Los Alamos remember as a born malaprop. Now, that combination of Oppenheimer and Groves

was remarkable because you would always tend to underestimate Groves, but he was the power behind Oppenheimer. That combination made the thing work.

Despite all the difficulties, the project got off to a fast start, all the more so considering that, at the beginning, they had no equipment. As Rabi drily observed, "There were no materials for the bomb that we could assemble, and their properties were not known." But as Richard Feynman, one of the youngest and most gifted graduate students observed, the theorists did not need a laboratory, let alone any apparatus, so they "could start working right away." Oppenheimer nominated Serber to give a series of lectures to the gathering, which was attended by thirty of the most distinguished scientists in the country. They assembled in the old school library, which was empty because none of the technical books and reports had arrived yet either. Serber stood before a small, rickety blackboard on wheels and tried to make himself heard above the banging of hammers and the carpenters' loud footsteps overhead. "The object of the project," he began, "is to produce a practical military weapon in the form of a bomb in which the energy is released by a fast neutron chain reaction in one or more of the materials known to show nuclear fission. . . ."

At first, Serber seemed like an odd choice to lead the conference. He was a slight, mild-mannered, almost puckish character, who spoke softly and with a pronounced lisp. A few minutes into his first talk, Oppenheimer had to dispatch Manley to go up front and tell him to please stop using the word "bomb" and substitute the term "gadget" instead. Serber was not a riveting speaker, but he held their attention despite the chaotic surroundings. At one point, an electrician's foot came through the ceiling, accompanied by a shower of sawdust, leaving a good-sized hole. It was not clear how the workman had breached Groves' awesome security, but had he been a spy, they all agreed he would have been treated to an earful. By the time Serber was finished, he had delivered five brilliant lectures, an hour or so each, summarizing what the project was about and the existing state of knowledge about the bomb. The tutorials became instant classics. Before Condon left, he compiled them, and mimeographed copies of *Los Alamos Report Number 1: The Los Alamos Primer* subsequently became required reading for every new physicist who joined the project. "He wasn't much of a speaker," one participant recalled. "But for ammunition he had everything Oppenheimer's theoretical group had uncovered during the last year. He knew it all cold and that was all he cared about."

While the lectures were proceeding, Oppenheimer began calling in the senior scientists one by one and informing them that, in order to facilitate the work, he was organizing the laboratory into divisions, each with overall responsibility for a number of subdivisions, or groups. Rabi had advised him that the MIT Rad Lab's structure had worked exceptionally well and that implementing it at Los Alamos would help create a sense of mission and camaraderie. Oppenheimer organized his staff accordingly around the various parts of the project: theoretical physics, experimental physics, chemistry, metallurgy, and ordnance. They would investigate two proposals for the bomb's design: the most straightforward was the gun-assembly method for the U-235 bomb, which involved firing a subcritical mass of fissionable material, in this case uranium, into a second subcritical mass of fissionable material, producing a supercritical mass, the chain reaction leading to an atomic explosion. The other method, the plutonium gun, was trickier, because theoretically the chain reaction of plutonium would be much faster, so consequently the gun would have to be much faster or it would predetonate. How much faster would have to be explored. Both assemblies would be investigated posthaste, although the uranium bomb was to be given priority.

Groves came in for much derision when he proposed what Teller later called "a very stupid way of assembly" for the plutonium weapon, which the scientists made fun of behind his back. But Oppenheimer always maintained that anyone from the lowliest employee could be the source of a good suggestion, and in the end Groves' absurd proposal planted the seed of an idea in someone in the audience. Later on, just as Serber was winding up his last lecture, a tall, gangly physicist named Seth Neddermeyer stood up and announced that he believed implosion was the way to go, by directing a blast of high explosives *inward* toward a quantity of fissionable material causing it to reach critical mass and detonate. While he faced considerable skepticism from Bethe, Fermi, and others, Neddermeyer's stubborn championing of this means of assembly led Oppenheimer to reluctantly assign him to convene an experimental study group on implosion. It was one more thing that would now have to be looked into, and they had little time and few resources. Neddermeyer later recalled thinking that Oppenheimer looked tired, and that the burden of responsibility for the success of the bomb project, for picking "the things that will work," was weighing heavily on his thin shoulders.

Rabi had also advised Oppenheimer that, saddled with all the duties of being scientific director, he could not also be responsible for heading the Theoretical Division. Rabi proposed Hans Bethe for the job, and Oppie agreed. When Teller learned that Oppenheimer had made Bethe head of the T Division, and Serber his group leader, he was furious. Oppenheimer had known full well that Teller would feel snubbed, but his primary responsibility was to keep everyone on track and the work progressing as quickly as possible. Teller persisted in talking about his own ideas and introducing obstacles in an effort to control the project and move it in a different direction.

One immediate benefit of the new laboratory structure was that it fell to Bethe to inform the intractable physicist that he was being further sidelined. He suggested to Teller that since he was interested in alternate ways the bomb could be made to work, he should go off and pursue fission calculations with his own small group, including his idea for a fusion weapon. Teller immediately protested, in part because he realized the task might take so long it would prevent him from making any real contribution to the atomic bomb. He had forced their hand with his consistent refusal to cooperate and in the end had to accede to their request that he work outside the Theoretical Division. While Teller never stopped raising questions and challenging ideas at meetings, Oppenheimer had effectively removed him from the main thrust of the activity at Los Alamos. He had brought the major source of disruption on the mesa into line, but at the cost of losing a first-rate mind. Oppenheimer made every effort to placate the brooding Hungarian by meeting with him privately, but after a while he told Greene to limit his audiences with Teller to once a week.

As soon as the opening session was over, Groves moved to clamp down on the scientists' freedom to discuss problems relating to different divisions of the project. He was alarmed by some of the free-wheeling exchanges he had heard during the first few weeks and strove to have the different sites strictly compartmentalized and to limit open discussion of the bomb. For the most part, Oppenheimer and Groves did not have serious disagreements about how to run the project, but in this area their agendas came into sharp conflict. "Groves wanted to partition everything," recalled Bethe. "He wanted each group to work by itself and no other group should know about it." Oppenheimer vehemently opposed any compartmentalization, arguing that progress was only made through interaction, that science was not possible without discussion, cross-fertilization, and collaboration. Groves only agreed after Conant, Rabi, Bacher, and much of the senior laboratory staff took the same line.

Oppenheimer established weekly colloquiums, where the division and group leaders could report their progress and problems in detail. The meetings, which became a Tuesday night tradition, came to be regarded as the linchpin of the laboratory's productivity and rapid progress. "Oppenheimer insisted that absolutely everyone should be interested, and should know, and should contribute," explained Bethe, adding that this open-handed approach encouraged even the lowliest of the scientists to take part and promoted a sense of enthusiasm and pulling together that became the heart and soul of their fledgling enterprise. He later wrote that Oppie's leadership in establishing the "democratic organization" of the laboratory was one of the key factors in its ultimate success:

> The governing board, where questions of general and technical laboratory policy were discussed, consisted of the division leaders (about eight of them). The coordinating council included all the group leaders, about 50 in number, and kept them all informed on the most important technical progress and problems of the various groups in the laboratory. All scientists having a B.A. degree were admitted to the colloquium in which specialized talks about laboratory problems were given. Each of these assemblies met once a week. In this manner everybody in the laboratory felt part of the whole and felt he should contribute to the success of the program. Very often a problem discussed in one of these meetings

> would intrigue a scientist in a completely different branch of the laboratory, and he would come up with an unexpected solution. . . . Oppenheimer had to fight hard for free discussion among all qualified members of the laboratory. But the free flow of information and discussion, together with Oppenheimer's personality, kept morale at its highest throughout the war.

Groves subscribed to the opinion of one of his former chief engineers—"[I have] no objection to any committee as long as I appoint them"—and established his own Review Committee after Conant counseled him that forming such a body would improve his relationship with the scientific community, regardless of whether or not he chose to pay attention to their reports.

> He [Conant] pointed out that these people were accustomed to making their views known to similar committees appointed by their university administrations, and that our adoption of this system would meet with their approbation. A further advantage which we both recognized was that a review committee, with its fresh outlook, might be able to make a suggestion that would be eagerly seized upon, whereas if the same suggestion came from me, it might be regarded as interference.

The Review Committee, which was composed of five scientists carefully selected by Groves for their familiarity with the project and sympathy with his views, approved the program laid out by the Los Alamos physicists and lauded Oppenheimer's performance as director.

Oppenheimer had always been quick to grasp technical problems, but his adept handling of mesa politics, and particular finesse when it came to Groves, far exceeded expectations. Bethe, who prior to Los Alamos had regarded Oppenheimer as "a difficult human being" and someone who could be very tactless and cutting when he wanted to be, was pleasantly surprised by the change in his demeanor. For all his initial doubts, Manley, too, was impressed by the "astonishingly rapid transformation of this theorist" into a highly effective leader and administrator.

Though he had become more aware of Oppenheimer's warmth and consideration as they worked together on the early stages of the project, he still had seen nothing to suggest the hidden talents that emerged at Los Alamos. Manley was most impressed by Oppenheimer's subtle orchestration of all the various players, which he likened to a "ballet," with each person knowing what was expected of him and playing his assigned part. "He had no great reluctance about using people," he observed. "But . . . it was an enjoyable experience because of the character of Robert to do it so adroitly." Manley often wondered if Groves, who was a very astute judge of people, "sensed the breadth of stature and capability of this man in areas which his previous activity had given so few objective hints for the future."

Greene, too, took note of this new, more mature turn in Oppenheimer's character. "It's a real mystery that he rose to this so fast, considering that he was a very diffident, shy person to begin with," she said. But he had the scientists' respect and their unstinting confidence—that he understood the problem and that it really could be done—and this gave him a new kind of strength and confidence. "He had that behind him," she said, adding that statesmanship was "something he learned, and perhaps not too well."

After the April gathering, Groves moved to tighten his hold over Los Alamos. In the wake of Condon's resignation, he rescinded Oppenheimer's authority to designate who came onto the Los Alamos site. From May 8, 1943, only he, and he alone, would have the power to authorize visitors to the secret city. Under his supervision, Oppenheimer would from then on periodically issue supplemental "Notes on Security," clearly designed to clarify and further amplify the policy of compartmentalization laid out in the original "Memorandum on the Los Alamos Project," which had been sent out to the scientists before they came and had served as an ominous introduction to their future life. The first memo had been stamped "Restricted" and warned the reader not to share the document with anyone but known project members—wives could be acquainted with the contents only if they were sworn to silence—and included a copy of the Espionage Act. It had provided the barest minimum of information, stated that limited movement outside Los Alamos and in the immediate area would be permitted, and concluded with a stern note from Oppenheimer:

> The extent to which we shall be able to maintain this comparative freedom will depend primarily on our success in keeping the affairs of the laboratory strictly within the confines of the Laboratory, on the cooperation which the project personnel affords us in its discretion on all project matters, and on our willingness to rupture completely our normal associations with those not on the project.

A subsequent note from Oppenheimer on May 22, 1943, which had Groves' fingerprints all over it, was far more severe in tone: "It is important that our personnel should maintain no social relations with people in the neighboring communities."

It was not enough that the scientists were shut away from prying eyes; they were to avoid any human society beyond the gates. The censorship regulations dictated that they never introduce themselves to local residents, identify their occupation in any way other than as "engineer," or mention where they were staying even in the most general terms lest the smallest detail betray their location. "Don't mention the topographical details which are essential to the Project," another security pamphlet warned. "Box 1663" was the only address they were allowed to give out—it went on their new IDs, forwarded mail, boxes containing their furniture, auto registrations, driver's licenses, insurance policies, bank accounts, and ration coupons. They could not sign their name to any form, even something as innocent as a library card, that might give away their identity or whereabouts. They were forbidden to travel more than one hundred miles from Los Alamos, circumscribing their lives within a tiny fragment of the map bordered by the New Mexican towns of Albuquerque, Cuba, Las Vegas, and Lamy. Exceptions would be made only for trips related to laboratory business and personal emergencies. If they passed people they knew on the street, they were to cut them dead. If they were recognized at any time, or happened upon an old friend from their former life, they were to keep the conversation short and would have to submit a lengthy report to the security force. They were RESTRICTED, as they were reminded again and again in large, bold letters in every new instruction booklet and directive.

Leaving the post for even a short time was not an easy proposition. The fence that enclosed Los Alamos was surrounded by an outer fence, separating the two-by-eight-mile-wide strip of land from the rest of the world. There were only two ways to access the closed city: the Main Gate, or East Gate Guard Tower, which was reached by the desert road from Santa Fe; and the less used Back Gate, or West Gate, which opened onto mountain country and woods and led down to the Valle Grande, an immense volcano basin now covered in pasture. Each guard station was posted with red-and-white warning signs that read:

U.S. GOVERNMENT PROPERTY
DANGER! PELIGROSO!
KEEP OUT

The uniformed sentries did not permit anyone to enter or leave the compound without proper authorization. The elaborate inspection process quickly became the bane of everyone's existence. The passes handed out to people by Dorothy McKibbin when they first arrived in Santa Fe were good for only twenty-four hours, which required newcomers to report with dispatch to the Pass Office on the Hill, located in the old stone pump house. There they would be photographed and fingerprinted, and subjected to the first of many strange and humiliating interrogations. Did they have any unusual birthmarks? Identifying scars? All this would be duly recorded in their files and noted on their permanent passes. They would then be informed of all the secrecy regulations and required to sign the Espionage Act. But even then they were not home free. The credentials had to be renewed every two weeks in the Pass Office, or the MPs would solemnly shake their heads and they would not be allowed through the gate. Many an irate physicist and his tearful wife were detained at the gate, or turned away completely, because their passes had expired. Hans Staub, one of the project's resident philosophers, was convinced he saw ominous implications in the steel fences. "Are those big, tough MPs, with their guns, here to keep us in or to keep the rest of the world out?" he demanded once, only half in jest. "There is an important distinction here and before I leave this place, I want to know the answer."

Telephone calls were routinely interrupted by overzealous monitors, who cut the connection every time a curious relative asked a probing question in order to caution personnel before they replied. Security was also steaming open their mail and perusing the contents. Adding insult to injury, Groves denied this was happening and maintained that a policy of self-censorship was adequate. But as an experiment, Fermi once slipped a strand of his own hair into a letter to his wife. She reported back to him that when she opened it, the hair was gone. As the months went on, the rumors persisted that G-2 was opening and reading letters, and the complaints increased. Groves determined that censorship was not such a bad idea, and Oppenheimer went along with it.

A list of regulations was then drawn up by army officials and circulated to the staff. Specifically, this meant that in addition to not revealing any details about their location, they were not to mention the names of colleagues at the lab or anything relating to their work or equipment. Sending photographs of Los Alamos, drawings, even doodles, was strictly against the rules. They were instructed to put their mail in the box unsealed, and the censors would collect it, and cart it off to the cramped offices next to Dorothy's on East Palace, where it would be read, sealed, and sent on its way. If the censors did not like what they read, they would blue-pencil the offending passages and send back the letter with suggested revisions. When Shirley Barnett sent her mother a sketch of how she had arranged the furniture in her new apartment, along with the dimensions of her tiny living room, the letter was returned with a note requesting she excise all the physical descriptions. Incoming mail was also read and resealed, and it arrived bearing the stamp "Opened by the Army Examiner."

The system was simple, but the psychological repercussions were complicated and profound. Letters from Los Alamos tended to be terse.

Bob Wilson's wife, Jane, recalled being too "painfully self-conscious," and altogether too terrified that she might in some way incriminate herself, to put pen to paper: "I couldn't write a letter without seeing a censor poring over it. I couldn't go to Santa Fe without being aware of hidden eyes upon me, watching, waiting to pounce on that inevitable miss-step. It wasn't a pleasant feeling." She wrote of a chance encounter with a college friend on the streets in Santa Fe that left her badly shaken. It had been thrilling to see someone "from the outside world, someone whose life wasn't mixed up in supersecret matters. But even this encounter was against the rules":

> "Come have a Coke with me," my friend suggested, little realizing the enormity of her proposition.
>
> I was numb with embarrassment. Woodenly, I accepted the invitation, although my conversation was a succession of fluid grunts. A moment's slip and I, by nature blabbermouthed, felt that I would find myself hurtling into the gaping entrance to hell. It was a relief to say goodbye. Then, like a child confessing she has been naughty, I reported my social engagement to the Security Officer. Everything had to be reported to the Security Officer. Living at Los Alamos was something like living in jail.

The scientists and their families tolerated these indignities and mourned their lost privacy and freedom. Accustomed to the lax atmosphere of university campuses, they reacted with varying degrees of resignation, anger, and dismay to their confinement. Even though they knew the importance of wartime secrecy, they could not help being disconcerted by the daily reality of being watched. On brief excursions into nearby Espanola or Santa Fe for supplies they might find themselves tailed by security. There were rumors that their apartments were bugged and that one or two physicists had been pulled aside and warned not to talk about their work at home. They felt security was always breathing down their necks, and Dorothy saw the toll it took on their behavior. They became increasingly stiff and unnatural in public, and at times the reply to the most casual question would suddenly freeze on their lips.

They were so plagued by G-2, she wrote, that project members would turn pale with fright if they thought their cloak of obscurity was about to be lifted. "When a shopkeeper automatically inquired, prompted more by way of western hospitality, 'Where are you from?' the answer was always a stammered Box 1663, as the speaker faded into the background. Security allowed them to say no more."

For the most part, the scientists and their families tried to be cooperative with their guardians. They wanted to safeguard the secrecy of the project, even if many of the rules seemed contradictory and vague. But to Dick Feynman, who was very young, very bright, and barely out of school, the censors seemed pointlessly arbitrary and high-handed, and he delighted in foiling them at every turn. Feynman was precocious, even by Los Alamos standards, and full of mischief. Despite his youth, he was included in all the high-level meetings with the senior scientists because he was regarded as unusually brilliant and innovative. He had only agreed to come to Los Alamos with the promise from Oppenheimer that he could regularly visit his wife, who had tuberculosis and was convalescing in a sanitarium in Albuquerque, a few hours away. Feynman had married his high school sweetheart and wrote to her almost daily. To make the correspondence more amusing for them both, he suggested she write to him in a code of her own devising that he would then decipher. When her letters arrived, written in what appeared to be gibberish—"TJXYWZTW1X3"—the censors naturally queried the content. Feynman would explain that the text was written in code and that he did not know what it said because he did not have the key. The censors did not approve of this game, and after much back and forth, it was finally agreed that she would enclose a key so they could read the letter and that they would then remove the key before forwarding the letter to him. But after trying to figure out a few more Feynman missives, Captain Peer de Silva, the chief of security at Los Alamos, changed his mind and declared, "No codes."

The fun did not end there. The Feynmans graduated to even more elaborate forms of cryptography. Letters from Feynman's wife began to arrive with words whited-out, whole passages missing, and in one instance "a hole cut out of the paper." One of the accepted rules of censorship on the post was that the censors only tampered with the outgoing mail. Everyone knew they also monitored the incoming mail, but "they were not supposed to take anything out." In his reminiscences of Los Alamos, Feynman described in detail the way he exploited every loophole in the system to drive the censors to distraction:

> There was always some kind of difficulty with the letters going back and forth. For example, my wife kept mentioning the fact that she felt uncomfortable writing with the feeling that a censor is looking over her shoulder. Now, as a rule, we aren't supposed to mention censorship. We aren't, but how can they tell *her?* So they keep sending me a note: "Your wife mentioned censorship." *Certainly,* my wife mentioned censorship. So finally they sent me a note that said, "Please inform your wife not to mention censorship in her letters." So I start my letter: "I have been instructed to inform you to not mention censorship in your letters." *Phoom, phoooom,* it comes right back!

At a certain point, Groves decided he had had enough of the budding genius's hijinks. This may have been when Feynman took to safecracking for his own entertainment and worked out how the Los Alamos combination locks functioned, which enabled him to open anyone's safe just by listening to the tiny clicks the knob made when turned backward and forward. He enjoyed embarrassing colleagues by leaving their safes open so they would be scolded by security, or by putting little notes inside signaling that he had struck again. The last straw may have been when he chose to demonstrate how easy it was to get into the classified facility by finding a hole that had been cut in the fence by some of the local workmen, who used it as a shortcut. Feynman went out the gate and in through the hole, and out and in again, round and round until he finally caught the attention of the sergeant at the gate. It is likely the general drew the line at having one of his scientific minions, a jumped-up graduate student no less, holding his security measures up to ridicule. Oppenheimer was forced to run interference for Feynman again and again, and had he not been so gifted, and his wife so desperately ill, there would have been hell to pay. As it was, his pranks quickly became a favorite topic at dinner, and if nothing else, the rapidity with which his more outrageous escapades made the rounds testified to how efficient the classified community's gossip network had become. "He caused a lot of trouble," said Greene. "But Dick always looked distraught, and Oppie made allowances."

All through the war, the mysterious little hole in the fence kept reappearing, and Indians from the nearby pueblos would climb through and come to the picture show, which cost twelve cents, and buy Cokes at the PX. There was a persistent rumor that Oppie was responsible for that hole.

For a very few, life behind Groves' tall fences was too claustrophobic to bear. Worst affected were the recent refugees from Europe, for whom the guards, dogs, and concertina wire brought back memories of the internment camps they had recently left behind. "A few people were absolutely incompatible with Los Alamos," acknowledged Greene. "One man arrived, his name I fortunately don't remember, and left within twenty-four hours. One girl was actually sent away. Then there were a couple of people who didn't get along very well but stuck it out. There always was the question [for Oppie], of whether you really should persuade someone to come because it was going to be such a close and difficult place to live and maybe they would be more trouble than they were worth." One of the few women physicists, Jane Roberg, who worked on calculations for the fusion weapon, asked to leave because of personal problems. The Swiss-born theoretician Felix Bloch also could not stand it. He had managed to escape Hitler's Germany and immigrate to America, only to find himself at the mercy of Groves' restrictive policies and Oppenheimer's equally autocratic management of the laboratory. The aristocratic Bloch made no secret of the fact that he was appalled by the experimental desert laboratory and felt the whole undertaking was a waste of time. In his opinion, radar, and not the bomb, would decide the war with Germany, and he asked to be released from the project. After weeks of simmering tension, Groves and Oppenheimer gave Bloch permission to depart, which they were loath to do given his abilities and how much he knew of the operation.

Teller always felt that his friendship with Bloch was a mark against him, and noted that the only time Oppenheimer invited him to join his regular poker games, a gathering of considerable status, was at the precise moment that Teller had promised to drive Bloch to the station. Teller viewed this as a typically manipulative Oppenheimer ploy and chose to see his friend off instead. As a parting gift, Bloch presented him with a plaque, purchased in one of the tourist shops, which fully expressed his contempt for the project: it depicted a car driving headlong into a tree.

SEVEN

Summer Camp

As THE CHIEF ARCHITECT of their misery, Groves was blamed for everything that disgruntled the scientific populace. He had made the mesa his de facto headquarters, and he walked around the place as if he owned it and rubbed almost everyone the wrong way. He would march past Priscilla Greene's desk heading straight for Oppie's office and, without breaking stride, would rebuke her for the dirty fingerprints on the door, demanding, "Don't you ever wash your hands?" Greene, who routinely listened in on all calls and took notes, as Oppenheimer could not possibly keep track of the dozens of conversations he had with people over the course of the day, would feel hurt and insulted when Groves bellowed, "Get her off the line." She complained bitterly to Oppenheimer, who shrugged and said that the general was tough on subordinates as a matter of policy. "I remember Robert sitting me down and telling me about Groves, that he could be mean, and sometimes had a nasty tongue," said Greene. "He said he did it on purpose, to keep people under his thumb, and to create an aura around himself."

Groves expected Los Alamos to operate like a military base, with the same discipline and austere living conditions. As a result of his tightfisted, no-frills policies, Los Alamos had none of the civilizing touches found in even the smallest towns, such as paved roads or sidewalks, forcing its inhabitants to slog through the spring mud like pig farmers. The scientists were further irritated by the fact that some aspects of life on the post seemed unduly primitive. In a typical requisitioning screwup, the army had ordered enormous cast-iron coal- and wood-burning stoves, known as "Black Beauties," which took hours to get started and which, once they got going, invariably turned the cramped quarters into saunas. Beyond the inconvenience and discomfort, there was the very real threat of fire. The flimsy wooden apartment buildings were veritable fire traps and, to make matters worse, came equipped with furnaces designed for buildings ten times the size. There was a constant danger that either the furnaces or the stoves would overheat. Every time the siren sounded, people rushed from their homes clutching their children for fear the whole block might be consumed in minutes. For many disillusioned wives, the Black Beauties were the last straw. When Jane Wilson complained that cooking on the smoke-belching monstrosities was an unnecessary hardship, Groves scoffed at her faintheartedness. The general insisted on coming over to her apartment and giving her a lesson on how it should be done, eventually getting down on all fours and blowing on the kindling for all he was worth. At 7,300 feet, however, a fire is slow to light. After an hour's exertion, the general succeeded in coaxing a flame from the box and departed, covered in soot. Shortly thereafter, the army issued everyone electric hot plates.

"People—I know I certainly did—used to blame everything on General Groves," said Greene. "When Hans Staub left, he wrote a letter to General Groves on the wall. I used to compose letters to General Groves in my head when things were particularly tough. I once almost set the whole apartment on fire [using] my neighbor's Black Beauty and got too big a fire going. . . . Looking back on it years later," she added, "it seems fairly adventuresome and not too difficult, but sometimes you felt a little put upon."

Despite all the technical matters he had to attend to, Oppenheimer tried to solve these domestic crises as they came up and to look after his staff's welfare. His concern for their happiness was perhaps less altruistic than practical, as he realized keeping the scientists and their families content was essential to securing a stable and productive workforce. He allayed tempers and injured feelings with his patience, an informal "we're-all-in-this-together" spirit, and warm first-name-basis relationships with everyone on the mesa. After Groves overheard some of the Italian-speaking physicists talking animatedly in their native language over lunch at Fuller Lodge, he went over to the table, which included Fermi, Bruno Rossi, and Segrè, and, as the latter indignantly recalled, informed them "he did not like us speaking Hungarian!" Groves went on to say that he expected English to be the only language spoken on the mesa. Understandably, that did not go over too well with the recent immigrants and their families—Teller among them—and the general's recommendation was roundly ignored. But when Groves complained that he continued to hear foreign languages spoken over dinner, Oppenheimer was forced to convene a half dozen top physicists and delicately suggest they refrain from speaking in their native tongues in public spaces. What they did in their homes or off campus, he added, was their own business.

From the beginning, Oppenheimer relied on Dorothy to deal with the more disgruntled wives—to listen to their troubles and sorrows, provide sound advice, and find ways to make their life at Los Alamos a little easier. She served as the project's unofficial den mother, soothing frayed nerves and alleviating the loneliness that came with the desert and great distances from friends and family. She reassured skittish young faculty wives, who had never spent a day in the wilderness before, that there was little chance they would find a rattlesnake in their shoe, and promised that the project's tall fences would keep out the coyotes they heard howling at night. Bears were occasionally sighted on the dry hills, but Dorothy advised people that the bears generally ran at the sight of humans and would not bother them if they did not bother their cubs. She told them the Rio Grande was usually too muddy to swim in, but the cool Santa Cruz Reservoir was less than an hour's drive and would make a nice Sunday outing. Dorothy would endeavor to find out what they were missing most, going to great lengths to procure a pony for a moping youngster, pumpernickel bread for homesick Austrians, and wines that had never been seen in those parts before. "All women brought their difficulties and their checks to Dorothy," wrote Laura Fermi in a memoir of the war years. "She endorsed the latter so they could be cashed at the bank, and smoothed out the first. . . . Women always came out of Dorothy's office with greater cheer than when going in."

Her door was always open. Despite all the comings and goings at 109, and the piles of boxes and packages that sometimes threatened to overwhelm her small office, she always had time for a chat and a considered appraisal of any problem. She never ceased to amaze people with her promptness and ingenuity in satisfying both their urgent needs and their more whimsical requests. She kept their secrets and, as their only contact with the outside world, often helped resolve personal problems requiring the greatest delicacy and discretion. She could be counted on for her common sense and sympathetic ear, except in the case of whiners. She had little patience for sissies, and said as much. So the altitude of 7,300 feet made some people woozy. So water boiled at 198 degrees Fahrenheit, foods took forever to cook, and a simple meal of meat and potatoes might take three hours to prepare. It wasn't the end of the world. She gave them a pressure cooker and told them to stop complaining. If she thought someone was indulging in a prolonged bout of self-pity, she would remind her that there was a war on and of the overriding importance of the work that had to be done. After all, Dorothy would scold, as a scientist's wife she was lucky to be together with her loved ones, with a husband who was coming home every night, and she should stop talking of leaving and start making herself useful. Oppenheimer, whose own wife could be a handful, was grateful to Dorothy for lightening his load, and he increasingly relied on her for help. He understood that Dorothy's so-called housekeeping activities encompassed hundreds of small tasks that were essential to so large an operation and that her good humor and kind ministrations went a long way toward making sure he did not lose any more employees, as he could not afford to lose a single one.

As they settled into life on the Hill, the scientists were helped by a feeling of pent-up excitement, which Bethe likened to the exhilarating first week of summer camp. It helped that the spring weather had turned fine and mild, and that the thin dry air tasted like champagne, making them feel giddy and feverish. On April 22, 1943, Kitty Oppenheimer threw a party to celebrate Oppie's fortieth birthday, and everyone discovered that at that altitude, a little liquor went a long way. Oppie worked his usual magic with martinis, and after just one cocktail, many of the revelers found their heads swimming and their balance seriously impaired. Records provided the music, and the drinking and dancing progressed. As was the case with most mesa parties, this was a chance to let off steam, and most of the revelers stayed until well after eleven, wandering home with the aid of flashlights to avoid falling into one of the many ditches by the side of the road and breaking an ankle. Given that everyone lived in virtually identical apartment complexes, which they called "greenhouses," there was much hilarity and blundering about in the dark. The occasion was marred only by the McMillans absence, and the fact that Oppie grew increasingly frantic as the evening wore on and they failed to turn up. Much later that night, they learned that there had been a car accident and the convertible the McMillans had been riding in had slid off one of the steep mountain roads and flipped over. The Jensens and the Weinekes, who had been leading the way in their own car, had gone for help. Ed McMillan escaped with only a cut on his forehead, but Elsie's skull was fractured, and she would require months of rest and hospital care. It was a cautionary tale that discouraged many Los Alamos wives from venturing down the mountain on their own, though the jolting ride up and back on the army bus, with the back end hanging over the edge on the turns, was almost as hair-raising.

Everyone took turns visiting Elsie McMillan at the tiny, barrackslike Los Alamos hospital, turning it into a regular meeting place during her long recuperation. A statuesque redheaded nurse named Harriet ("Petey") Peterson presided over the five-room facility, and her dog, an enormous Irish wolfhound named Timoshenko, sleepily kept guard outside the front door. The post's lawyer, Smitty Carlisle, made regular excuses to stop by and flirt with Peterson, and despite giving him a hard time, she would marry him many years later.

The project's very tall, very blond medical director, Louis Hempelmann, was also in love, and he spent much of his free time conducting a telephone courtship with Elinor Pulitzer, the beautiful daughter of the newspaper magnate Joseph Pulitzer II, the editor and publisher of the *St. Louis Post Dispatch.* He finally worked up the courage to propose, and

Oppenheimer gave him permission to slip away to get married. When he returned shortly thereafter with his glamorous bride, whose debutante ball eighteen months earlier had made all the society columns, her lavish wardrobe was the talk of the mesa—particularly after she let it be known that one of her Palm Beach ensembles cost $500, which was more than many of the physicists earned in a month. She came from a completely different world than everyone else on the Hill and had no idea how to cook or clean or cope without staff. But she was very funny about her wartime predicament and willing to learn, and everyone admired her gumption. Not long after she arrived, she decided to throw a big dinner party. Unaware that it would take hours for a roast to cook at that altitude, she neglected to put the meat in the oven until her guests arrived. Everyone got so drunk waiting for the food that she earned a reputation as an excellent hostess. The last thing anyone could remember that night was Groves trying to negotiate the stairs of the duplex, saying, "I'll go down first," and Kitty Oppenheimer following right behind him, saying, "That's just fine, General. I'll fall on you."

To the younger members of the project, it seemed as if they had been invited to take part in a grand adventure, along with many of the most famous scientists in the world. They were awed by the collection of talent around them and happily put up with the inconveniences and exhausting hours for the chance to sit at the feet of great physicists like Oppenheimer, Bethe, Rabi, the brilliant mathematician John von Neumann, and the incomparable Enrico Fermi, who would visit from Chicago. "We had a sense of patriotic duty, but also a sense of being privileged to be in that company," said Harold Agnew. "That they would pay any attention to us was amazing. It was really a thrill to be there."

By the time Beverly Agnew caught up with her husband, the worst of the mud season was over and the town was beginning to take shape. Los Alamos struck her as an amazing cross between an alpine resort and a mining camp. She had a fleeting impression of a ski lodge, inspired by the wooden chalets and distant snowcapped peaks, and the spectacle of so many people sportily attired in Western clothes and boots. "I remember when I stepped off the train at Lamy, I was wearing a proper tweed suit, and I wore gloves, heels, a hat and carried a purse," she said. "And here was this woman waiting to pick me up, and she had on blue jeans! At that point in my life, I had never seen anything like it. I will always remember how very shocked I was. Of course, after that, I wore them all the time."

Beverly rolled up her sleeves and took on the peculiarities of life on the utilitarian post as a sporting challenge. In the beginning, she and Harold more or less "camped out" in a small, two-bedroom apartment with another couple and six men whose families had not arrived yet. Each couple had their own bedroom, and six cots were set up in the living room. When they finally got their own place, she found the cramped apartment no worse than most graduate student housing. The Commissary shelves were stacked with canned goods, albeit many of them unheard-of brands and unheard-of sizes, including industrial-sized jugs of mustard and vanilla extract and two-gallon tins of peas and corn. But the meat was better and more plentiful than what was available at home because of the wartime shortages. "Those were chaotic days, but for us they were fun," she said. "I cooked a lot, and it wasn't unusual to have three or four Nobel Prize winners to dinner."

Beverly went straight to work in Oppie's office as promised, helping Priscilla Greene get all the new arrivals sorted. By then, Oppenheimer's original plan for thirty scientists and their dependants had long since fallen by the wayside, and the second rough estimate of a hundred or more was already a distant dream as the number needed for the project continued to snowball. "I guess Oppie and I both had this amazed feeling—everyone did," recalled Greene. "You didn't know what was going to happen. It just happened, and then all of a sudden you'd say, 'Wow, here we are 150. Wow, here we are 750.' Once in a while you had to really make a big plan because there wasn't any place for people to live. Nobody sat down and said, 'Gee, we must have made a mistake; we didn't plan it right in the first place.' It just happened to us all. There was nothing you could do about it."

They soon resigned themselves to the fact that the housing construction would never stop. Huge army bulldozers continuously tore away at the rock cliffs, leveling everything in their path. One morning, Mici Teller had settled on a blanket under a shady clump of pine trees near her apartment and was preparing to feed her baby son his bottle, when a young soldier politely asked her to move. He had orders to clear the surrounding area. Determined to preserve her shade trees from the plow, she refused to budge. The soldier left, but he returned the next day insisting that he had orders "to finish this neck of the woods." Every bit as stubborn as her husband, Mici rallied a group of wives and the women all set up chairs under the trees. There was nothing the soldier could do but report back to his commanding officer. When advised of the standoff, Groves relented and said, "Leave the trees." As Edward Teller later observed, even their general, who was not accustomed to deferring to civilians, seems to have possessed "a little knowledge about human engineering."

As the operation swelled in size, the office became so desperate for support staff that Oppenheimer had to send Priscilla Greene back to Berkeley to recruit secretaries. "Finding appropriate people was very difficult," she said. "People called friends, and friends of friends. If you knew somebody who might be useful, they would be brought in." The Serbers suggested Priscilla's twenty-one-year-old kid brother, DeMotte, who had mastoids and was 4F, and the next thing she knew he was joining the project as a technician. Sam Allison came into the office from the Met Lab in Chicago and brought a secretary. Eventually, Bob Bacher called on his brother-in-law, David Dow, a New York lawyer, who arrived one day in a topcoat and bowler hat and took over a lot of the administrative tasks from Oppenheimer. "It became a proper office for a director of a large laboratory instead of just one fellow and a girl typing," said Greene. In the meantime, Dorothy screened local employees and managed to hire a fetching blonde named Mary McCauley to help with the office work. She almost immediately caught the eye of the physicist Art Wahl, and everyone, including Oppie, enjoyed the sport of watching Los Alamos's first budding romance.

New people were arriving daily: electricians, engineers, plumbers, carpenters, machinists, clerks, soldiers, schoolteachers, nurses, doctors, even a much-needed post pediatrician. The post's medical director, Louis Hempelmann, had been recruited from the Mallinckrodt Institute of Radiology in St. Louis, where he had worked in nuclear medicine using a cyclotron. He brought with him several classmates from the Washington

University School of Medicine in St. Louis, including Henry Barnett, James Nolan, and Paul Hagemann. Henry Barnett had trained as a pediatrician and had only recently completed his degree when he was asked to enlist and issued his travel orders. Barnett and Nolan, the post's ob-gyn, arrived in the nick of time. The story on the Hill was that when Hempelmann went to deliver his first baby, the young radiologist passed out at the sight of all the blood. When he woke up, he was recovering on a bed next to the new mother.

When Henry Barnett and his twenty-four-year-old wife, Shirley, arrived at 109 East Palace Avenue in June, all they knew was that he had been temporarily stationed in Santa Fe, presumably before traveling on to Europe. Dorothy McKibbin gave them their passes and took them to lunch at La Fonda, where she gently broke the news that they were headed to Los Alamos and that once the gates closed behind them, they would be there until the war was over. When Shirley Barnett heard that she would be allowed no visitors, and no visits home, she burst into tears. "It was scary as hell," she said. "The moment Henry and I were alone for a moment I said, 'I don't want to stay here.' Of course, there was no choice. I wasn't scared of the bomb because I didn't know about it then. I was scared that I wouldn't be able to leave, or to see my family or home for months, even years. Henry was a pediatrician, but he was in the army and involved in who knows what. The whole picture was very frightening. I couldn't believe it was happening."

Dorothy was warm, steady, and understanding, and she helped Shirley come around to the idea that as a wartime assignment, Los Alamos was not the worst place to be. Before lunch was over, Dorothy had her laughing over the maddening idiosyncrasies and foibles of army life, and the two women parted fast friends. In times of great uncertainty, friendships can flower in a matter of moments, and strangers can become inseparable in the blink of an eye. In their small, inward-turned society, it was only natural that like-minded people bonded quickly. Pretty, petite, and sharp as a tack, Shirley Barnett immediately hit it off with Priscilla Greene and Charlotte Serber, and on their recommendation, soon found herself working for Oppenheimer. (Beverly Agnew was pregnant, and Oppenheimer was once again in need of another assistant.) Thanks to an unforeseen baby boom, encouraged no doubt by the lack of distractions after dark, Henry Barnett became the most popular man on the mesa, and between the two of them, the Barnetts were privy to most of the unofficial secrets at Los Alamos.

Working in Oppie's office, Shirley Barnett should have been subject to a rigorous security clearance, but Oppie did not care and asked her to start before her clearance came through. All Los Alamos personnel, along with the family members accompanying them to the site, were expected to fill out the detailed questionnaires. At first glance, the government forms seemed standard, requesting information on family background, relatives, schools, military service, employment since 1935, and travel to foreign countries since 1935 and the reasons for each visit. But the most telling question, according to Barnett, was the one asking for a list of memberships in organizations since 1930. Three character references were also required, along with their addresses. "You could tell the clearance forms the G-2 filled out on people were mostly concerned with Communist connections—had they ever been Communists, did they have friends who were Communists, and so on," she said. The security checks, supervised by the War Department counterintelligence operations, could take upward of a month, though during heavy recruiting periods they could take much longer. Shirley worried about this for a while, but just sitting around waiting was driving her crazy, so she went to work.

Every potential employee was investigated to some degree, but because of the thousands recruited to work at the Manhattan Project laboratories, support staff—truck drivers, cleaning staff, cafeteria workers—with no access to classified material were given only limited checks. By contrast, the background investigations of top physicists, particularly those who were foreign-born, were, in Groves' words, "most thorough and might go back to infancy." Everyone was fingerprinted, and the files forwarded to the FBI. The main barrier to clearance was any attempt to conceal an arrest record, although given the demand for certain highly skilled laborers, exceptions were made in cases of traffic violations and drunkenness. No one convicted of rape, arson, or narcotics charges was hired because of "demonstrated weakness in moral fiber." The objective, according to Groves, was "to find out before employing anyone whether there was anything in his background that would make him a possible source of danger, paying particular attention to his vulnerability to blackmail, arising from some prior indiscretion."

A new person was supposed to be kept on "nonsecret phases" of the work until the hurried investigation could be completed, but in practice the employee's lack of clearance was overlooked. Security officers never approved Shirley Barnett, even though she worked in Oppenheimer's office and handled highly sensitive correspondence and classified documents. "Many people, including me, were never cleared because the standards were so ridiculous," she said. "If you had any kind of leftist background—and anybody worth their salt would have had some connections during the 1930s when anti-fascism was popular among intellectuals—you were never finally cleared. We all sort of ignored it because we were all there, and they wanted us to work." During the war, many clearance forms crossed Oppenheimer's desk, and Shirley vividly remembered seeing one where Groves had impatiently scrawled: "The Communists are our allies in this war. It's the Nazis we are fighting. Go look at them." Thereafter, she had a soft spot for the general. "Groves did things his own way," she said. "He wasn't an idealogue. He cared about security, but he could be unorthodox when it suited him."

One way or another, the security arrangements continued to preoccupy their thoughts, particularly as the mesa's population continued to mushroom, and the many new employees thronged the cafés and curio shops in Santa Fe, and uniformed WACs and navy officers began to outnumber the tourists on the streets. Rumors circulated about what was going on up at Los Alamos, where so many people, packages, and delivery trucks kept disappearing. Oppenheimer was worried that people would start asking questions about all the scientists, and about the loud booms that had begun to ring out from the canyons and could be heard in Santa Fe on fine mornings. He asked Dorothy to try to discourage all the talk by reaching out to her wide network of friends and contacts, but despite her best efforts the gossip continued unabated. One popular story had it that Los Alamos was a wartime plant that made windshield wipers for submarines. Others insisted that workers were actually assembling submarines in a factory. This theory persisted even though there was no deep water for hundreds of miles around: against all reason, people actually believed the army had cut a secret passage to float the subs down the Rio Grande.

A vacationing reporter from the Cleveland *Plain Dealer* followed up on the rumors and got as far as the first guard station before being turned back. Refusing to be deterred, he published an article entitled "The Forbidden City" and speculated excitedly about all the secret doings. The article was accompanied by a cartoon parodying the gate and the MPs in their battle helmets. "The Mr. Big in the city is a college professor, Dr. J. Robert Oppenheimer, called the 'Second Einstein,' " he reported. "[A] widespread belief is that he is developing ordnance and explosives. Supporters of this guess argue that it accounts for the number of mechanics working on the production of a single device." When the *Plain Dealer* break was picked up by *The Santa Fe New Mexican*, it sent security into a tizzy, and the Manhattan District's Counter-Intelligence Corps (CIC) came down hard on both papers.

After that, the security boys really cracked down on the local press, patrols were increased, and everything tightened up. Even if keeping Los Alamos an absolute secret was a logistical impossibility—innumerable day workers, delivery truck drivers, Indian housemaids, and assorted cowboys passed through on a daily basis—if the army conceded the effort altogether, the project would have to close up shop. Instead, Oppenheimer decided it would be advisable to put out a "story" about the project that, he wrote in a memo to Groves, "if disseminated in the right way, might serve somewhat to reduce the curiosity of the local population":

> We propose that it be let known that the Los Alamos Project is working on a new type of rocket and that the detail be added that this is a largely electrical device. We feel that the story will have a certain credibility; that the loud noises which we will soon be making here will fit in with the subject; and the fact, unfortunately not kept completely secret, that we are installing a good deal of electrical equipment, and the further fact that we have a large group of civilian specialists would fit in quite well. We fur-

> ther believe that the remoteness of the site for such a development and the secrecy which has surrounded the project would be appropriate, and that the circumstances that a good deal of work is in fact being done on rockets, together with the appeal of the word, makes this story one which is both exciting and credible.

After gaining the approval of the governing board of the laboratory, Oppenheimer informed his staff of this "official" smoke story and, in an unnecessarily elaborate touch, further instructed them "neither to contradict nor to support a story of this kind should they run into it." Oppenheimer and the lieutenant in charge of security decided that it would be a good idea to launch their rumor counteroffensive by organizing a small party of staffers to visit a few local watering holes in Santa Fe "for the purpose of deliberately spilling something about an electromagnetic gun." Oppie's first choice for the mission was Charlotte Serber, who it was generally agreed could talk the ear off anyone. Escorting her would be the always gregarious and charming John Manley.

One afternoon, Oppenheimer summoned them into the director's office, where Priscilla Greene and the G-2 lieutenant were already seated, and informed them of their mission. "No one is to be told of this assignment," Oppenheimer said. "If you are successful, you will be reported on by G-2 in Santa Fe and by other Los Alamosites who overhear you. You will be protected if you get into trouble, but for the moment it is a secret mission." The pair reluctantly agreed to go, but Charlotte requested that her husband be allowed to come along for moral support. Manley said he would take Priscilla and make their "spy ring" a foursome. With orders to spread false rumors all over town, they made a tentative date for the following evening. The four of them went down to La Cantina, the cocktail lounge at La Fonda, which was the main place in town and usually packed with locals as well as tourists, though on this particular evening business was rather slow. They got a table and drinks and began talking loudly among themselves in the hopes of ensnaring eavesdroppers, but their conversation was a bit awkward, and Charlotte could not help feeling a little silly and self-conscious. As Bob Serber recalled, it was "difficult to work electric rockets into the conversation":

> Nobody paid any attention that we could see, so after a while we left there and decided to go to a low-down bar. The bar was jumping, jammed and crowded and full of Spanish-Americans. We got a booth and drinks, and John and Priscilla started to dance and talk about electric rockets. Pretty soon a Spanish-American kid came over and asked Charlotte to dance. It turned out that he'd had a job as a construction worker on the site for a while and couldn't care less, and the only thing he wanted to talk about was his ambition to own a horse ranch.

None of the patrons of Joe King's Blue Ribbon Bar seemed the least bit interested in who they were or what they were saying. As the night wore on, they could see that their efforts were falling flat. Finally, Bob Serber announced, "Well, I'm just going to do it!" He bellied up to the bar and tried to draw the men on either side of him into conversation with the opening gambit, "Did you ever hear what they're doing up there?" As Serber was usually a rather retiring person, Greene looked on with amazement as he practically seized a man by the coat lapels. "He couldn't really get anyone to talk to him, and he was sort of a little bit drunk at that point—we all were—and he sort of shook one of them by the shoulders and shouted at him that we were building rockets," said Greene. "The only problem was that this fellow was so drunk, we were sure there was no way he would ever remember." It was 3 A.M. by the time they made it back to the post. On the long drive back, they all agreed they were flops at the spy business.

In the end, the rocket rumor never took. The foursome made several more forays to the Blue Ribbon Bar over the next two months, but those outings proved no more successful. On one of their last outings, Charlotte attempted to spread false stories about the project at a Santa Fe beauty parlor, while John Manley worked the patrons at the local barber shop. They came away feeling embarrassed and decided to retire as counterespionage agents. While the army was never too happy about it, probably the most enduring rumor about Los Alamos, no doubt prompted by Dorothy's scavenging scarce baby clothes and cribs for new mothers on the Hill, was that it was a home for pregnant WACs.

By August, people had relaxed and let down their guard a little. "The Santa Feans soon became accustomed to the queer ways of the scientists," wrote Dorothy. "They claimed they could spot them from a great distance." Laboratory personnel were permitted one day a month to do their shopping, and they streamed into town on weekends to hunt for luxuries like stockings and whiskey, which were always in short supply, and to fill up on Mexican food and take part in the local festivals. On Sundays, several couples would share their monthly government-issued C-coupons to buy gas, which was rationed during the war, and organize picnic outings to the Valle Grande, Jemez River, or Rio Frijoles, where there was good hiking and fishing and they could go for a dip in the river if the water level was high enough. Dorothy's office, adjacent to the shady courtyard bordered by tall hollyhocks and zinnias, was a busy hub of activity for project members and their families on their prized visits to civilization. Dogs were tied outside. Babies napped in strollers under the huge elm tree outside her door while their parents dashed around doing errands. Bachelors loitered by her desk and asked her advice on their love lives or on the best place to eat or to buy turquoise trinkets and cheap silver.

Dorothy invited them all to make her office their headquarters, a place where they could leave parcels, meet friends, or just stop by for a chat. "She had an air of handling people easily," wrote Bernice Brode, who was married to the physicist Robert Brode. "Only later did I come to know the serious difficulties she avoided for us all":

> She was very lovely, with shining hair and dressed in blue tweed to match her eyes. She had a quiet grace in the midst of all the hubbub. She was a hostess rather than a chargé d'affaires. . . . So 109 East Palace, and Dorothy, our only link with Santa Fe, became our private, secret club in the capital of New Mexico. There we could talk and make plans and have no fear of being overheard.

Dorothy gradually became more accustomed to dealing with her important charges and grew bolder within the secure confines of her office.

She flirted with the handsome physicists and, between a wink and a nod, cultivated a breezy familiarity that endeared her to one and all. This was clever company after all, and with the world in such a terrible state, and the work so unrelentingly serious, she felt the least she could do was provide witty conversation and a spot of relief. In an unguarded moment, she might even be so forward as to blithely violate the security injunctions that regulated their every move and communication, and to tempt them to follow suit. It was a game they all played, and it made the long hours easier and their hearts lighter. "If there was no stranger around and I was feeling very wicked," she wrote, "I would glance in all directions, examine the empty air, raise an eyebrow and whisper tensely, blowing through my teeth like a suppressed wind instrument, 'Are you a phhh ht?' And the young man would nod, and we would say no more, but smoulder within with shared excitement."

E I G H T

Lost Almost

AFTER THE CHAOS of the first few months, life at Los Alamos eased into a more predictable, deliberate pace. The summer sun was hot, and the sky, porcelain blue with white clouds hugging the mountains, as in a postcard. The Jemez hills were topped with skullcaps of pale green, creating the illusion of grass, though they were really only the thick carpet of pine trees. The mud had dried to a fine brown dust that coated everything and which the trucks and cars kicked up into great choking clouds as they tore through town. The deafening grind of construction had quieted, if only sporadically. Ashley Pond sparkled in the sunshine, and for the first time since the scientists and their families had arrived, a certain harmony and beauty returned to the mesa. Little girls played hopscotch in front of their houses, boys played ball in the dirt road, and everywhere you looked there were people on bicycles.

Oppenheimer, who enjoyed the rarified setting more than most, was at pains to point out the advantages of their splendid isolation. Sunburned as a native and clad in his uniform of blue jeans, checked shirt, and silver-studded New Mexican belt, Oppie urged everyone to get off the post on Sundays, their one day off, and wander out into the silent wilderness of aspens, blue spruce, and ponderosas. The hills were full of old trails established by generations of schoolboys, shady and sweet with the smell of pine. Oppie encouraged people to buy horses, and asked

Dorothy to recommend some reputable local horse dealers. The army also had saddle ponies that could be rented by the hour, and provided stables and corrals for a small fee. A number of the scientists were skilled alpinists and shared Oppie's romantic fascination with the mountains. The metallurgist Cyril Smith took to hiking up and down Lake Peak (12,500 feet), as did Fermi and Bethe, who claimed they did some of their best thinking that way. People took to calling Los Alamos "Shangri-La," a joking reference to the idyllic city hidden in the mountains of Tibet featured in James Hilton's popular novel *Lost Horizon,* which more than a few scientists had read before dropping out of sight themselves. They could not help but see their own community as a similar kind of social experiment, a separate culture, obscured from view, whose very existence was vigorously denied, which would either flourish or founder cut off from the outside world.

Down below, in the real world, the slaughter in the Pacific continued unabated. The victory at Midway that had been so heartening in June 1942 seemed like a distant memory. Though the German Wehrmacht was driving deep into Russia, it had been checked in the desert. Rommel had been stopped at the border of Egypt. The Italian campaign had begun. Allied bombs rained on Sicily, and the invasion of the European mainland seemed inevitable. The possibility that one of the three Axis powers might fall was cause for hope. President Roosevelt and Prime Minister Winston Churchill declared that the only terms the Allies would accept were those of "unconditional surrender."

The Los Alamos participants were starved for uncensored news on their mountaintop. Newcomers were thoroughly debriefed, and senior scientists returning from meetings in Washington or Chicago would find themselves surrounded at dinner at Fuller Lodge and inundated with questions. Almost everyone got *The New York Times* and devoured *The Denver Post.* People always fought over the local Santa Fe papers while waiting in line at the PX because there were never enough to go around. Dorothy scrounged all the extra copies she could and sent them up. She knew that many of her friends on the Hill were émigrés who had escaped Nazi persecution in their own countries. "Many had families and relatives still in Europe, living in poverty, concentration camps, subject to starvation or vicious death," she recalled. "Anxiety and fear for their people haunted them day and night."

Bad news from the front cast a shadow across the entire mesa. At the same time, a victorious battle could be counted on to put the junior military personnel in a particularly foul temper. It was widely known that they had all applied for overseas service and were "burned to the ground," as Smitty Carlisle put it, to find themselves stranded in the desert and missing all the excitement. One of the earliest bits of mesa lore had to do with a WAC who had never been farther west than Albany, New York. Just as the Hill bus rounded the highest point of the ascent and the jagged Jemez Mountains peeked out, she fainted dead away, though whether from amazement or sheer disappointment no one knew. Back in April 1943, when the first of the recently organized WAC detachments was ordered off the train in Lamy, Dorothy had had her hands full coping with the outpouring of grief and fury. "Of all the incoming personnel the WACs and some of the soldiers were at the lowest ebb in this office," she wrote. "They had not been told what was going to happen to them":

> One WAC told me she was not allowed to tell her closest buddy she was going overseas, and was whisked out of her bed at two A.M. and sent silently on her way. When the train ran west and stopped at Lamy she thought it was all a big mistake; the sand and piñon trees didn't look like any ocean she had ever seen. The nightmare continued when she walked into the old mud building which had nothing marine about it and was told she had forty-five miles yet to go farther into that "beat up old land." Many WACs were tearful in those early days, not that they weren't good soldiers, but the shock was too much for them.

Quite a few she ushered still weeping into army cars and sent up to the Hill. For weeks afterward, they could be seen moping around the mesa and were unnecessarily surly behind the checkout counters of the Commissary. Soldiers, who arrived with no further orders than Santa Fe, were also frustrated to learn they would be sitting out the war holed up in the desert. They did not even try to hide their contempt for the scientists, disdainfully referred to as "longhairs," they had been assigned to babysit. To them, Los Alamos was an interminable limbo—"Lost Almost," in the words of one dejected GI.

Theirs was a community turned powerfully in on itself, ironically in ways that, a century earlier, would have been true of people isolated by long weeks of travel from the rest of the world and forced to make what amounted to a civilization of their own. In this case, however, they were not inventing their own culture, so much as being asked to accommodate to the most unusual and arbitrary of arrangements, and discontent festered despite the magnificent surroundings. Segrè never failed to be impressed by the view, but it did nothing to ease his doubts about their internment. The mess hall in Fuller Lodge had a porch that opened onto the lawn, and he often stood there after meals silently contemplating the Sangre de Cristos, dominated by Truchas Peak, and miles and miles of the Rio Grande Valley. One afternoon after admiring the vista with Rabi, Segrè observed that in all likelihood, "after ten years of looking at it, we would have had enough of the view." Reminded of the comment by Rabi years later, Segrè said that it revealed "what we thought about the possible duration of our enterprise and of the war."

"It's hard to believe we could be unhappy in such a beautiful, gorgeous part of the world, but we hated it so," said Shirley Barnett. "All we could think about was home, what we missed—New England green. It was difficult because we did not know how long we were going to be there, we didn't know how long the war was going to go on, and it had been going on a long time. There were times when all I wanted to do was pull down the shade and never see another mountain again. It was hard not to feel trapped."

The altitude and isolation worked on everyone's nerves. Los Alamos ran on bells, military-style: the first whistle blew at 7 A.M., summoning the scientists to work in an hour's time. For a population that was not used to obedience as a way of life, the shrill sirens only seemed to magnify the immense pressure of impossible deadlines and heavy responsibilities. Oppenheimer, who had never been known to keep regular hours at Berkeley, was now the first to arrive in the morning and often stayed in his office long after everyone else had left for the day. Sam Allison, an experimental physicist who temporarily shared an office with him when he first came to the site from Chicago, said his one ambition was to be already sitting at his desk when Oppie walked through the door.

The laboratory personnel were keenly aware of Oppenheimer's nervous, chain-smoking presence, and his sense of urgency and determination drove them to extend themselves as never before. When security, following Groves' orders, locked up the equipment each evening following the five o'clock siren, the physicists took to sawing off the padlocks on the stockroom doors, and stayed at the lab and worked into the night. They set up army cots and snatched a few hours sleep in their offices, not stopping to go home at all. The men were consumed with the grim task of beating the Germans to the draw in the development of a diabolically powerful weapon, and the even grimmer prospect of what their failure might mean. First and foremost on their minds was the idea, as Conant emphatically put it on one of his periodic visits to the site, that "whoever gets this first will win the war."

The heady experience of coming together on the mesa to create a unique scientific enterprise had given way to the harsh reality of everything that had to be accomplished. Above all, recalled Manley, the knowledge that the German scientists had a head start was a constant hand at their backs, a reminder of the need to push ahead and notch some progress:

> Just before Los Alamos really got going, the last measurements on how much uranium 235 might be needed for a weapon had increased over the previous low estimate by almost a factor of 2; it was about 5 kilograms in absolute amounts. These 5 kilograms meant nearly two months extra production for each weapon from the electromagnetic separator which had been authorized at a hundred grams a day. Since we had no idea where the Germans were in this whole business—whether they had isotope separation plants going, whether they had a chain reaction going and were making plutonium, or were almost ready to drop bombs—these two months could mean we'd lose. However, there was a

> chance we could recover some of this apparent loss. Maybe, if we were really clever and got an extremely good material we could get back most of that factor of 2 that we had just lost. We were playing that kind of game continually.

The scientists were hardly alone in feeling the pressure. The physicists' families also lived with it, lived with the feeling that the work took precedence over everything, that the work had to get done. "It was always hurry, hurry, hurry!" wrote Elsie McMillan. "Work this morning, work this afternoon, work until 4:00 A.M. Work, work, work." To most of the physicists' wives, it seemed as if the men disappeared into the Tech Area each morning as if into the belly of the beast, leaving them to fend for themselves. "The Tech Area was a great pit which swallowed our scientist husbands out of sight, almost out of our lives," recalled Ruth Marshak, the wife of Robert Marshak, the deputy head of the Theoretical Division. Lonely, confused, and anxious, many of the women felt alienated from their husbands as never before. Physicists who for years had come home and bored their wives with endless hours of shop talk now never spoke of their work, which consumed their days, and sometimes their nights and weekends. They sat through family dinners in an exhausted stupor, staring at the ceiling, or persevered in making polite small talk, mentioning the leaky faucet or latest mesa traffic accident, which sounded unnatural and forced. They would suddenly disappear for days at a time, working around the clock on their experiments, only to return looking gray and worried. Cautioned not to share their burden, they let the silence build walls between them and their spouses. "Secrecy becomes a habit," said Rose Bethe, who had known about the bomb project from its inception and had once engaged in long ethical arguments about it with her husband. "Hans stopped talking about his work. We just stopped talking."

Los Alamos was a community of walls within walls. To Ruth Marshak, the fence penning the compound "had a real and tangible effect on the psychology of the people behind it. It was a tangible barrier," she said, "a symbol of our isolated lives. Within it lay the most secret part of the atomic bomb project." The Tech Area, with its ominous third security fence segregating the scientists from their wives, represented an even more profound emotional and physical divide. This was the forbidden zone. It was impossible to enter the Tech Area without a badge: a white badge indicated the senior cadre of scientific personnel with full clearance and granted access to all Tech Area buildings; blue provided limited access, specific to one work station; orange was for the support staff—the secretaries, clerks, and typists—composed of scientists' wives with skills and time and willingness to work.

Inevitably, the badges came to connote a certain social status. Los Alamos was not a casteless society any more than the laboratory was. Social lines were drawn according to the importance of one's position in the Tech Area hierarchy, with special distinction accorded to the handpicked crew who were first to arrive on the mesa with Oppie, facetiously referred to as "the Mayflower crowd." Women who had badges were in the know, and they enjoyed a shared sense of excitement and purpose with the project leaders that some of the men's own wives did not. Charlotte Serber, whom Oppie had made head of the Tech Area library and classified document room, had the honor of being the sole female group leader. Charlotte was privy to most of the work that was going on, as was Oppie's assistant, Priscilla Greene. "They were major-domos, no doubt about it," said Harold Agnew. "They more or less ran the place." More important in the pecking order of Los Alamos, they were invited to the Oppenheimers' home for dinner parties, where they traded stories about the latest Feynman safecracking incident and cryptic office gossip that made the other women present feel left out.

Only a few physicists dared to violate Groves' admonitions and confide in their wives the exact nature of what they were doing in the Tech Area. Some couples were so nervous that the army might be eavesdropping on their conversations, they would save private talks for long hikes or rides into the open countryside. Despite their suspicions and all too many slips of the tongue, most of the wives at Los Alamos existed in a peculiar state of deliberate ignorance. Many took the view that it was easier not to know, because that way they had no secrets to guard, though Elsie McMillan was glad when she finally learned the truth. "In a way I was fortunate to know what they were actually doing," she recalled, "a fact I discovered from a high-up leak." (Her sister, Mary, was married to Ernest

Lawrence, the director of the top-secret Hanford site, which was producing the plutonium for the bomb.) "Ed nearly fell out of bed the night I admitted I knew," Elsie said, "but it was a relief to be able to talk to each other freely in private. A relief that few other wives shared."

The pressures were immense, and insidious. "No wonder the inhabitants became touchy and restless," observed Segrè, who had more perspective than most as he enjoyed the rare luxury of commuting back and forth between Los Alamos and Berkeley for the first few months of the project:

> Often they resented petty things to which they would never have paid attention under normal circumstances. Rank, housing assignments, the part of town in which one lived, social invitations, administrative assignments, everything became important, occasionally in a childish way. The fact that one willy-nilly always saw the same people added to the difficulties. The wives, displaced from their usual surroundings, only added to the problems. Without the absorbing technical work of the husbands, and unavoidably in the dark about what went on in the laboratories, they became depressed, quarrelsome and gossipy.

There was no question that life at Los Alamos was hardest on the young mothers. "They turned to Henry, who had broad shoulders and dispensed as much wisdom as he could," said Shirley Barnett. "Pediatricians fall into the category of confidante, and he was very tuned in to their psychological problems." Henry Barnett, just out of medical school, had braced himself for wartime emergencies, but instead found himself dealing with an extremely high-strung group of women and children suffering from a host of stress-related problems, from headaches, insomnia, and fatigue to acute anxiety and depression. He and his colleagues diagnosed some of the problems as "Los Alamositis," the result of so many people from so many different parts of the world coming together on the mesa and pooling their germs. It was not long before "Doc" Barnett realized that much of the dizziness, exhaustion, and nausea he was seeing was due to morning sickness, as one young wife after another turned up pregnant. Because medical care at Los Alamos was completely free and very good, many of the young couples felt encouraged to get a start on their families. The hospital was soon busy delivering so many babies, it was dubbed RFD, for "rural free delivery."

For most women on the mesa, the chief indignity was the expectation that they would raise their families next to one of the world's most advanced laboratories, while at the same time putting up with conditions straight out of the pioneer days. They complained to Dorothy about everything: it was a constant struggle to obtain fresh milk for babies, eggs were rotten on arrival, and what little fruit and produce the Commissary carried was shriveled and almost inedible after the long haul from Texas. There was never enough water—memos alerting them that "the water situation in this camp is critical" were frequent—and what little came sputtering out of the faucet was often accompanied by algae, sediment, and, on a bad day, worms. At other times, the water was so overchlorinated, it dissolved their precious hosiery, which was almost impossible to replace in Santa Fe. "The water went off when one had soaped for a shower but not entered it yet," recalled Dorothy, reciting from a long list of grievances. "The power went off regularly at 5:30, just at dinner cooking time." Tech Area experiments routinely drained all the juice from town, rendering hot plates and electric ovens useless. There were periods when the electricity stayed off for hours, making it necessary to eat by candlelight. If the meal was not ready, families often skipped dinner altogether. Bread never rose, cakes fell, and nerves frayed. The contrast between the latest accelerator in the laboratory and the hand-operated mangle in the post's laundry room was almost more than the women could bear.

"In the mountains of New Mexico," Jane Wilson wrote, "the women aged":

> We aged day to day. Our electric power was uncertain. Our water supply ran out. Crisis succeeded crisis. Everything went wrong. We had few of the conveniences which most of us had taken for granted in the past. No mailman, no milkman, no laundryman, no paper boy knocked at our doors. There were no telephones in

> our homes. We shared unique difficulties of living with our husbands without sharing the recompensing thrill or sometimes even the knowledge of the great scientific experiment which was in progress.

It was painfully clear that some of the young wives had not taken well to being transplanted onto a military post, particularly when they were also deprived of many of the basic resources they were accustomed to, and the strain was beginning to show. In a rare concession to the opposite sex the army had installed a beauty parlor when the town first got going, but that did not compensate for the lack of a reliable dry cleaners or laundry service. With dirty diapers piling up, and typhoid endemic to the area, more than one desperate new mother came close to burning down the house while boiling diapers on the stove. The problems got to be so serious, according to Segrè, that Oppenheimer consulted a psychiatrist on how to cope with them. The doctor advised him to "find work to keep the women busy and to pay them so that they would have a tangible proof of their usefulness." By then, the women had already come to much the same conclusion and were taking steps to remedy the situation themselves. Women who had never held jobs before, and had little or no training, went to work either full-time or part-time in the Tech Area as "human computers," or adding machine operators, working on long, complicated sums. They were given a three-month crash course in computing by Joseph Hirschfelder, a balding chemist and ballistics expert, and then put to work in the Theoretical wing. Just to make things more complicated, most of the Tech Area jobs were for three-eighths-time work—not half-time or even three-quarters-time positions—as though a solid grounding in fractions were requisite. Those who could not stomach the pressure or factorylike grind took jobs in the community as teachers, administrators, and medical technicians.

After putting up with as much as they could, the women waged their own private war with Groves. The general, who ruled the outpost from the relative comfort of Washington, had dictated that they make do with absurdly backward conditions. Tired of feeling powerless, and determined to improve their living standards, they rebelled. They deluged Oppenheimer with complaints, laid siege to the post commander, and organized meetings in which they articulated their demands. If for no other reason than to restore peace and to stop them wasting so much of Oppenheimer's time, Groves gave in on a number of issues. After that, Dorothy observed that things at Los Alamos began to change for the better, as the scientists and their wives "pitched in and started schools, churches and a library."

The wives arranged for church services in one of the project's two theaters, importing priests and ministers from Santa Fe and rising at dawn to clear the floor of cigarette butts and bottles left over from the Saturday night dances. Although Groves once remarked that he thought physicists were a "godless bunch," they strove to achieve a pious frame of mind in their makeshift house of worship, though the odor of stale beer made that almost impossible. The same theaters with their rows of hard wooden benches were used to stage amateur orchestra performances, choral recitals, and theatricals, all part of the wives' desperate attempts to civilize life on the dreary army post they were forced to call home. They organized community laundries and instructed the Commissary that better vegetables could be had locally in the villages. They started a mesa paper, the *Daily Bulletin*, a mimeographed sheet that covered all the community events the army saw fit to print. It bore the admonition, "This paper is for the site—keep it here."

The scientists and their wives also created a community radio station, KRS, which was accomplished by hijacking the public address system the military used to communicate with the town residents in the absence of telephones. The control room was on the ground floor of the Big House, and the post's power lines doubled as an antenna. A capacitor cut off the signal, which limited the broadcast to within a five-mile radius. Volunteers manning the station played jazz and classical records culled from the residents' own collections. Occasionally, one of the more talented in their number would give a "live" concert. The announcement that Edward [Teller] was playing the piano almost always signaled a moody selection from Wagner. When they achieved a public address system in the technical buildings, some prankster decided to have some fun with the new toy: for two days the operator, by request, paged "Werner Heisenberg, Werner Heisenberg" over and over again in dulcet tones. Finally, a kindly physicist told her she was being had, and that Herr Heisenberg was in Berlin running the Nazi bomb project.

They also talked Groves into starting Los Alamos's four-room school, which was built on the slope west of the water tower. Initially, Groves had refused to allow a high school on the post and had only yielded to the argument that several essential physicists with older children would refuse to come to Los Alamos unless college preparation was available. They hired a principal and induced the wives who had graduate degrees to serve as instructors to the sixty students in the combined junior and senior high school. Alice Smith, who had a Ph.D. in English history, was talked into teaching social studies; Jane Wilson volunteered to teach English; David Inglis' wife, Betty, agreed to teach math; and a young chemist named Barbara Long arranged to take time out from her job in the Tech Area to teach science. To make it easier for the wives with small children to devote their days to community jobs, a nursery school was established. Such was their enthusiasm and idealism at the time, the organizers assumed the offspring of so many Nobel Prize winners would naturally be addicted to study, so they extended the academic year to eleven months, with only a brief respite in August. As it turned out, the children in Los Alamos were like children everywhere else, and after enduring their howling protests all that first summer, the school authorities adopted a normal schedule.

Groves had declared himself all in favor of the scientists' wives working to "keep them out of mischief." But, as Elsie McMillan observed, more was needed: "Even poor General Groves realized that with so many women working, and babies being born in spite of him, and people getting sick, we had to have some domestic help." Oppenheimer appealed to Dorothy for assistance, and together with Vera Williams, the wife of the physicist John Williams, she organized a maid service to do the heavy housework and watch over small children. Dorothy recruited Anita Martinez, whose mother-in-law, Maria Martinez, was an acclaimed New Mexican pueblo artist known as "Maria, the Potter of San Ildefonso." Several of those lucky enough to retain Anita's services were invited to visit her family's adobe home and began collecting Maria's beautiful blackware pottery, which unbeknownst to them would later be worth a fortune. Martinez went all over the valley recruiting Spanish and Indian women to work at Los Alamos and told them to report to Dorothy at 109 East Palace.

Dorothy screened, fingerprinted, and issued passes to the dozens of pueblo women and men who served on the Hill as maids and janitors, waitresses and cooks. Though they said little in answer to her questions, most were fluent in three languages and communicated in a mishmash of English, Spanish, and their native Tewa. Throngs of them gathered outside her office at 109 each morning—the women dressed in colorful pueblo shawl-like mantas and high, white, deerskin boots, their glossy black hair pulled back in chignons or two thick braids—their native dignity undaunted by this latest Yankee incursion. The army buses would pick them up and haul them to the mesa, depositing them outside the post Housing Office in the old Ranch School garage, where they would be assigned jobs, and then bring them back to town at the end of the day. It was hard to say how impressed they were by General Groves' strict security bans, but it seemed to Dorothy that the Indians thoroughly enjoyed their excursions to this strange other world, and they regularly returned loaded down with goods purchased at the army PX.

Like everything else at Los Alamos, maid service was strictly rationed, with working women getting first dibs. As the nonworking wives complained they needed more help and haunted the pseudo-employment office trying to pick up any extra Indian maids, working women like Charlotte Serber realized they were losing ground. When it finally dawned on them that the system was no longer working in their favor, they pulled rank and instituted reforms. "When demand got so far ahead of supply that things were thoroughly out of hand and hair-pulling arguments seemed a likely prospect," she wrote, "the Housing Office inaugurated a priority system. Illness and pregnancy were the highest caste, full-time working wives came next, then part-time working wives with children, nonworking wives with children, part-time working wives without children, and lastly, the nonworking childless wives." Elaborate as this system seemed, it by no means put an end to the squabbling between the working wives and those in the "leisure class."

In setting up their novel community, Oppenheimer, Groves, and his army engineers had seen to the practical necessities of providing places to live, work, eat, and gather, but not a lot of attention was given to the social needs of the population. Not surprisingly, social problems began to crop up almost immediately. Whether they wanted to or not, as the tensions mounted, "Oppie and General Groves had to talk about problems concerning the community," said Greene. "It had a strong effect on morale and how people got along."

For one thing, Los Alamos was a very young community, with the majority being more or less college age. Single men and women were packed like sardines in dormitories, with the WACs sharing bunk beds placed only two feet apart, seventy-five to eighty to a barrack. A goodly number of the young people had never been far from home before and did not necessarily handle their newfound freedom well. Given the close quarters, and the fact that there was not much to do after dark, a fair amount of carousing went on at the two PXs, which sold cigarettes, burgers, and warm Cokes to the scratchy tunes of a jukebox. They were the closest thing to a singles' scene at Los Alamos and were always smoky, noisy, and crowded at night.

There were also frequent "dorm parties," held in one of the large dormitory lounges, that featured a makeshift dance band and a lot of dancing. The main focus of the parties was always the mystery punch, served in a five-foot glass chemical reagent jar pinched from one of the laboratories and spiked with whatever booze could be purchased for the occasion in Santa Fe and enough Tech Area 200 proof to guarantee a good time. After most of the dancers were exhausted, Dick Feynman usually did a drum solo, and if Rabi was around, he performed on his comb. The raucous mixers, which were attended by everyone from Nobel laureates to the lowliest graduate students, usually went on until the early hours of Sunday morning.

The parties ranged from casual to formal affairs held at Fuller Lodge, when all the scientists and their wives would make an effort to dress up in their finest clothes. Square dancing became a popular pastime for the older set. Sometimes invitations would be issued, though mostly hosts relied on word of mouth. One of the first big blowouts was called the

"Necktie Party," and a mimeographed summons requested gentlemen to "wear a necktie if you have one." Shirley Barnett recalled a "Suppressed Desire" costume party hosted by her husband and the mesa's overworked medical staff. The three much-in-demand doctors came dressed in their pajamas with pillows strapped to their heads, a testament to their desire for a good night's sleep. Others came dressed as well-known screen actresses, pin-ups, historic figures, and, not a stretch for this crowd, college professors. The highlight of the night was when Harold Agnew won first prize for his uncanny impersonation of Jezebel. One fellow came in roller skates and just did loops around the room. Single women were a rare commodity on the post, and there was always a long stag line. Dorothy never lacked for dance partners. Determined not to miss out on the fun, she drove all the way from Santa Fe dressed in a leopard-skin number, with just a coat slung over her shoulders. She said later that she prayed all the way she would not have an accident or any car trouble, and end up trying to explain to the local police what she was doing on the lonely back roads late at night in such a crazy getup. While someone fetched her a stiff drink, she sighed that she had her hands full as it was trying "not to explain" what she was doing all day on her crazy job.

"We were awfully busy, we worked all the time, so the parties were naturally pretty wild," said Shirley Barnett. "We were all young and liked to have fun, and it took the sting out of all the restrictions. Everyone would go, and Oppie always put in an appearance." There were a lot of romances, but given their age and the extenuating circumstances, Barnett remembered being surprised at the time at how little "hanky-panky" there was among the married couples. "There was some, but much less than you might expect," she said. "Partly because we had no time, and because there was no privacy, no back stair. Anything you engaged in ran the risk of exposure pretty soon."

They all "drank like fish," added Barnett, and "going to work the next day with a hangover was perfectly acceptable." Alcohol was regarded as the nearest anesthetic to hand, and they all needed to occasionally blot out the reality of what they were really doing there. Whiskey and gin were in short supply, but cheap rum and Mexican vodka were easily purchased in Santa Fe. Taking their cue from Oppie, they made vodka martinis their drink of choice, and John von Neumann, who was great fun at parties, once drank fifteen in a single evening as a kind of experiment. The next day he remarked, "I know my stomach has a cast iron lining, but it must have developed a crack." Tins of tomato juice, which had long ago disappeared off the supermarket shelves in most parts of America, could be bought at the Commissary, and everyone swapped home-made cures for benders.

While the scientists and their wives looked like college students, and for the most part acted like them, the army had the added difficulty that the laboratory employees were in fact independent working adults over whom it could not exercise the same control as over soldiers. There were the inevitable broken curfews, alcoholic binges, fist fights, incidents of "co-mingling" in the single-sex dorms, and outbreaks of the clap. Complaints that a number of WACs were soliciting men outside the PX, and at a price, caused quite a flap. While all this provoked the wrath of the authorities, nothing particularly untoward ever happened, though stories about the wild goings-on would later confirm the suspicions of some outsiders that Los Alamos was a place of licentiousness and loose morals. At one point, the post commander attempted to crack down on the late-night carousing and threatened to place one of the WAC shacks, or women's dorms, off-limits. The women tearfully protested the MPs' action. The bachelors argued even more forcefully against closing the dorm, the community took up their cause, and the whole matter threatened to escalate into a full-blown mesa scandal. "The post commander got quite upset and said, 'Women were not to be allowed in the men's dormitories after ten o'clock at night and visa versa,' " said Greene, who was once caught necking with a young chemist in his room. "People were outraged at the suggestion—these bright young fellows were used to doing what they wanted to do. Oppie took it really rather calmly and wrote them a polite note back saying that they really were quite mature and it ought to be that one did not interfere with this."

By late spring, it was already apparent to the scientists that they needed to have a greater say in the rules that governed their lives. As Alice Smith put it, "The American preference for being governed by almost any guy in tweeds or a sack suit rather than by an efficient chap in uniform soon led to agitation for a civilian representative body." Since their fractious little community clearly required some kind of orderly forum through which they could negotiate compromises and clear the air, Robert Wilson proposed a Town Council be formed. The council would be a civilian governing body, its members democratically elected, and could serve as an advisory committee to the army administration on domestic problems. Oppenheimer was cautiously supportive, as was Whitney Ashbridge, the new CO, who had come in with the mandate to secure greater civilian cooperation. In the beginning they both agreed it would be best to proceed with appointed members, lest it appear the council was trying to subvert military authority.

The army, however, was loath to take this constituency seriously and did its best to ignore it. "Robert Wilson was seen as quite a hot-headed young man for doing this," said Greene. Wilson had already earned a reputation as a singularly energetic and resourceful individual for organizing his own barbershop after noticing members of his division wasted valuable time standing on line for a haircut. After his request for additional barbers was rejected, he requisitioned a barbershop chair and other supplies through the laboratory and set up his own shop with a technician who was handy with a straight razor. After Groves had spotted Wilson's impromptu Tech Area salon, he ordered that the post's barbershop be expanded and its staff enlarged. He was less tolerant of Wilson's Town Council, which allowed the civilians to second-guess the post command—awkward to say the least—and which became a thorn in the administration's side. Greene remembered Deke Parsons, who enjoyed being the only naval officer in an army camp, drolly noting at a Town Council meeting that such a board was not standard. "You know, usually on a military post, the commander is the social arbiter and top dog. It's really sort of hard for the military here because everybody looks down on them."

Having won this small concession to democracy, however, the scientists would not be denied, and by June 1943 a Town Council with six members drawn from the community was duly elected. Bob Wilson became the first elected chairman. The council faced its first major crisis in the dog days of summer. The instigator was quite literally a dog, a large Airedale belonging to the physicist Bob Davis, which had been observed foaming at the mouth. Reports of a sick dog menacing pedestrians spread like wildfire, touching off the mesa's first rabies scare. Parents were so frightened they kept their children at home and immediately demanded that all the dogs that roamed freely around the post be banished for health and safety reasons. The army responded by issuing the order—the dogs had to go. But for many of the workaholic scientists, whose dogs were very nearly their best friends, this order was tantamount to treason. Ed McMillan promptly announced he would leave Los Alamos before he let anyone confiscate his cocker spaniel. As the rhetoric grew hotter, the pro-dog and anti-dog forces stormed the Big House, where at the end of a cranky, emotionally overwrought meeting it was clear that the vast majority of the population was in favor of keeping pets on the mesa. The army agreed to rescind its order of removal. After all the dust settled, it appeared that Davis' Airedale had probably just eaten some soap flakes and he, too, would be spared.

The Town Council weighed in on everything concerning day-to-day life at Los Alamos—from the hazards of having no traffic lights to slow the army jeeps and no sidewalks to protect pedestrians, to the problems of noise, flies, bad food, no food, and so on. It also dealt with those issues where army and civilian life collided, as was the case with the rifle range. The practice range had been carefully laid out on the highest ground on the mesa in the early days of the project, but as the town spread it was perfectly situated to spray bullets all over the makeshift golf course, baseball field, horse corrals, and picnic areas, all places frequented by small children. The chronic problem was finally resolved when the range went up in flames after a month of blistering heat. The council met on Monday evenings in the private dining room in Fuller Lodge, and as time went on, people took to drifting in after dinner to lodge some protest or listen in on local justice. One young British physicist recalled gently being fined five dollars for speeding, by Victor Weisskopf, a Viennese physicist who was one of the town's most popular "mayors."

While it is debatable how much governing the Town Council ever did, trying to impose discipline on what Oppenheimer once called their "odd community" was certainly an eye-opening experience for the innocent academics, who until recently had been living in quiet campus towns. The council received recurring reports of "goings-on" in the women's dorms—the culprit usually turned out to be a soldier-husband who had been billeted in the men's barracks and was bent on a conjugal visit—and the investigations required butting into a multitude of areas that were nobody's business, but they proved vastly entertaining to the growing throngs who packed the balconies during the meetings. As Wilson later recalled, by the time the council got to the bottom of the WAC shack controversy, and the "flourishing business" the girls were doing servicing the boys, "I was a considerably more learned physicist than I had intended to be a few years earlier when going into physics was not all that different from taking the cloth."

It still fell to Oppenheimer, as the laboratory's administrator, to arbitrate all manner of mundane disputes over living quarters, construction, mail censorship, seniority, promotions, and salaries, almost all of which were well beyond his experience and purview at Berkeley. He turned out to be surprisingly good at this, but unfortunately the nature of the problems was such that he often had to choose between a number of equally bad possibilities. Los Alamos's pay scale was a particularly sore subject. It had been settled at the beginning of the project that all the established scientists would receive twelve-tenths of their university salary, as they would be carrying a heavier load than their usual ten months of teaching; younger scientists would receive less, based on the OSRD scale; and graduate students even less than that. Meanwhile, the technicians and construction crews were paid the prevailing commercial rate. This meant that a doctoral candidate like Harold Agnew was earning $125 a week, while the electrician or carpenter working alongside him was pulling down $500 a week. The grad students were willing to make sacrifices for their country, but this was above and beyond their endurance. "We all did pretty much the same things," said Agnew, who was not happy when he discovered the disparity. "If the scientists needed us to hook up a motor or build a bench, you just did it. There were no union rules or anything like that. So a bunch of us got together and decided this was unfair and we would go to the director."

Agnew approached Oppenheimer and explained their grievance. Oppenheimer agreed to hear them out. Later that day, he ambled over to the Z building wearing an indulgent expression and chain-smoked his way through the meeting. "About twenty of us had gathered," recalled Agnew. "He stood in the hall and listened while we explained our beef. Then he smiled and said, 'Well, there's a difference. You know why you're here, and what you're doing, and they don't.' Then he turned around and walked out. And we all stood there smiling and nodding in agreement, and said, 'Yeah!' " Agnew shook his head admiringly and laughed. "That was almost four hundred bucks we lost on our special knowledge. But the way he did it—with just those two sentences—he had us."

At times like that, Oppenheimer radiated power. With his grave, almost priestly manner, he could electrify a crowd, or with one masterful gesture silence opposition. On the rare occasions he lost patience, his usual warmth would be displaced by the infamous "blue glare," an icy stare he leveled at those who crossed him. But ultimately, those who were there say he was so compelling because he led by example. In September, he sent a letter to Robert Sproul, the president of the University of California, which officially oversaw the Los Alamos contract, requesting a reduction in his salary. He explained that his current War Department salary of $10,000 a year actually exceeded by $200 his peacetime pay as "a professor of physics and not a director of anything." Noting that as he did not regard "work done for the Government of the United States in time of war as the occasion for any essential increase in income," he suggested his future salary should be brought into line using "the procedure we usually follow." Of course, Oppenheimer could well afford such high principles, but his peers found it no less commendable. He copied the letter to Groves, who wrote back that he heartily approved of his attitude.

In one way or another, everyone became caught up in the Oppenheimer charisma. He established the tone, and it followed that they would do what needed to be done and know how to invent what did not yet exist. "Oppenheimer stretched me," recalled Bob Wilson. "His style, the poetic vision of what we were doing, of life, of a relationship to people, inflamed me. In his presence, I became more intelligent, more vocal, more intense, more prescient, more poetic myself."

No detail of their lives, inside or outside the laboratory, seemed to escape Oppie's notice. No matter how busy he was, he would take the time to stop by their office with a suggestion, a word of encouragement, a lit cigarette, or an expression of confidence that left them feeling flattered and validated. In typical know-it-all fashion, he once stopped a physicist's wife and congratulated her on her pregnancy—she swore she had not yet told a soul—and politely inquired after her health. Then, before she could utter a single complaint, he proceeded to list all her ailments in rapid order and strode off down the road. When the woman later told a friend of the encounter, she said, "How the devil does Oppie know how a pregnant woman feels?" Then grinning in spite of herself, she added, "You know, he was dead right."

With such small insights and kindnesses Oppenheimer succeeded in charming the mesa and holding sway over his community of crackpots. He thrived in his role as president of his tiny republic, and for a man who was once a shy loner, he became a gregarious campaigner, shaking hands and kissing babies like a seasoned politician. On Sundays, Oppie could be seen riding his handsome chestnut stallion across town on his way to the mountain trails, wrote Joseph Hirschfelder, "greeting each one he passed with a wave of his pork-pie hat and a friendly remark":

> He knew everyone who lived in Los Alamos, from the top scientists to the children of the Spanish-American janitors—they were all Oppenheimer's family. . . . Each of us could walk in, sit on his desk, and tell him how we thought something could be improved. Oppy would listen attentively, argue with us, and sometimes dress us down with a clever cutting sarcasm. At all times he knew exactly what each of us was working on, sometimes having a better grasp of what we were doing than we did ourselves. Needless to say, we all adored and worshipped him.

NINE

Welcome Distractions

To boost his own spirits, along with the mesa's slumping morale, Oppenheimer began taking small groups of friends to dinner at Miss Warner's small teahouse by the Otowi Bridge, which in earlier times had been frequented by the Ranch School boys and their families and by tourists on their way to see the ancient cliff dwellings in the Frijoles Canyon. Edith Warner had transformed a little shack by the train tracks of the old "chili line," which once connected Antonito, Colorado, with Santa Fe. Supplies for the Los Alamos Ranch School were hauled along the line as far as the Otowi Switch, where Miss Warner served as station mistress from 1928 to 1941, when the line was discontinued. Miss Warner was a Pennsylvania Quaker who had come out west because of her health and had made her home on the mesa back when the Indians of the San Ildefonso Pueblo were her only neighbors. She had caught a bad case of "New Mexicoitis," as Peggy Pond Church dubbed the affliction by which people succumbed to the spell of the area. Warner was a thin, pixieish woman, with a faded beauty and regretful air, who wrote unpublished poetry and had taken a Hopi named Tilano as her lover. She was exactly the sort of self-styled individualist Oppie liked to collect and that he invested with great importance in his life. Soon after he was married, he took his wife, Kitty, over to meet Miss Warner, and sample her legendary cooking. Later, when he invited small groups of scientists to go down for delicious home-cooked dinners at her little house by the river, they could not help wondering if he had planned it that way all along.

In truth, Miss Warner's was almost the first casualty of the military occupation of the Pajarito Plateau. By late 1942, the closing of the chili line had reduced Miss Warner's business to a slow trickle, and the news that the Ranch School would also be shutting down, and the army blockading the road to Frijoles, that area's major tourist attraction, seemed to doom her small establishment. One evening in the spring of '43, Oppenheimer and Groves stopped by "to reassure her," as Oppenheimer later recalled, and Groves suggested that Miss Warner might consider running the dining facility at the school lodge, or perhaps a cafeteria on the post. With her livelihood threatened, Miss Warner had little choice but to agree to consider the general's proposal. But moments after they left, Oppenheimer was back, standing on her stoop, the blue eyes plaintive. "Don't do it," he told her, before climbing back into the car and disappearing into the night.

After that, Oppenheimer had stopped by on a regular basis and soon hit on a compromise. He asked Miss Warner if she would turn her dining room, known for its savory chicken and the fabulous chocolate cake that was beloved by so many schoolboys, into a private restaurant for the laboratory personnel. She agreed, and Groves approved the plan. For security purposes, her house would be closed to tourists. Soon Oppenheimer was bringing down groups of eight or ten for small, intimate dinners once or twice a week and introducing them to his old friend Edith, who never joined in, but padded quietly back and forth between the table and stove in her soft buckskin moccasins. Visiting dignitaries like Fermi, Compton, and Conant were always treated to a wonderful dinner at Miss Warner's, where the simple, nourishing stews arrived steaming hot on big terra-cotta plates and the fresh corn, salads, sweet relishes, and five varieties of squash all came from her garden. Her companion, Tilano, with his weathered face and long, graying braids, acted as butler, and at the end of every meal he served the strong black coffee in big pottery cups.

Somehow, Oppenheimer had sensed that in stealing away to Miss Warner's tiny, welcoming home, the Los Alamos scientists would find, as he did, some measure of relief from life on the barren military post and the relentless six-day-a-week grind. It was a badly needed escape and did much to restore everyone's outlook on life. By late summer, Miss Warner's reputation was such that her one large table was booked for months in advance, though old friends like Oppie and Dorothy could always get a reservation. The Agnews, McMillans, and Serbers, along with the other couples who had made Miss Warner's acquaintance early on, were privileged to claim one night each week for their own party, and the less fortunate had to wait until notified that it was their turn. A new adobe dining room was added to accommodate the demand, and Miss Warner charged only the modest sum of two dollars a head and refused to accept tips. Dorothy fondly recalled those cozy, candlelit dinners in her dining room, scented with the aroma of freshly baked bread, where the food was always indescribably delicious, particularly the chocolate cake with fresh raspberries, which she served in hand-carved cottonwood bowls. "That cake was famous around the country," Dorothy recalled. "There was no secret about its recipe, and many people tried to make it but it never tasted as good as Edith's."

At the end of the evening, fortified by the cooking and warmed by the fire and conversation, they would all step out into the quiet darkness of the Pajarito Plateau, where the only sound was the river, which Edith called "the song of the Rio Grande." Miss Warner looked down on alcohol and did not allow it in her dining room, so guests usually shared a bottle on the twenty-mile drive down the winding dirt road and polished it off during the return journey. Thanks to Groves' economies, the post never had any streetlights, so as they made their way home, the night sky over Los Alamos was always carpeted in stars that looked impossibly bright and close enough to touch. That they could be grateful to the general for this was something. "Robert well realized what these dinners and Miss Warner's presence would do for our morale," wrote Elsie McMillan. "The moment one walked into her home, one felt the beauty, peace, dedication, and love that existed there."

Just as Miss Warner had once been an example to Dorothy when as a young widow she had struggled to make a new life for herself in New Mexico, she now taught the scientists to appreciate their isolated surroundings. She was a deeply spiritual woman, and her reverence for the land was expressed in everything from her well-tended garden to the fireplace brooms of hand-picked, dried grass and the neatly bundled piñon wood that she used for kindling, and which scented every room with its pungent fragrance. She made her home a lovely respite without the benefit of electricity or indoor plumbing. The old pine floors of the log cabin were smooth and gray from years of scrubbing. She worked miracles on her slow wood stove and firmly put the Los Alamos wives in their place when they complained about their awful Black Beauties, stating that in her experience, "a coal or wood stove was the only proper way to cook." She also did a great deal to help alleviate their sense of deprivation by allowing first Kitty, and then Elsie, to help themselves to all the surplus fruits, vegetables, eggs, and chickens from her garden. Miss Warner had little money of her own, barely making enough to cover her simple needs, and Elsie worried that they could not begin to repay her kindness with the meager profits from what they earned selling her harvest to the housewives on the Hill. "I am ashamed to say it was like sale day in a bargain basement," Elsie wrote of the greedy mob who would overrun her kitchen and make off with the coveted produce, "as so-called 'ladies' would fight over a bunch of fresh carrots or a dozen wonderful Warner eggs."

Miss Warner wore herself out cooking and waiting on the Los Alamos crowd, and Dorothy fretted that the work was too much for her. She had always been thin, and she grew frailer over the months, but all the available help in San Ildefonso was working for good wages on the Hill, and there was no one to help Edith and Tilano with the backbreaking chores. The Agnews took to stopping by and helping Edith haul water from the well, which was drawn by bucket using a rope and pulley and then carried to the kitchen. As time went on, Miss Warner also let them come down and help pick fruit, as she and Tilano were getting too old and unsteady to climb trees. After Harold Agnew noticed she needed batteries to run the old radio in her kitchen, he took to hoarding the used dry cells from the laboratory, which were discarded when they were low, and bringing them to her so she could keep up with the news and listen to the symphony on Sunday nights. That was the least they could do for the one friend—other than Dorothy, who was part of the project—they would be allowed to make outside of Los Alamos during the war years.

Miss Warner was a special case, and one of the few exceptions to Groves' maximum security laws. Officially, she was not supposed to know anything about the project, but she was far too canny a woman not to come to her own conclusions. As she wrote in one of her annual Christmas letters after the war, "I had not known what was being done up there, though in the beginning I had suspected atomic research." She never took part in their discussions before or after dinner, or inquired about the secret city on the Hill. She regarded feeding the "hungry scientists" as her duty, though some of her guests knew that such a peaceful woman would deplore weapons research and sensed an unspoken reproach. "It would have been easy for you to reject our problem," Phil Morrison, one of the physicists who frequented her "candlelit table," wrote her in gratitude in 1945. "You could have drawn away from the Hill people and their concerns and remained in the compact life of the valley. But you did not."

Morrison's letter is a hymn to the small teahouse at the quiet Otowi crossing and the woman who helped make their time on the Hill, a time made of "long night hours and of critical discussions, of busy desert days and patient waiting in the laboratory," both more bearable and memorable:

> We lived in a community. We grew to know each other. But that was hardly novel; most of us had been friends long before Los Alamos. What was new was the life around us we began to share. We learned to watch the snow on the Sangres and to look for deer in Water canyon. We found that on the mesas and in the valley there was an old and strange culture; there were our neighbors, the people of the pueblos, and there were the caves in Otowi to remind us that other men had sought water in this dry land. Not the smallest part of the life we came to lead, Miss Warner, was you. Evenings in your place by the river, by the table so neatly set, before the fireplace so carefully contrived, gave us a little of your reassurance, allowed us to belong, took us from the green temporary houses and the bulldozed roads. We shall not forget. . . .

In the midst of everything that summer, Oppenheimer shared with his staff a letter from President Roosevelt that further lifted their spirits. In early July, Roosevelt had written asking that Oppie assure the scientists on the project that their efforts in the face of considerable danger and personal sacrifice counted and that the country was grateful. "I am sure we can rely on their continued wholehearted and unselfish labors," wrote the president. "Whatever the enemy may be planning, American science will be equal to the challenge. With this thought in mind, I send this note of confidence and appreciation." Oppenheimer read the letter out loud at the weekly staff colloquium held in the theater each Tuesday evening. When Dorothy heard the president's words, it was one of the proudest days of her life.

By autumn, everyone was distracted by all the mesa romances. As they were more or less living in each other's pockets, it was common knowledge that Priscilla Greene had fallen in love with the quiet young chemist named Robert Duffield, known to everyone as "Duff" because there were too many Roberts at Los Alamos. She had first met him back in Berkeley when he came into the office because his employment form had the wrong middle name—his was Brokaw, apparently quite a distinguished moniker in the part of New Jersey he came from—and they had argued. It was not a good start, but when they met again on the Hill a few months later, they hit it off. They began spending quite a lot of time together, playing tennis and going to the movies on Sundays. In July, when Dorothy called to say she had found Priscilla a puppy, an English bulldog named Truchas, Duff offered to accompany Priscilla into town to pick it up. They had lunch with Dorothy in the outdoor café La Placita, at La Fonda, and little Truchas slept in a planter that held one of the patio's shady trees. Not long after that, Priscilla ran down the hall of the T building telling everyone that she was getting married.

At first, Oppenheimer did not approve of the union and tried to talk Priscilla out of it. He worried about "shipboard romances" like theirs, brought on by close proximity and the pressures of war. He also argued that Duff was too junior, though Priscilla suspected that was just because Oppenheimer had someone else in mind for her, a scientist whom he thought of as a more suitable match. When a last-minute meeting with security officers called by Groves in Cheyenne, Wyoming, meant he would have to miss the ceremony, Oppenheimer insisted Priscilla drive with him to Santa Fe so he could give her the benefit of some fatherly advice. "I can remember sitting in the Buick with him waiting for Groves," recalled Greene. "I had the feeling he thought I was marrying the wrong man, and that our courtship had been too quick. But then we did everything quickly in those days."

Dorothy approved of Duff because he was from Princeton, and she insisted the couple get married at her home. She told them how to go about getting a license, found a trustworthy local judge to do the honors, and made all the arrangements. Because of the secrecy surrounding project employees, she could not tell the presiding judge the surnames of the couple, so the ceremony had a rather informal air, as everyone was obliged to be on a first-name basis. She and Priscilla went into the mountains and collected armfuls of glorious yellow aspen, and filled the house with flowers. The rosebushes were in bloom, and the garden looked beautiful. The wedding took place on September 5, on a Sunday morning, and all their friends from the Hill came, including the Serbers, Wilsons, Manleys, Williamses, Agnews, and, of course, Kitty. The ceremony was held in the farmhouse's living room, and the judge and prospective groom stood nervously before the pueblo-style fireplace while Priscilla and her bridesmaid made what Dorothy described as "quite a dramatic entrance" through one of the large windows. No family could attend for security reasons, but Priscilla's brother DeMotte, who was already working on the Hill, was on hand to give her away. Dorothy's son, Kevin, popped the champagne and balanced glasses of bubbly on the back of his huge pet tortoise, which crawled slowly around the courtyard transporting drinks to the guests.

It was the Fiesta weekend, and after Priscilla and Duff drove off for their three-day honeymoon in Taos, the rest of the wedding party joined the traditional local celebration already in progress and carried on well into the night. "We were the first people to get married at Los Alamos, and we had a marvelous wedding," Greene recalled. "Actually, getting married in Los Alamos was a much less tense thing than getting married anywhere else because you only had your friends. You didn't have any family who had to be appeased or worried about."

Three weeks later, the festivities were repeated when Priscilla's bridesmaid, a young secretary from Berkeley named Marjorie Hall, married Hugh Bradner, the young Berkeley physicist who was one of the first to arrive on the mesa with Oppie. Dorothy had played matchmaker, turning to them after lunch at La Fonda one day and observing pointedly that her house was "a lovely place for a wedding." Bradner took the hint and proposed a few days later. "We were in love, and she saw it," said Marge. "Hugh was the first person I met when I came through. I walked into her Santa Fe office, and this good-looking guy said, 'Come in,' and I fell for him. She knew it. She didn't miss a thing." As usual, Dorothy saw to everything. This time, Priscilla served as matron of honor, and Henry Barnett as the best man. Oppenheimer stood in as father of the bride, and his armed guards stood watch at the end of Dorothy's drive, much to the consternation of her neighbors.

After that, several more weddings at Dorothy's house followed in quick succession, and people began jokingly referring to her living room as "the little chapel around the corner." Oppenheimer came whenever he could, standing in for missing family and, on more than one occasion, giving the bride away. That Oppenheimer took an active interest in their lives, and cared enough to share in their anxieties and problems as well as in the joys of another wedding or a new baby, instilled in people a feeling that they were pulling together, and renewed their faith and dedication.

In September, when tragedy struck the mesa with the sudden death of Barbara Long, the popular science teacher and wife of a group leader, Oppenheimer's calm direction helped steady their nerves. Long's mysterious illness touched off a widespread panic that the unidentified form of paralysis that felled her might be polio, and she had been quickly transferred from the post infirmary to Bruns Hospital in Santa Fe. To avoid a widespread outbreak, the post's doctors were forced to lock down the site under almost complete quarantine. Laboratory personnel were not allowed to come and go from Santa Fe. The post school was closed for a week, and parents were ordered to keep their children inside. Henry Barnett kept a close eye on any toddlers who showed symptoms of fever. Jittery staffers huddled in small groups to exchange scraps of information and news. When word of Long's death came a few days later, Alice Smith observed that for the shocked and frightened community, held captive behind the barbed-wire fence, it was "reassuring to know that Oppenheimer himself had been the first to visit the bereaved husband."

"For those who never saw Oppenheimer at work inside the Tech Area fence," wrote Smith, "the mystique of his leadership included an element of his personal concern":

> He seemed to understand the uprooted feeling that afflicted newcomers, many of whom had left homes as pleasant as the Oppenheimers' own house in Berkeley. Dismayed by lack of privacy and recurrent milk, water, and power shortages, they were somewhat appeased by the knowledge that it was Oppenheimer who had included fireplaces and large closets in the original house plans. He no longer came to dinner bearing bouquets of flowers, the gesture for which he was famous among Berkeley hostesses, but he gave both employed and nonworking wives a sense that their presence and participation in the collective enterprise was important.

Even for Dorothy, whose life in Santa Fe was not nearly as contained and stifling as that of the Hill dwellers, Oppenheimer had become a beacon, a source of strength and inspiration, and the central focus of her life. She had been lonelier after her husband's death than she dared acknowledge, and had buried those feelings deep inside her, afraid that they would betray a certain weakness. All consumptives were schooled in determination and denial, and Dorothy had held fast to those hard tenets of survival. She had never given in to bitterness and had more than contented herself with her health, her son, and their simple life in Santa Fe. But now she found herself caught up in a vital and thrilling society, and for reasons she could not begin to explain, she felt unaccountably at home, as if she had finally found a place where she belonged. Not since Sunmount had she felt a part of something larger than herself, and the intense commonality of effort was seductive. Oppie had brought her in and made her not only a part of the project, but a part of his intimate circle. Their friendship, forged in the wake of innumerable dramas and intrigues, meant everything to her.

She was deeply touched when, worried by the solitary nature of her assignment in Santa Fe, Oppenheimer again suggested she move up to the Hill with the rest of them. Had she been on her own, Dorothy would have accepted in a second. All the excitement was on the Hill and in the mysterious production going on in the Tech Area, whatever it was. "I was really tempted to get in on the action," she admitted years later. But after giving it some serious thought, she declined his offer. "Kevin had his pigs, goats, a pony, even a jalopy," she reasoned. "If we moved, it would mean two rooms and no yard." Penelope, the McKibbin pig, was a particular problem. She was always escaping the fenced-in garden and raising a hue and cry with the neighbors. She shuddered to think of the havoc the pig would wreak in the close quarters of Los Alamos. Besides, the scientists all counted on her as their anchor in the real world and on her house as a safe haven, a "place to relax, to spend the night, to have weddings." Oppenheimer would renew the offer again and again in the months to come, but each time she would turn him down. She knew she was more valuable to the project, and to Oppie, on the outside. It filled her with pride that she was in a position to help him, to be his eyes and ears in town, and no one, not even Oppie himself, could convince her otherwise.

Yet, being with Oppenheimer—the precious few hours she spent in his company—was irresistible to Dorothy. She looked forward all week to Saturday, when she would steal away from town and drive the winding dirt road up to the post to buy huge chocolate bars for Kevin at the PX and attend a party given by her new friends. As time went on, she sometimes took Kevin along, and they would stay overnight with anyone who had a room to spare. Since Oppie had mandated that Sundays be devoted to outdoor recreation and relaxation, her days off were taken up with exploring the area with her industrious new companions. Just a few miles east of Los Alamos, the scientists discovered the ruins of Tsankawi and visited the ruins of Otowi, and Navawi, in Mortandad Canyon, and to the north, the ancient cliff city of Puye. They hunted for arrowheads and collected samples of all the minerals, finding amethyst, quartz, malachite, microlite, calcite, lepidolite, silver, fluorite, mica, and turquoise. Some even caught the fever of the Old West and took up mining with a passion, filling empty soup cans with chunks of copper. It entertained her no end to see such great minds gulping dust as they dry-panned for tiny grains of gold. "It seemed to me," she wrote, "that the Los Alamos people found out more about the mineralogy, geology, rare metals and stones, streams and camping possibilities of this country than we had ever known in the many years of living here."

Like the Indians who shared the Pajarito Plateau with the scientists, Oppenheimer believed it was still an enchanted land, and despite all the hardships and confusion, Dorothy could see that he had succeeded in converting almost everyone to his way of thinking. From their vantage point at Los Alamos, raised high above the world, even something as simple as the sunset behind the Jemez peaks became terrifyingly beautiful in its great sweep of color and sky. When they considered the scale of the towering dark mountains and limitless heavens, all their problems were reduced in importance to that of any of the earth's tiny creatures. Perhaps Oppie, who had known what a perilous journey awaited them, had hoped they would look down with wonder at the world and feel uplifted and inspired even in their darkest hour. Perhaps that was what drew him back to the mountains time and again. If it was not quite paradise lost, Los Alamos had become for Dorothy a magical place, a mystical place, and to remind herself of how lucky she was, she needed only to see the extraordinary man she was working for, or to look out at the vast desert broken only by majestic palisades where, as D. H. Lawrence wrote, "only the tawny eagle could really sail out into the splendour of it all."

The project was scarcely six months old, but already it was apparent to her how they all looked to Oppenheimer, and depended on him to get them through the uncertain days ahead. Only he seemed to see the way to the end of the war and to the thing that would save American boys' lives, European boys' lives, countless lives. "He commanded the greatest respect and gave us all the ability to do things we didn't think we could do," she told a reporter years later. "Of course, at first we didn't know what was going on, we didn't know anything, and then it began to unfold like a book just being written."

TEN

Nothing Dangerous

For Oppenheimer personally, the warm weather had brought with it a growing sense of unease and self-doubt. While he projected confidence in the formal meetings and colloquiums, and continued to demonstrate the same dazzling ability to zero in on a problem—if not necessarily the solution—during technical discussions, he felt increasingly in over his head and pulled in different directions. For all that he might have imagined that by rooting his work in the familiar wilderness setting he would re-create the stimulating experience of freedom and adventurousness he had always enjoyed at Perro Caliente, he could not have been more wrong. Instead, he found himself chained to a desk and faced with a crushing load of administrative difficulties and decisions.

As the size and scale of the project continued to escalate monthly, with costs doubling and quadrupling, and concerns about housing, construction, salaries, and procurement of supplies taking up more and more of his time, Oppenheimer began to feel unequal to the task of managing the project's large and complex corporate machinery. To Priscilla Greene, he looked increasing thin, haggard, and worried. Much later, Bob Bacher told Alice Smith that several times during that first summer Oppenheimer confided that he felt overwhelmed by the responsibility of leading the laboratory and worried that he would not be able to see it through. Bacher's advice to Oppie was short and sweet: he had no choice but to carry on as director, for "No one else could do the job."

Contributing to Oppenheimer's sense of vulnerability were the regular debriefings with army intelligence forces, who, incredibly, had yet to approve his security clearance. Consequently, he found himself in the absurd position of being ordered to waste no time in producing the new weapon, while being undercut at every turn with trivial and distracting questions about his past. To make matters worse, the questioning had intensified in recent weeks, following a trip to San Francisco he took that June to visit his former fiancée, Jean Tatlock.

Oppenheimer had stayed in touch with Tatlock after his marriage to Kitty. Though she had continued to call him and wrote with some regularity, in the years between 1939 and 1943 they had seen little of one another. Tatlock had drifted further and further into religious and political fringe groups, and her bouts of melancholia had become more frequent and severe. Late in the summer of '42, just after the Berkeley weapons conference, Serber had been surprised to see Oppie and his old flame pacing along the sidewalk below Oppie's house at One Eagle Hill Road. It turned out that Tatlock, the daughter of a noted English professor at the university, was visiting her father, who lived around the corner. As Kitty later told Serber, "Whenever Jean was hit by a bad depression, she would appeal to Oppie for support."

During the hectic winter and spring of '43, while Oppenheimer had been preoccupied with the planning and launch of Los Alamos, Tatlock had taken a turn for the worse. Although a psychologist herself, she had not been able to rid her mind of thoughts about Oppenheimer, whom she told friends she regretted not marrying. She had put herself in the care of a psychiatrist and was receiving treatment for depression at Mt. Zion Hospital in San Francisco. Oppenheimer had received word on several occasions that she needed to see him, but the press of events had made that impossible. But by June, after receiving a message from her passed on by his former landlady, Oppenheimer felt he could no longer ignore her urgent appeals. They had been "twice close enough to marriage" to think of themselves as engaged, he would explain years later during his loyalty hearings, when his visit to a former girlfriend who was a Communist activist was used to incriminate him. Whether out of pity, or in deference to their long and complicated history, Oppenheimer maintained that he had felt obliged to comply with her request. "I almost had to," he stated. "She was not much of a Communist, but she was certainly a member of the party. There was nothing dangerous about that. Nothing potentially dangerous about that."

Oppenheimer traveled to Berkeley on Saturday, June 12, and busied himself with meetings at the university. He caught up with Lawrence, dining with him on Sunday night. He had planned to spend some time with Tatlock on Monday at her apartment on Telegraph Hill and take the train back to Los Alamos that night, but he found her condition more fragile than he had expected and could not easily take his leave. Tatlock was "extremely unhappy," he later testified, and confessed she was still in love with him. Oppenheimer ended up missing his train and stayed the night in her apartment. He must have known that he was being tailed by G-2 agents and had certainly read enough FBI files to realize that his visit would become part of his dossier, but he did not know enough to try to insulate himself from any appearance of impropriety. Compounding matters, he agreed to meet with her again the following evening, and after dinner she drove him to the airport. Late for important meetings at Los Alamos, Oppenheimer attempted to make up for lost time by catching the first plane back to New Mexico, breaking Groves' moratorium against key project members flying. All of this, of course, was duly reported by his round-the-clock surveillance team and submitted in a comprehensive report to Lieutenant Colonel Boris Pash, who was chief of Counter-Intelligence for the Ninth Army Corps on the West Coast.

Predictably, Pash and G-2 were greatly interested in what they regarded as Oppenheimer's reckless decision to meet privately with a card-carrying Communist. They already had sufficient reason to be concerned about Robert's close connections with party members and insistence on hiring left-wing professors to work at Los Alamos, but the report that he had traveled from the highly classified atomic bomb laboratory for a reunion with someone as suspect and unstable as Tatlock provided fresh grounds for doubt. The question that confounded them was, Why would anyone in a position as highly sensitive as Oppenheimer's continue to pursue associations with Communist adherents, especially in time of war? Why would he openly court suspicion, or even allow himself to be drawn into a situation that might possibly be construed as suspicious?

While these considerations might seem obvious, it is impossible to know what Oppenheimer, already bound in a fantastic web of secrecy and lies and false identities as proscribed by the project, actually viewed as risky behavior. In his arrogance, he may have believed that as director of the Los Alamos laboratory he was held to a different standard, that in effect his brilliance and importance to the project rendered him both irreplaceable and beyond reproach. "Look, I have had a lot of secrets in my head a long time," he testified during the hearing. "It does not matter who I associate with. I don't talk about those secrets. Only a very skillful guy might pick up a trace of information as to where I had been or what I was up to."

In the same way that his egotism and air of superiority had once alienated many of his academic colleagues, they again worked against him with his military counterparts. Pash formed an immediate dislike of the haughty New York physicist and was far from convinced of his infallibility. As far as he was concerned, Oppenheimer's meeting with a Communist "contact" like Tatlock provided all the ammunition he needed to alert Washington that they were dealing with a potential traitor. In late June, he forwarded a summary of the surveillance data to Lieutenant Colonel John Lansdale, his chief at the Pentagon, along with his memorandum stating in unequivocal terms his judgment that "the subject," his preferred term for Oppenheimer, was not to be trusted: "In view of the fact that this office believes that the subject still is or may be connected with the Communist Party in the Project . . ." He then spun his own elaborate conspiracy theories, including "the possibility of his developing a scientific work to a certain extent, then turning it over to the Party without submitting any phase of it to the US Government." A hard-liner on security risks, Pash recommended that Oppenheimer be "removed completely from the project and dismissed from employment by the US Government."

As a result of Oppenheimer's flagrant flouting of the rules, the army security officers in charge of the Los Alamos site did not like or trust him.

This all but guaranteed that their file on him, first opened by the FBI, would never be closed. Not only would the investigation into his past continue, he would be forced to submit to round after round of probing interviews. Much of the questioning was done by Lansdale, a young army intelligence officer and Harvard Law graduate, who was Groves' chief aide on security matters. The thirty-one-year-old Lansdale had originally been recruited by Conant in early 1942 to keep tabs on the political activities of the Berkeley physicists. At the time, Conant, who was all too aware of how gossipy academics could be, worried that the notoriously left-wing Berkeley contingent could not be trusted to safeguard their atomic research with the zealousness the situation required and might allow information to fall into the wrong hands. His worst fears were confirmed when Lansdale reported back a few weeks later that while on Berkeley's campus he had filled a notebook with snippets of conversations relating to atomic research, all of which Conant regarded as serious breaches of security. Conant, who feared that whatever the Nazis gleaned of the United States' project would galvanize their efforts to build the first atomic weapon, immediately sent Lansdale back to Berkeley with instructions to put the fear of God into the physicists when it came to national security. In the months that followed, Lansdale had continued his monitoring of Berkeley's security lapses and, during the early phase of the project, had focused his attention on the character of the newly appointed Oppenheimer, his Communist wife, and their many "pink" associations.

Lansdale had made a point of getting to know Kitty personally and later recalled that on one afternoon when he went to interview her, she offered him a martini. "Not the kind to serve tea," he noted, implying that Oppenheimer's wife took social drinking to a new level. Kitty made no bones about the fact that she knew exactly why Lansdale was snooping around. But at the same time, she made it clear that she was now totally committed to Oppenheimer and his career, and the young intelligence agent was struck by how passionately she had taken up the role of dutiful wife:

> As we say in the lingo, she was trying to rope me, just as I was trying to rope her. The thing that impressed me was how hard she

> was trying. Intensely, emotionally, with everything she had. She struck me as a curious personality, at once frail and very strong. I felt she'd go to any lengths for what she believed in.

Lansdale, an affable, blue-blooded lawyer with little experience in spying, decided to play on her ambition for her husband to draw her out and tried to present himself as "a person of balance, honestly wanting to evaluate Oppenheimer's position." He believed she had been in sympathy with her first husband, who had been a member of the Communist Party, and doubted that her "abstract opinions" had altered much with time. As he explained years later, "feelings were her source of belief":

> I got the impression of a woman who'd craved some sort of quality or distinction of character she could attach herself to, who'd had to find it in order to live. She didn't care how much I knew of what she'd done before she met Oppenheimer or how it looked to me. Gradually I began to see that nothing in her past and nothing in her other husbands meant anything to her compared to him.

Lansdale came away convinced that Kitty was never the least bit fooled by his tactics: "[She] hated me and everything I stand for." While she was no conventional patriot, he had left with a grudging admiration for her sense of conviction, and he reported to Groves that her protectiveness of her husband might actually serve them well in the long run, as her "strength of will was a powerful influence in keeping Dr. Oppenheimer away from what we would regard as dangerous associations."

In his long, rambling conversations with Oppenheimer at Los Alamos, Lansdale tried to explore the physicist's various friendships with Berkeley Communists. At one point, he even let Oppenheimer know that G-2 had doubts about the union activities of several of his former students. Their suspicions centered on Giovanni Rossi Lomanitz, a gifted twenty-one-year-old physicist, who had been working for Oppie at Berkeley since the spring of '42 and was being considered for promotion by Ernest Lawrence. Because of his position at the lab, Lomanitz was fully aware of the work being done and the kind of weapon it indicated. Toward the end of July, Lawrence, who had never had any interest in politics, went ahead and made Lomanitz a group leader. Security did not want to challenge Lawrence's decision outright, but unhappy about having a campus radical with so much access to classified material, G-2 arranged for Lomanitz to be drafted. Instead of steering clear of this hot potato, and letting Lawrence fight his own personnel battles, Oppenheimer interceded on Lomanitz's behalf, sending a telegram pleading his case to project headquarters. With nuclear physicists in short supply, Oppenheimer argued, Lomanitz was indispensable.

On August 10, Lansdale, during another of his lengthy interviews with Oppenheimer at Los Alamos, counseled him against getting involved in such sensitive issues. Among his many doubts about Oppenheimer, Lansadale worried that the Los Alamos director's extreme competence in his field led him to falsely believe he was also competent in other fields, whether it was playing at politics or matching wits with suspected Communists. Like many brilliant men, he could be surprisingly dense. "For goodness' sake," he told Oppenheimer, "lay off Lomanitz and stop raising questions."

In the midst of the session, Oppenheimer turned to Lansdale and pointedly observed that when it came to his own staff, although he would not be concerned about a physicist's past affiliations if his present frame of mind was constructive, he definitely did not want any *current* members of the Communist Party working at Los Alamos because there would always be a "question of divided loyalties." The strong statement was surprising coming from Oppenheimer, and to Lansdale seemed out of character. It was one of many statements Oppenheimer would make that would later be hard for security officials to know how to interpret and would leave them wondering if he was simply trying to be conciliatory or was attempting to lead them astray.

Oppenheimer not only ignored Lansdale's warning, he interjected himself even further into the Lomanitz mess. On August 25, he again traveled to Berkeley on business, and while he was there, he stopped by the campus security office in Durant Hall and asked the local chief, Lieutenant Lyall Johnson, if it would be all right if he met briefly with Lomanitz. Johnson discouraged the idea, stating that he thought the young physicist was trouble, but allowed the Los Alamos director to do as he saw fit. Oppenheimer went on to have a brief but unsatisfactory meeting with Lomanitz, who admitted that he was still a party activist and claimed he was being framed. Annoyed and fed up, Oppenheimer, according to his own account, parted on bad terms with his ex-student. That might have been the end of it, but before leaving Johnson's security office in Durant Hall that day, Oppenheimer mentioned in passing that he had been told by a friend that a certain British engineer named George C. Eltenton, who was employed by the Shell Development Corporation, had offered to supply technical data to the Russians. It is not clear why Oppenheimer chose that moment to help security by volunteering the tip about Eltenton, particularly since the conversation with his friend had taken place some eight months earlier. It is almost certain that he had no idea how inflammatory this action would prove to be.

As soon as Oppenheimer left, Johnson relayed the Eltenton tip to Colonel Pash, his superior, whose jurisdiction over the West Coast atomic facilities included Lawrence's Berkeley laboratory. Pash, who was already investigating leads that the Soviets had infiltrated the Manhattan Project, called Oppenheimer and asked him to come back the next day for a friendly chat.

On August 26, 1943, Pash and Johnson sat down with Oppenheimer in the security office on campus and asked him to elaborate on the Eltenton tip. Unbeknownst to Oppenheimer, Pash had arranged for an officer to surreptitiously tape the interview. Pash quickly got down to business and asked for the identity of his contact, but Oppenheimer balked at naming names. Oppenheimer was reluctant to drag in his old friend Haakon Chevalier, a lecturer in Romance languages at the University of California, whom he had met through the Teachers' Union during his early dabblings in left-wing politics. Whether Oppenheimer was apprehensive about what his past might yield, or greatly underestimated his interrogators and the importance they attached to his Communist associations, is difficult to gauge. Certainly he wanted to protect Chevalier, who had stood by him after his scandalous elopement with Kitty and was so trusted a friend that the Oppenheimers had left their two-month-old baby son in the care of Chevalier and his wife while they took a badly needed vacation to New Mexico in the spring of 1941. Tired and distracted, he may not have seen why he should waste time and energy on something he regarded as a trifling matter.

In any case, under several hours of Pash's persistent questioning, Oppenheimer refused to give up Eltenton's contact and failed to give a coherent account of how he came to be approached by someone soliciting technical information about the bomb project for the Russians. He told Pash that the go-between was acting on Eltenton's behalf and knew a man proficient at "microfilming" who could pass the information on to the Soviet consulate. He also added, according to the transcript of their conversation, that three other approaches had been made and that these three scientists "were troubled by them, and sometimes came and discussed them" with him. When Pash insisted again that Oppenheimer had a duty to furnish the name of the man involved in such subversive activity, Oppenheimer refused, saying that he had rebuffed the approach. Since no information was leaked, and the mission had failed, he did not see the need to impeach his friend.

A week after their meeting, Pash fired off a memo to Groves summarizing the interview. Drawing a direct connection between the Lomanitz situation and the Eltenton tip, Pash concluded that the reason Oppenheimer had suddenly spoken up about an incident that had occurred eight months earlier was that G-2 agents were closing in on his secret contacts, and he wanted to throw them off the trail and distract them from where his true sympathies lay.

It did not help matters that Los Alamos's brash young security officer, Captain Peer de Silva, a handsome twenty-six-year-old who was two years out of West Point, had also taken an instant dislike to Oppenheimer. Cold, correct, and a stickler for details, de Silva, who was under Pash's command, was one of several agents investigating whether Oppenheimer should be given final clearance, an issue that security was reluctant to sign off on. De Silva was appalled by the inconsistencies in Oppenheimer's file and wrote Pash that the physicist "must either be incredibly naïve and almost childlike in his sense of reality, or he himself is extremely clever and disloyal." Clearly de Silva inclined toward the latter. No doubt prompted by Pash, he sent his own memo to Washington on September 2 charging that Oppenheimer was involved in treachery:

> The writer wishes to go on record as saying J. R. Oppenheimer is playing a key part in the attempt of the Soviet Union to secure, by espionage, highly secret information which is vital to the United States.

All of these charges cast rather a dark shadow over the man Groves had chosen to head the Los Alamos laboratory. Moreover, their timing could not have been more awkward. The nervous sleuths in army intelligence had been proceeding at such a snail's pace in granting clearances to the top physicists that Conant had worried this might "seriously delay" the bomb project and had pushed Vannevar Bush to recommend quick approval for Oppenheimer and other vital scientists despite their "unusual" backgrounds. Goaded into action, Groves ordered the directive on July 20, only five weeks before his top security officers would accuse the Los Alamos director of treason:

> In accordance with my verbal directions of July 15, it is desired that clearance be issued for the employment of Julius Robert Oppenheimer without delay, irrespective of the information which you have concerning Mr. Oppenheimer. He is absolutely essential to the project.

Despite the blemishes already on Oppenheimer's record by that July, Groves had decided to back him. There was much not to like about Oppenheimer's politics, and his lapses in judgment and occasionally undiplomatic comportment caused them all to worry if they had selected the right man to be the leader. But Groves had come to the same inescapable conclusion as Bob Bacher: there was no one else. Not one to take unnecessary chances, however, Groves decided to hedge his bet. Intent on assuring that Oppenheimer behaved and the project proceeded smoothly, Groves installed de Silva as head of security at Los Alamos. That way, his man could keep an eye on the restless laboratory director's movements for the duration of the project and make sure he did not wander too far off the reservation.

If G-2's intrusions into his life left Oppenheimer feeling he was no longer his own man, a letter from Groves on July 29 sealed his fate. Not only was he to be held accountable for his past indiscretions, Groves made it clear he was now a prisoner in a jail of his own making:

> In view of the nature of the work on which you are engaged, the knowledge of it which is possessed by you and the dependence which rests upon you for its successful accomplishment, it seems necessary to ask you to take certain practical precautions with respect to your personal safety.
>
> It is requested that:
>
> (a) You refrain from flying in airplanes of any description; the time saved is not worth the risk. (If emergency demands their use my prior consent should be requested.)
>
> (b) You refrain from driving an automobile for any appreciable distance (above a few miles) and from being without suitable protection on any lonely road, such as the road from Los Alamos to Santa Fe. On such trips you should be accompanied by a competent, able bodied, armed guard. There is no objection to the guard serving as chauffeur.
>
> (c) Your cars be driven with due regard to safety and that in driving about town a guard of some kind should be used, particularly during the hours of darkness. The cost of such a guard is a proper charge against the United States.
>
> I realize that some of these precautions may be personally burdensome and that they may appear to you to be unduly restrictive but I am asking you to bear with them until our work is successfully completed.

Groves also ordered special MPs to stand guard outside the Oppenheimers' house on Bathtub Row. The additional surveillance was uncomfortable, and Kitty found the lack of privacy particularly grating. It also caused no end of problems when she dashed out of the house without her badge, only to find on her return that the uniformed men outside the door would not allow her to enter her own home. No matter how many times she insisted she was the laboratory director's wife, the guards would refuse her entry. On more than one occasion, she had to appeal to a neighbor to prove she was who she claimed to be, and only then would the guards let her in to retrieve her pass and put the matter to rest. Never one to be easily intimidated, Kitty soon found a way to turn the situation to her advantage. She put the MPs to work as babysitters, relying on them to watch little Peter, who would be asleep in his crib, while she went out to do errands. Elsie McMillan recalled returning home one day from the Commissary to find a grinning young officer at her back door. He saluted and said, "Good afternoon, Mrs. Oppenheimer, the baby you left in the bedroom is quite all right." She replied, "Thank you very much, but I am not Mrs. Oppenheimer and I didn't leave a baby in my house." He said, "My God, I'm guarding the wrong house!" and took off at a run. Shortly thereafter, a fence was put around the Oppenheimers' home, though she could not help wondering if it was meant to ward off spies or keep future guards from making the same mistake.

Kitty Oppenheimer was the unofficial hostess of Los Alamos, but while she enjoyed the status and perks of the role, and was not above occasionally lording it over others, she remained for the most part a reluctant member of the community. Not everyone thrived on the enforced togetherness and small-town clubbiness that characterized life on an isolated army post, and Kitty, brittle and taciturn by nature, was less suited to it than most. She never really seemed to fit in, and she failed to rise to the occasion and become a social force in the small colony as might have been expected of the laboratory director's wife. Instead, she ceded that role to more outgoing women, such as Martha Parsons, who did most of the large entertaining, and Rose Bethe, a beautiful Ingrid Bergman lookalike, who tackled the Housing Office and was a great organizer and cheerleader in the early days of Los Alamos. Kitty was imperious, unconventional, and, as demonstrated by her rapid succession of marriages, generally indifferent to the middle class mores that prevailed on the mesa. Every bit as quick and cutting as Oppie, she had the same strong opinions on matters of style and comportment, and it was never hard to tell whom she favored and whom she did not. Unlike her husband, however, she did not worry about staff morale and made little effort to temper her frankness or sarcasm.

To visiting VIPs, she appeared clever, vivacious, and charming, and she threw delightful tipsy parties. She was a wonderful cook, and her dinners were made all the more memorable by a guest list studded with Nobel laureates. To newly arriving friends and eminent male physicists, she could be solicitous, showing up at their door that first disconcerting morning with a ready-made "welcome basket" of essentials, including milk, bread, assorted dishes, flatware, and a can opener to tide them over until they were unpacked and settled. On a day-to-day basis, however, she was a difficult and divisive presence. Despite the fact that the Indian and Spanish American maids who were bused in daily were in short supply and allotted strictly on the basis of need, she felt entitled to the services of a daily housekeeper, inspiring considerable resentment in the other wives. And at a time when rationing also dictated that Hill wives had to save C-coupons and carpool to have enough gas for the eighty-mile round-trip drive to Santa Fe once a month—and that was with coasting downhill—she could often be seen tearing off to town in her pickup truck. A fearless driver, she thought nothing of tackling the winding, treacherous road on her own.

All this might have mattered less had she not been the director's wife. But in their narrow little world, she wielded enormous power and, according to Priscilla Greene, often wielded it quite mercilessly. If she had no use for someone, it showed. She was coldly dismissive of Rose Bethe, who was as congenial as her husband and well liked by everyone. But Rose was also a favorite of Oppie's, which may have been all the excuse Kitty needed to turn against her. "Kitty was impossible," said Priscilla Greene. "She was not friendly, she was the boss's wife, and she could really be mean. She could also cause trouble for you, so you had to be very careful."

Kitty warily regarded her husband's endearing way with the opposite sex and jealously guarded his affections. All of Oppenheimer's assistants quickly learned to give her a wide berth. Kitty taught Priscilla Greene that lesson upon her arrival in Santa Fe. Priscilla had accompanied baby Peter and his nurse out west and had settled into the room adjoining the family's suite at La Fonda. Joe Stevenson had booked her the room next door for convenience, but Kitty was not pleased with the arrangement. She marched straight into Priscilla's room, locked the connecting door, and walked out without so much as a word. The next day, Priscilla discovered she had been removed to the opposite side of the hotel.

Regardless of how most people felt about Kitty, no woman dared turn down an invitation to one of her dinners or tea parties for fear of the repercussions, not just for herself, but also for her husband. Kitty had an unkind word for everyone, and the women worried her comments might reach Oppenheimer's ears. "Kitty was a very strange woman," said Emily Morrison, the wife of Phil Morrison, one of Oppie's protégés. "She would pick a pet, one of the wives, and be extraordinarily friendly with her, and then drop her for no reason. She had temporary favorites. That's the way she was. She did it to one person after another." They all saw how she went through people, picking fights with old friends and then snubbing them publicly. "She could be a very bewitching person," said Morrison, "but she was someone to be wary of."

At one point, Kitty's wrath turned on Charlotte Serber, one of the most prominent and popular women at Los Alamos. "Kitty threw a hate on Charlotte," said Shirley Barnett. "She would boycott her and throw parties and not invite the Serbers, which was ridiculous. Everybody was aware of it, and it was very hurtful. But Kitty was capable of that." Oppie knew better, she said, but "went along with it; it was probably a matter of keeping the peace at home."

When it came to Kitty, Oppenheimer was so passive as to be complicit. Even when she was at her worst, he would look the other way. He was an intensely private person, and while it was always hard to read his feelings, her sabotaging of the Serbers had to have been particularly awkward and embarrassing, as they were among his oldest and closest friends and he continued to see them daily in the Tech Area. Nevertheless, he did not intervene, and the Serbers were dropped from their cocktail crowd. This followed a pattern that had emerged shortly after their marriage, when Oppie broke off from many of his old Berkeley pals, either because they did not get on with his new wife or because they failed to meet with her approval. "He was very aware, but what could he do?" said Priscilla Greene. "It was such a bad, stupid marriage in the first place. I don't think he had terribly good taste in women. And in a way, he felt responsible for Kitty."

At Los Alamos, Kitty kept to herself. She brooded, and it showed in her face, which was pretty and animated when she smiled, but was increasingly closed and drawn down into a scowl. She wanted desperately to be a good wife to Oppenheimer, but he was not an easy man at the best of times, and at Los Alamos he was tense, distracted, and remote. No longer free to make the leisurely overnight pack trips they used to take into the mountains, he left their five horses to her to look after. In his whole time at Los Alamos, Oppenheimer found time for only one weekend excursion. Lonely and bored, Kitty reached out to Shirley Barnett, who had begun helping out in Oppie's office and whose husband was already her trusted pediatrician. "Kitty needed companionship, and I was young and less threatening than the others," she recalled. "She would invite me to spend my days off with her, and we'd go for lunch and shopping trips to Santa Fe, or as far as Albuquerque. She always had a bottle of something with her when she was driving, and you could always tell when she was getting drunk because she would talk more freely."

On those long, alcohol-fueled drives, Kitty would often reminisce about when she had first visited New Mexico with Oppie, when they would ride over to the Ranch School and Miss Warner's on horseback, and it was apparent that those were fond memories for her. When she had been drinking more than usual, she would touch on her past, though she rarely mentioned her family and remained circumspect. "She was fascinating, but not very nice," reflected Shirley. "She was not very happy, and you got the sense she never really had been. I think it stemmed from the fact that she'd had a complicated life and had been married before. The great love of her life was her first husband [*sic*], the one who was killed in the Spanish Civil War, and she never really got over it."

Kitty had indeed had a complicated life, and that made her particularly vulnerable to the emotional and intellectual pressure cooker that was Los Alamos. She was born Kathryn Puening in Germany in 1910, was brought to the United States at the age of three by her parents, Franz and Kathe Vissering Puening, and became a citizen by virtue of her father's naturalization in 1922. Her father was a mining engineer, and her family settled in Pittsburgh, where Franz Puening made a comfortable living in the steel industry. Kitty had wanted for nothing. But from the bright, attractive, and indulged only child of successful immigrants, Kitty grew to be a willful and rebellious young woman, whose first essays at independence were marred by an impetuousness that bordered on recklessness. After graduating from high school in 1928, Kitty entered the University of Pittsburgh, but, restless and unfocused, she left after only a year and headed for Paris. She drifted through courses at the Sorbonne and then did a stint at the University of Grenoble, but she was by her own admission more interested in the nightlife than her studies. Her European sojourn reportedly climaxed in a brief, unhappy alliance with a young musician from Boston, Frank Wells Ramseyer, Jr., in the spring of 1933. She never spoke of it and later grudgingly revealed to the FBI that the union was annulled in September after only a few months. Kitty clearly considered the matter closed and not worthy of mention. That fall, she enrolled in the University of Wisconsin, only to abandon her studies again a few months later to run off with the man whom she would always refer to as her first husband.

Joseph Dallet, the tall, good-looking son of a Boston investment banker, had spurned his family to dedicate himself to, of all things, unionizing steel workers. Kitty probably could not have chosen anyone more unsettling to her upright German father. Dallet had dropped out of Dartmouth in his junior year and had distanced himself from his well-to-do family, which was "conservative in political and social outlook," according to the brief biography that accompanies a published collection of his letters. After traveling in Europe and trying his hand at business, he finally went to work as a longshoreman "not because of economic pressure, but because he felt then that failure to earn his living by productive labor was to be a parasite." By the time he and Kitty met at a New Year's Eve party, he had joined the Communist Party and was a steel union organizer in Youngstown, Ohio. While he may have struck Kitty as a dashing and romantic revolutionary, leading strikes and trying to persuade workers to join the ranks of organized labor was not child's play. It was a hard and dangerous occupation. Dallet and his militant comrades were repeatedly arrested, and they regularly squared off with the steel trust gunmen and sluggers.

In the first throes of love, Kitty left college for Youngstown and impulsively joined the party. She tried to immerse herself in his union work, typing letters, mimeographing leaflets, and doing odd jobs around the office. To prove her commitment to the class struggle, however, she was also expected to get her hands dirty distributing pamphlets and hawking *The Daily Worker,* the Communist Party newspaper, in the streets around the mills. She and Dallet had no money, shared a rundown communal house with other party members, and lived in the kind of squalor she had never in her life countenanced. After two years, poverty got the better of her politics, and she had had enough. In June 1936, she told Dallet she wanted a separation. "I felt I didn't want to attend party meetings or do the kind of work that I was doing in the office," she would state later at Oppenheimer's loyalty hearing. "That made him unhappy. We agreed that we couldn't go on that way." Kitty's mother, who had never approved of the marriage, welcomed her back with open arms. But after several months of brooding in England with her parents, who had relocated abroad, she wrote Dallet saying she would rejoin him. Months went by with no word from him, and the waiting left her dejected and depressed. It was then she learned to her horror that her mother had been hiding Dallet's letters. "When she finally discovered this, Kitty was heartbroken," recalled Anne Wilson, who knew Kitty better than most. "Her mother was a real dragon, a very hard, repressive woman. She disappeared one day over the side of a transatlantic ship, and nobody missed her. That says it all."

Furious at her mother's treachery, Kitty ran straight back to her estranged husband, whose tender missives were all addressed to "Kitty Darling" or "Dearest Love." Kitty was briefly reunited with Dallet in early March 1937, meeting him in Cherbourg, where the *Queen Mary* docked. They took the train to Paris, where they had ten days together before he left for Spain to join the large contingent of idealistic young American volunteers fighting for the Loyalist forces in the Spanish Civil War. Infected by his enthusiasm for the cause, she wrote begging him to allow her to come fight by his side in Spain. He wrote back that he was "compelled to say no," as wives were not welcome on the front lines, though he added, "Personally, I think you would make a first-rate tank-driver in addition to the other things you've listed." Kitty, with no choice but to wait for him, returned to England. She wrote him faithfully several times a week and sent dozens of "snaps" of herself for him to show around. In July, Dallet hastily wrote, "Wonderful news. You can come." They made plans to meet in Spain, but a burst appendix delayed her trip, and it was weeks before she was fit to travel.

In October, still very much in love, she was en route to Spain to spend several days leave with him when a telegram reached her in Paris that Dallet had been shot and killed during an advance of the International Brigade. Devastated, she remained in Paris for a week, where she was comforted by Steve Nelson, her husband's close friend and comrade in arms, who brought her news of his bravery. To Nelson, Dallet died a hero, falling in battle against the fascists "without flinching an eyelash." He convinced Kitty to contribute to a monument to his memory and leading role in the party by allowing the Workers Library to publish Dallet's private letters to her from Spain, for which Nelson penned a laudatory introduction. To Kitty, though, it must have been small consolation that her husband gave his life to the cause before theirs had had a chance to begin.

A year later, in November 1938, Kitty married again on the rebound, this time to the British doctor Richard Stewart Harrison, whom she had known off and on for years. She later described her third marriage as "singularly unsuccessful from the start" and told Shirley Barnett it was over almost before it began. On returning to America that fall, Kitty went back to college, majoring in biology at the University of Pennsylvania, while Harrison remained on the opposite coast, completing his internship at a California hospital. She had only just rejoined Harrison in

Pasadena, moving into his apartment for the first time following her graduation, when she was introduced to Oppenheimer at an August faculty party. "I fell in love with Robert that day," she wrote, "but hoped to conceal it. I had agreed to stay with Dick Harrison because of his conviction that a divorce might ruin a rising doctor." But she was twenty-nine, and as impatient as ever. Fifteen months later, she was pregnant by her new lover and filing for a divorce in Reno.

A shotgun wedding was not Oppenheimer's style, and all his old friends immediately suspected him of doing the honorable thing. The rumor was confirmed when it turned out that she was already "showing" on her wedding day. Kitty admitted the truth to Anne Wilson. "She'd set her hat for him. She did it the old-fashioned way, she got pregnant, and Robert was just innocent enough to go for it."

They were two of a piece, from their early flirtation, and eventual disillusionment, with radicalism to their tormented past love affairs and the deep insecurities that lay barely concealed beneath the sophisticated facades. If people were moved to judge her more harshly, however, it was because of the way she threw herself across his path and forced his hand in marriage. Their relationship was, in the eyes of almost everyone who knew them, a *coup de foudre*—yet another destructive force in the life of a gifted and sensitive man fated, as Diana Trilling observed after the hearings, "to suffer as much for his virtue as his error," since they were so interchangeable. Yet, that Kitty adored Oppenheimer, identified herself totally with him, and in her own way struggled to fortify and buttress him during the worst times at the end of the war and afterward, there could be no doubt. Her protectiveness brought out the tiger in her, and she would go to great lengths to keep him safe or subdue a perceived threat. In his wartime memoir, the British physicist Rudolf Peierls paid tribute to her mettle, noting that she was "a person of great courage, both in the saddle of a horse and when facing a hostile authority," a reference to the later Oppenheimer hearings.

Kitty could not have foreseen when she married the Berkeley physicist that she would be thrust into the middle of such a vicious environment and subjected to the kind of intense personal scrutiny that she and Oppenheimer were to endure from his appointment as director of Los

Alamos to the day, almost a decade later, when his security clearance was revoked in 1954. If she began to withdraw into herself at Los Alamos, it was partly because of her personality, but it may also have been partly because of a new wariness, a concern for her position and worries about her past. It was not an easy time to be German in this country, and it created conflicting feelings in Kitty, who alternated between defiance and defensiveness. She had been raised to be proud of her German ancestry, once boasting to the Serbers that she was descended from royalty and that in Europe she had been treated with deference because of her formidable family. She managed to permanently alienate her sister-in-law, Jackie Oppenheimer—Frank's wife—by once too often playing the aristocrat to her working girl. The two women thoroughly disliked each other, and after Oppie's marriage to Kitty, Frank saw less of his brother.

But at the same time, Kitty was burdened with the knowledge that she was closely related to Field Marshal Wilhelm Keitel, chief of staff of the Wehrmacht. Keitel was with Hitler when the German army marched into Poland in September 1939 and was prominently by his side again at the Führer's triumphant victory over France. It was Keitel who personally presented the humiliating terms of the armistice to the defeated French government at the railway coach at Rethondes, near Compiègne, on June 21, 1940, where twenty-two years earlier Germany had been similarly beaten and humiliated at the hands of the French. Kitty had told friends that when the war broke out, she had taken quite a punishing from some of her college classmates. With anti-Nazi propaganda flooding the airwaves and American popular opinion, so recently isolationist, now whipping itself into a fever pitch in its campaign against the "murderous Huns," it would have been no small wonder if at times she felt embattled and alone.

Dorothy considered Kitty to be as trying as everyone else did, but for Oppie's sake, she exercised patience. She knew he worried about Kitty, who had suffered through a difficult first pregnancy and was expecting a second child. It touched her that their director, who was such a towering figure in his field and had such powers of persuasion, was at such a loss as to what to do or say when it came to his own wife. That humanized him in her eyes. He was a man who appeared trapped in a difficult marriage, and there were times it hurt her to see him so troubled. Everyone drank at Los Alamos, but Kitty drank with an abandon that was disturbing to watch and which clearly destined her for the alcoholism that took hold as the war came to a close. She hardly bothered to eat, and judging by the way clothes hung loosely off Oppie's bony frame, Dorothy hated to think what passed for meals in their household. More than one Los Alamos wife who could recall staggering home from Kitty's bungalow blind drunk would later comment on the meager fare that accompanied the liquid lunch.

Dorothy, well aware of the strain in the Oppenheimers' domestic life, was far too discreet to talk about it. She made it her business to be useful, to be there when she was needed, and to quickly absent herself when the crisis had passed. As the project consumed more of Oppenheimer's time, and the pressure and isolation took its toll, she increasingly functioned as Kitty's confidante, and a crutch to lean on. Though trained as a biologist, Kitty never worked more than sporadically at the lab. She filled the days by going on long rides with her neighbor Martha Parsons, but, apart from a handful of avid horsewomen, she had few friends. Dorothy was direct and down to earth, and her sensible, lowbrow advice must have been a good antidote to the elaborate, overwrought debates Kitty had with Oppie. Dorothy's independence, along with her bohemian air, had a certain appeal for Kitty, who fancied herself something of a free spirit. Not one to be easily flattered, Dorothy understood the relationship was largely one of convenience and, within limits, tried to be accommodating. Kitty was in desperate need of diversion, and Dorothy would organize Sunday hikes or take her off to visit small villages to look for Indian pots and Spanish rugs. But she stopped short of being her nursemaid. "Kitty made Dorothy very nervous," said Shirley Barnett. "She had to walk on eggshells all the time around her. Everyone did. Dorothy was extremely fond of Oppie, and her whole feeling about him and the project was starry-eyed. She would have done anything for him. She was absolutely loyal. But the closeness was something Kitty decided on. Dorothy was never really Kitty's friend."

Kitty would take pains to praise Dorothy in front of Oppie and the others and say how much she liked her, but her comments were always barbed, as they were whenever she talked about any woman who was close to her husband. According to Shirley Barnett, Kitty would point out that Dorothy, given her fine Smith education, had never fulfilled her potential and was stuck performing "a journeyman's job." On other occasions, she would object to inviting Dorothy to a dinner party, complaining that her position was really a secondary one and she did not belong at a gathering of senior staff. Kitty never lost an opportunity to slight Dorothy or voice some minor complaint. "Kitty looked down on Dorothy," said Shirley Barnett. "But then she looked down on most people. She was a snob."

In the end, Kitty was her own worst enemy. Though her marriage to Oppenheimer offered her some measure of respect and protection while she was at Los Alamos, her callous treatment of so many of his friends and colleagues earned her their lasting enmity. Even independent of her drinking, many people remember her as a cold and manipulative person. After the war, the Dutch physicist Abraham Pais recalled seeing Kitty at Princeton at the spring dance she and Oppie were hosting for the Institute of Advanced Study. Kitty was berating a young secretary, coldly instructing the poor girl that for her next evening dress she should choose blue because pink did not "suit [her] at all." The encounter left Pais, one of the world's leading theoreticians, shaking with anger and led him to describe Kitty as "the most despicable female I have ever known, because of her cruelty."

Despite all his skirmishes with G-2, by early September, Oppenheimer comforted himself that he had finally disposed of his security problems. During a sixteen-hour train trip with Groves and Lansdale, the three of them had reviewed a variety of security issues in detail and Oppenheimer had expressed regret at being drawn into the political quagmire created by his former student Lomanitz. They had also discussed Oppie's meeting with Pash and Oppenheimer's discomfort in giving up the name of his contact, and thereby dragging an old friend into a situation into which he was sure his friend had no involvement. Oppenheimer wanted to behave honorably, he told Groves, but he said he would divulge the name only if he were ordered to. Not wanting to stir up a hornet's nest, and reassured that Oppenheimer had meant well when he reported the espionage attempt to security in the first place, Groves had let the issue go.

Shortly after the train trip, however, the two highly critical memos from Pash and de Silva landed on Groves' desk. The general realized he now had little choice but to pursue the matter to the bitter end. Groves authorized Lansdale, whom he had promoted to colonel and made chief of security for the entire atomic project, to conduct his own investigation. While only thirty-five, Lansdale was a lawyer by training, and consequently a cut above Pash, and would be a far more skilled and subtle interrogator.

On September 12, 1943, Oppenheimer was summoned to Washington, where he was formally interviewed by Lansdale in Groves' office. Only this time, the questioning took the form of a cross-examination, with Lansdale systematically going through the list of Oppenheimer's friends and associates who were suspected to be Communist sympathizers in hopes of flushing out Eltenton's mysterious contact. Oppenheimer again declined to give Lansdale the name he was looking for, but in the course of being questioned volunteered that Lomanitz was probably a party member and named another former student, Joseph Weinberg, along with Serber's wife, Charlotte, and, last but not least, Kitty. By that time, Oppenheimer was probably confident that the FBI had open files on all of them and he was not telling Groves anything the FBI did not already know. Lansdale brought up Steve Nelson, Joe Dallet's former comrade, who had renewed his friendship with Kitty in 1941 and visited their home at Eagle Hill. He also asked about Haakon Chevalier, inquiring if he was a member of the party. Oppenheimer again chose not to disclose everything he knew, stating somewhat cagily, "He's a member of the faculty and I know him well. I wouldn't be surprised if he were a member. He is quite Red." They continued to dance around the matter for two hours, with Lansdale using every trick in the book to see if he could get Oppenheimer to crack and reveal the name of the intermediary and the three scientists he had approached. At one point, Oppenheimer asked Lansdale, "Why do you look so worried?"

"Because I'm not getting anywhere," he said.

"Well," Oppenheimer replied sanguinely, "except on that one point, I think you're getting everywhere that I can get you."

The questioning continued, with Lansdale repeatedly circling back to the Eltenton affair in an effort to pressure Oppenheimer to reveal his source, but to no avail. At the end, Lansdale told Oppenheimer that he liked him personally and had "no suspicions whatsoever," but at the same time advised him that his investigation of the matter was far from over. "Don't think it's the last time I'm going to ask you, because it isn't," he added.

Before taking his leave, Oppenheimer seemed to put forward a theory for his refusal to cooperate. "Well, I know where I stand on these things," he told Lansdale. "At least I'm not worried about that. It is, however, as you have asked me, a question of some past loyalties. . . . I would regard it as a low trick to involve someone where I would bet dollars to doughnuts he wasn't involved."

Lansdale replied, "OK, sir," and brought the grueling session to an end. But it was not OK, not by any measure, and in his subsequent memorandum to Groves, Lansdale indicated his firm belief that the director of Los Alamos should be made to reveal the name of the man suspected of a serious espionage attempt on behalf of a foreign power. Lansdale liked Oppenheimer, as he had commented more than once during the interview, but the physicist's obsession with retaining his own self-respect and private code of honor struck him as a kind of inverse vanity, and quite trivial in a time of war.

Even Oppenheimer, with his enormous gift for opacity, and a fastidious intellectualism that often left him completely blind to the political realities facing him, must have realized that he had only been stalling and that he could be compelled at any time to tell the army exactly what they wanted to know. What he could not have anticipated, however, was that his stubborn impeding of the Eltenton investigation, together with a number of concurrent security investigations into his activities and those of several former students, would raise so many troubling questions about his own loyalty.

On December 12, Groves returned to Los Alamos. During his visit, he once again took up the matter of security risks with Oppenheimer, only this time he was determined to get to the bottom of the matter. It was almost Christmas, three months had passed since Groves had confronted Oppenheimer in their shared Pullman compartment, and far from going away, the doubts and concerns about Oppenheimer's unwillingness to cooperate had festered and spread. This time, Groves was not going to be put off, however noble Oppenheimer's intentions toward his friend. Before he left Los Alamos, Groves ordered Oppenheimer to make a full disclosure. The next day, Colonel Nichols sent out three telegrams to the chief security officers of the Manhattan Project. The wire to Peer de Silva at Los Alamos read:

> HAAKON CHEVALIER TO BE REPORTED BY OPPENHEIMER TO BE PROFESSOR AT RAD LAB WHO MADE CONTACTS FOR ELTENTON. CLASSIFIED SECRET. OPPENHEIMER BELIEVED CHEVALIER ENGAGED IN NO FURTHER ACTIVITY OTHER THAN THREE ORIGINAL ATTEMPTS.

It is impossible to calculate what it must have cost Oppenheimer to give up the name of his old friend, knowing full well that implicating him in Eltenton's espionage plot would at the very least turn his life upside down and damage his career. As for the names of the three other scientists, he reportedly admitted to Groves that he had embroidered the story somewhat, and there were not three but one. (Although accounts differ, Lansdale later maintained that in exchange for Groves' word that he would not tell security, Oppenheimer admitted that Chevalier had tried to approach him through his brother, Frank, whom he hoped to enlist as a spy. But three years later, in 1946, Oppenheimer would tell the FBI yet another story, claiming the tale he had told Pash was a complete fabrication, and that Chevalier had never solicited anyone but him. At the security board hearing in 1954, he would stick to this line, saying that it would not have made sense for Chevalier to go through his brother since Chevalier was his friend and this would have been "a peculiarly roundabout and unnatural thing." Which version, if any, is the truth remains unclear.) But one thing is certain: by the end of 1943, Oppenheimer's own security file contained a tangle of lies. And as Oppie well knew, scientists were being investigated and dismissed from service for far less than the accusations that had been raised against him.

Groves was no fool, and it did not escape him that Oppenheimer was being evasive. The physicist's memory, which for myriad small details on the all-encompassing bomb project was always vivid and reliable, suddenly seemed to be playing tricks on him. His answers were often vague and inconsistent, and inevitably created the impression that he was being deliberately misleading. But it is likely Groves assured himself that the worst Oppenheimer could be accused of was dissembling about an espionage attempt that he had foiled, and that his actions had probably been motivated by a desire to protect his brother, his friend, or the lot of them. His reluctance to cooperate with security was understandable under the circumstances, if not admirable. The whole truth was impossible to determine, and what was known did not seem to warrant extreme action. Overriding security's objections, Groves did not withdraw his support of Los Alamos's director. The two men parted cordially, but the episode left the door open for security to wreak havoc on Oppenheimer's life in the years to come.

If Oppenheimer had once relished power, he was now learning that it could be a hard and lonely position. Burdened with the weight of the bomb project, and the guilt of handing Chevalier to the FBI on a silver platter,* he passed his first grim Christmas at Los Alamos. Not long afterward, on a cold blue January morning, Charlotte Serber walked into her husband's office in the T building and informed him she had just heard from old friends in Berkeley that Jean Tatlock had committed suicide the night before. She had taken a fistful of sleeping pills and drowned herself in the bathtub of her studio apartment on Telegraph Hill. By the time her father had discovered her body, she had been dead several days. She had left a hastily scrawled note, her shaky writing already showing the effect of the drugs:

*Chevalier was dismissed from his teaching post and left Berkeley. Years later, he maintained he was just trying to warn Oppenheimer about Eltenton, and never understood why Oppie turned the incident into such a "fantastic lie."

> . . . To those who loved me and helped me, all love and courage. I wanted to live and give and I got paralyzed somehow. I tried like hell to understand and couldn't. . . . At least, I could take away the burden of a paralyzed soul from a fighting world. . . .

Oppie needed to be told, and Charlotte asked her husband if he would break the news. By the time Bob Serber got to Oppie's office, he realized at once that he was too late. "I saw by his face that he had already heard. He was deeply grieved." Tatlock's untimely death was a sad coda to those idealistic days when she had first introduced Oppie to political activism and the principles of communism, and to many of the left-wing colleagues who were now listed in his FBI file as fellow travelers and possible collaborators. As Oppenheimer walked out of his office and headed for the silent wilderness, he must have mourned the loss of an old flame, and of a former innocence—both his and the country's.

ELEVEN

The Big Shot

THE LAST MONTHS of 1943 had brought a great rush of activity at Los Alamos with the much-anticipated arrival of the British Technical Mission. As if there were not enough to do already, Dorothy had to see to many of the preparations for the newcomers, who would need instructions, passes, housing, furnishings, and transport ready and waiting. It would also mean that there would be more mail, telegrams, phone calls, laboratory equipment, and lost luggage. Everything would be marked "Urgent" and required yesterday. While the authoritarian Groves was a great believer in the chain of command, Oppie was much more informal, and thought nothing of picking up the phone to make additional requests or last-minute suggestions himself. On top of this, Dorothy had to attend to all the regular business that naturally emanated from the Housing Office, daily dispatching all the new arrivals up to the mesa, along with two vans of furniture and at least one truckload of freight and laboratory equipment. Friends from the Hill stopped by to drop off packages bearing advertisements like "Send a Salami to Your Boy in the Army," which she would promise to post. There were also all the security passes that needed to be collected from, or reissued to, traveling staff members. At any given moment, the project had enough people on the go that they kept at least one double berth reserved daily on both the east- and westbound trains, and a drawing room twice a week.

Dorothy, now universally recognized as indispensable, operated on a grander scale from her humble command post than the Spanish viceroys who had occupied the adobe fortress centuries before. She was given an assistant to help answer her phone, which had been known to ring in over a hundred calls in a single day, and assigned her own small battalion of WAC couriers and uniformed drivers, anything she needed to get the job done. She marshaled her forces and prepared for the incoming. Because of the general lack of organization at Los Alamos, which was further compounded by wartime contingencies, nothing ever happened on schedule. Everything was flexible. The British mission's arrival dates and the numbers expected changed from one hour to the next. Groves' refusal to allow key project members to travel by air meant that the British had to come by train, and the Super Chief, which would be bringing them west after they changed trains in Chicago, was so chronically late that when it actually pulled into Lamy on time one afternoon no one was surprised to discover it was the previous day's train. All Dorothy knew with any certainty was that roughly two dozen scientists would be coming from the United Kingdom in December, and that among them was someone who warranted all kinds of special attention and far more than the usual security protocol. Whoever it was, he must be a big shot.

The cooperation between the British and American scientists was an event of international significance and represented the culmination of two years of negotiations between Churchill and Roosevelt. In 1939 and 1940, when fission work in the United States was still moving at a snail's pace, the British scientists were making considerable progress and were confident enough of atomic energy's potential usefulness as an explosive weapon to impress both Lawrence and Oppenheimer during the Berkeley conference in the summer of '42. With Groves' approval, Oppenheimer had reciprocated by sending the British a report on the conference's conclusions to Rudolf Peierls, the director of the British bomb project, which was code-named Tube Alloys. Peierls, a Dutch Jew, had been rescued by the Academic Assistance Committee and had joined the British effort. By then, the constant bombing and threat of invasion had forced Churchill to concede that they needed the United States' help in researching and developing an atomic bomb if it were to be completed in time to help save their island from Nazi domination. The Americans, however, had remained wary and were reluctant to share classified military weapons and techniques. After the fall of France, the British, in a bold move, had sent a delegation headed by Sir Henry Tizard across the ocean to try to initiate the exchange of secret military and scientific research. In a series of meetings in Washington, D.C., and Tuxedo Park, New York, the British scientists had overwhelmed the suspicious American army and navy officers with tangible evidence of their crucial breakthrough in the field of radar, proving to their red-faced hosts that the United States was, as an astounded Vannevar Bush had reported to Conant, "five years behind on the detection of planes." After that, the Americans and British entered into an unprecedented partnership to start a secret wartime radar laboratory to develop powerful new detection devices that would help England in its life-and-death struggle, and speed the day of victory.

But when it came time to pool their valuable atomic information, both countries played their cards close to the vest. The prime minister's advisors were reluctant to surrender what they believed was the one thing that could continue to guarantee Britain's status as the preeminent imperial power. The Americans, galvanized by the Pearl Harbor disaster, were happy for all the technical assistance the British could provide but, as soon as the project was poised to move from the research stage to large-scale production, became skittish about who would control the fruits of their collaboration. Talks between Churchill and Roosevelt had become increasingly tense as the two nations bickered over who would gain more military and diplomatic advantages in the postwar world, and it was another eight months before they resolved their differences. Finally, by the summer of '43, Britain, realizing it was in no position to compete with the United States' huge commitment of money and resources, agreed to the resumption of a "full and effective collaboration," with the production plants based in the United States. In the fall, a Combined Policy Committee was established, and by late 1943, British atomic scientists began arriving in the United States and were assigned to the various Manhattan Project laboratories.

Earlier that fall, during his last visit to Los Alamos, Groves had briefed Oppenheimer on the select list of people whom he could expect at Los Alamos. The eminent British physicist Sir James Chadwick, discoverer of the neutron, would be leading the mission, which to Dorothy's surprise contained only a handful of actual Englishmen and instead included German, Austrian, Swiss, and Polish scientists, all sporting brand-new British passports. "There were around twenty members," Dorothy recalled, ticking off the list with her encyclopedic memory for names and faces: Otto Frisch, the nephew of Lise Meitner and one of the first to arrive; Rudolf Peierls; William Penney; Ernest Titterton; Philip Moon; James Tuck; Joseph Rotblat; Egon Bretscher; and Klaus Fuchs, to name a few.

One name on the list stood out from all the others: Niels Bohr. The Danish physicist was a living legend, and his narrow escape to Sweden in October had been the talk of the mesa for weeks afterward. The story had been told and retold: Bohr, despite being half Jewish, had defiantly remained in occupied Copenhagen, and on the day he learned that he had been slated for arrest by the Germans, he and his wife slipped away to the seaside. They hid in a gardener's shed until nightfall, when the underground had arranged for a small fishing boat to take them across the sound to safety in Sweden. Just before leaving, Bohr received a tip that the following day all Jews and other "undesirables" were to be rounded up and deported. At great personal risk, he traveled immediately to Stockholm, which was thick with German spies, and campaigned relentlessly to help secure the asylum in Sweden of more than seven thousand Danish Jews.

After receiving a formal invitation to come to England from Lord Cherwell, Churchill's scientific advisor, the fifty-eight-year-old physicist made the last leg of his trip huddled in the bay of a Mosquito bomber, losing consciousness during the flight when the plane climbed in altitude and he failed to hear the pilot's instructions to put on his oxygen mask. Bohr remained in London for several weeks, where he was reunited with his son Aage, who along with another of Bohr's sons had crossed into Sweden separately. Bohr was briefed by Chadwick on the progress of the secret Tube Alloys project and agreed to join the delegation that was being sent to America to help build the bomb. The British were delighted at the prospect of having the Danish Nobel laureate and his son as part of their team and considered this quite a feather in their cap. The Bohrs were to travel to America separately, and there was so much subterfuge and mystery attached to their plans that one member of the mission asked why "they hadn't been packed and sent in a crate; it would have been so much simpler."

By the time the British began trickling into Los Alamos in groups of twos and threes, there was already a frosty chill in the air. There was no room on Bathtub Row for the distinguished guests—even neighboring Snob Hollow was packed to overflowing—so they were apologetically informed that the bachelors would be put up in the Big House, but married scientists would have to make do with houses in the newer, less desirable part of town. Most had come on their own, though in some cases their families would follow later, so they easily accommodated themselves to dorm living. The British were impressed by the speed of the construction and watched in amazement as workmen laid a foundation of a few concrete blocks and then, using wooden planks nailed to a framework of timber, completed a new cabin in only a couple of days. While the Americans complained about the shoddy buildings, to the war-weary British, who had endured the blackouts and bombardments of wartime England, life at Los Alamos seemed safe and relatively comfortable. There was no nightly wail of the air-raid sirens and no waking up to find a well-known building had been reduced to an empty shell. Just the chance to eat fresh fruit and fried eggs after several years of wartime austerity was an unheard-of luxury.

For many of them, this was also their first introduction to the United States, and the dizzying altitude and dry climate could not have been more different from that of their soggy island. Otto Frisch, who arrived with the first contingent, thought that Los Alamos, with its alien landscape and eccentric inhabitants, was the most exceptional "small town" he had ever seen: "I had the pleasant notion that if I struck out on any evening in an arbitrary direction and knocked on the first door that I saw I would find interesting people inside, engaged in music making or in stimulating conversation."

The British were also pleased and surprised to find themselves reunited with so many of their European cohorts at Los Alamos and, despite the fractious negotiations that preceded their coming, by the warm and collegial atmosphere that greeted them. Two old school buildings behind Fuller Lodge were renovated for their use, and they were quickly made to feel at home. Oppenheimer had originally planned to have Chadwick head up his own team at the laboratory, but the mission brought badly needed specialists in nuclear physics, electronics, and explosives, and he ended up plugging them into the existing divisions where they could do the most good. The hardest thing for them to adjust to was the distinguished American director's informal style, which extended from the laboratory uniform of jeans and open-neck shirts to calling everyone by his first name. Oppenheimer apparently startled more than one proper Englishman by strolling over, his thumbs dangling from his big Mexican silver belt buckle, and greeting him with the words, "Welcome to Los Alamos, and who the devil are you?" But the new arrivals were very good sports, particularly during the first few meetings when, as Frisch recalled, it seemed that almost everyone answered to the name Bob.

The British scientists' firsthand experience of war had a sobering effect on many of the young American scientists. Shortly after their arrival, Bill Penney, a British mathematician who was an expert on the effects of blast waves, gave a talk at one of the colloquiums on the damage caused by the German bombardment of England, and his cool recitation of the facts and figures left the audience in a shocked silence. "His presentation was in the scientific matter-of-fact style, with his usual brightly smiling face," recalled Peierls, who was only able to stay at the site for a few days of conferences and meetings, but would be rejoining the mission later. "Many of the Americans had not been exposed to such a detailed and realistic discussion of casualties." After that, his impressed American colleagues dubbed him "the smiling killer." (Months later, people were stunned by the news that Penney's wife was among those killed in one of the worst bombing raids on London.) The refugee scientists brought with them their personal experiences of exile and persecution, and the hardships and tragedies that followed. Rudolf Peierls and his Russian wife, Eugenia, had fled Austria for England, but had grown so afraid of the pummeling blitz and what would happen if the Nazis invaded that they had packed their two children, aged four and six, off to Canada. They had endured a separation of four long years and were finally reunited again in America, and Peierls was looking forward to their joining him on the Hill in the summer. Joseph Rotblat's wife and family were missing in Poland, and he had no idea if they were alive or dead. (It was not until after the war that Rotblat learned she had been killed in Lublin.) There were no words for the overwhelming anger and sadness the Americans felt at hearing these stories, or the gnawing sense that time was working against them. The atomic weapon that could stop the death and destruction was still many months away from being a reality, and was new and unproven. The uncertainty and frustration were almost unbearable at times, but it made them all the more resolved to get on with the job and complete the awful task they had been assigned.

The first Christmas on the mesa was hard on everyone, particularly the members of the new British contingent, as it was punctuated by thoughts of the hard times at home and families and relatives in harm's way. But there was nothing to do but make the best of it, and people decorated their homes and threw holiday parties in an effort to make the season as festive as possible. There were many formal dinners, and to mark the occasion the men wore black tie and the women got dressed up in their finest evening clothes and carefully picked their away across boards laid along the side of the muddy roads so as not to ruin their shoes. The weather cooperated by blanketing Los Alamos in white, hiding the bulldozed fields and construction scars under deep drifts of snow. Foot-long icicles hung from the shingled water tower. The town resembled an old-fashioned Christmas card, with a thick frosting of powder transforming the lodge and rustic log cabins into gingerbread houses. Each evening, the sunset turned the snowcapped mountains in the distance a delicious pink, like mounds of strawberry ice cream. The air was cold, dry, and crisp. Apart from the occasional blizzard, the winter sky was cloudless, and the sun was hot on their faces. Everyone took up sledding and skiing, raiding the supply of old equipment left behind by the schoolboys. Ice skating became a favorite weekend activity. The skating pond was down in Los Alamos Canyon, and the soldiers built a shelter and bonfire pit, where people could warm their hands on bitter cold nights. The pond became a popular gathering spot and was crowded with families and courting couples on Sundays.

The army did its best to get into the holiday spirit, and one of the ubiquitous memorandums designated one hillside as an official "tree-hunting area." Soldiers cut down Christmas trees of every size and shape for the laboratory personnel to choose from. A huge blue spruce was chopped down and erected in Fuller Lodge and strung with lights. Dorothy cornered the market on tinsel and trimmings at the Woolworth's in Santa Fe and did her best to fill all the orders for ribbon, paper, and candles. She rushed parcels up to the Hill as soon as they arrived, loading them onto the buses. She asked local merchants to stock assorted seasonal delicacies she thought some of the foreign scientists might be missing, including good brandy and cigars. She advised a few ambitious physicists where they might hunt for wild turkeys and instructed more than one urban housewife on how to pluck and dress the birds for cooking. She also warned them to be sure to remove as much of the lead shot as they could or they would be in for a rude surprise at dinner. Two days before Christmas, Kitty called down and asked Dorothy to get her a big goose for their holiday meal. It was such short notice, that Dorothy practically turned the town upside down until she finally located one and had it sent straight up. She later grumbled that all she got for her trouble was Kitty's complaint that it had "a crooked breast."

On Christmas Eve, the community chorus piled into the back of an army truck and sang carols as they slowly drove around town. It was a strange and lovely Christmas, but the true cause for cheer that holiday season arrived in the person of Niels Bohr and with his son Aage, who quietly materialized on the mesa in late December. Security mandated that Bohr's comings and goings be as stealthy as possible, so even the *Daily Bulletin*, which barraged them with announcements and orders, made no mention of their celebrated visitor. Oppenheimer had taken the added precaution of assigning both Bohrs, as well as Chadwick, pseudonyms in advance as, he wrote Groves, they would doubtless be receiving all kinds of important long-distance calls and mail, and "it would be preferable if such well known names were not put in circulation." As soon as they set foot on American soil, Groves' security force presented them their new identities: Niels Bohr became "Nicholas Baker," and his son Aage was "Jim Baker." Following the complicated instructions given to them before they left New York, the Bohrs/Bakers got off the train at the stop before Lamy, where they were met by an army car. They were then driven to an isolated stretch of road, transferred to another vehicle, and then whisked up to the site.

Bohr's arrival at Los Alamos buoyed everyone's spirits. He was famous as one of the fathers of quantum theory and had taught a whole generation of physicists how to change their way of thinking to accommodate the uncertainty principle and the dual particle-wave nature of matter. Many of the project members had studied with him and had made the pilgrimage to his Institute of Theoretical Physics in Copenhagen. But above all, he was the man who had first reported to American scientists the news of fission—that the Germans had produced an atomic fission which could lead to the vast release of atomic energy. It was, in effect, the warning shot in the arms race that had soon consumed physicists and changed their profession, and their lives, forever. He was not very old, but to them he seemed ancient, the great pioneer of quantum theory and the embodiment of wisdom. Caught like all of them in the grip of history, he had come to the desert to preside over the building of the first atomic bomb. Fresh from his frightening experience at the hands of the Nazis, Bohr, whose sad face seemed to have borne witness to a world of experience beyond that of most of the mesa's young scientists, appeared to them as a kind of spiritual leader.

"Bohr at Los Alamos was marvelous," Oppenheimer recalled in a lecture after the war. "He made the enterprise seem hopeful when many were not free of misgiving." Bohr reenergized them with his passionate words about Hitler and reassured them that no one would succeed in enslaving Europe in the same way again. "He said nothing like that would ever happen again," said Oppenheimer. "And his own high hope that the outcome would be good, and that in this the role of objectivity, the cooperation which he had experienced among scientists would play a helpful part; all this, all of us wanted very much to believe."

"It was the first time we became aware of the sense in all these terrible things," recalled Victor Weisskopf. "Because Bohr right away participated not only in the work, but in our discussions. Every great and deep difficulty bears in itself its own solution. . . . This we learned from him."

As it turned out, Bohr had not escaped Europe empty-handed. He had brought with him a drawing that the German physicist Werner Heisenberg, a brilliant former protégé, had secretly passed to him after a troubling conversation about the military applications of atomic energy and the moral implications of doing such work. Afterward, Bohr suspected Heisenberg was on an elaborate fishing expedition and was attempting to find out what his famous mentor knew of the Allies' progress in building a bomb. While no two accounts of their clandestine meeting agree, it ended on a bitter note, and Heisenberg left Bohr with more questions than answers. The biggest riddle of all was the drawing itself. Bohr could not be sure what he was looking at: Was it a sketch of an experimental heavy-water reactor the Nazis were working on, or was it misinformation meant to confuse the enemy? In any event, Bohr knew he had no choice but to alert the Allies of its existence, and he did so immediately after reaching London.

On New Year's Eve day, Oppenheimer called a handful of senior staff to his office. When Serber got there, he found both Bohrs were present, as well as Bethe, Teller, and Weisskopf.

> Oppie handed me a scrap of paper that looked like it had been carelessly ripped from a note pad. It bore a sketch, and he asked me what I thought the sketch represented. After a minute I handed it back and said it looked like a heavy water moderated nuclear reactor. He then told me Bohr had gotten it from Heisenberg. The question was whether it could be interpreted as a weapon. The Los Alamos experts gathered in that room all agreed that it was useless as an explosive.

Although Bohr had already raised the question with Groves, Oppenheimer promptly sent Groves a summary of the meeting, along with a memorandum prepared by Bethe and Teller.

1.

Dorothy with her husband, Joseph McKibbin, and baby, Kevin, shortly after his birth on December 6, 1930. Before Kevin was a year old, Joe McKibbin died of Hodgkin's disease. In spring 1932, the young widow packed up her belongings and baby and moved to Santa Fe, where she built a rustic adobe farmhouse.

2.

3.

4.

When Robert Oppenheimer arrived in Santa Fe in March 1943, the Los Alamos Ranch School had been closed by the War Department and three thousand army engineers had turned it into a muddy construction site. The arriving physicists were appalled by the steep, twisting dirt road up the mountain and by the narrow Otowi Bridge, which was "too fragile" for heavy trucks.

5.

6.

No one could go to Los Alamos without first reporting to 109 East Palace, the small Santa Fe office that was a front for the classified laboratory (*below*). Dorothy's job was to issue the security passes that had to be presented at the Main Gate (*left*) before anyone was permitted to enter the heavily guarded site.

7.

General Leslie R. Groves (*left*) ran Los Alamos as a "no frills" military post, while Robert Oppenheimer fought to preserve a few civilizing touches.

8.

9.

Security at Los Alamos was designed to keep information and personnel from getting out. Military police patrolled the barbed-wire perimeter twenty-four hours a day, and the elaborate inspection process required for coming and going became the bane of the scientists' and their families' existence. Within the barbed wire fence penning the Tech Area (*below*) lay "the most secret part of the atomic bomb project."

10.

Los Alamos looked like a frontier boom town with row upon row of prefabricated apartment complexes all painted o.d.—"olive drab." The old school trading post (*below*) sold an odd assortment of supplies, including little boys T-shirts and moccasins. The wooden water tower was the only landmark by which newcomers could orient themselves among the maze of unnamed streets.

12.

13.

14.

Dorothy, Oppenheimer, and Victor Weisskopf at a party in Los Alamos, circa 1944.

15.

Oppenheimer's wife, Kitty, enjoyed the perks of being the "First Lady" of Los Alamos, but she was moody and withdrawn and did not participate in the mesa's active social life.

6.

George Kistiakowsky (*below*) and Enrico Fermi (*left*), both initially consultants at Los Alamos, were persuaded to move to the Hill during the project's second year and helped with the big push to perfect the implosion bomb.

17.

9.

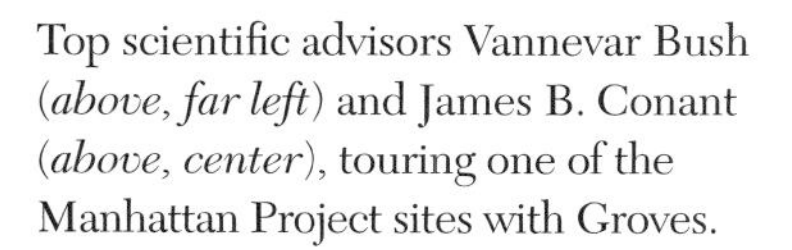

Top scientific advisors Vannevar Bush (*above, far left*) and James B. Conant (*above, center*), touring one of the Manhattan Project sites with Groves.

18.

(*Right*) Frank Oppenheimer, a physicist and Robert's brother, came to Los Alamos in 1945 and worked at the Trinity site.

20.

21

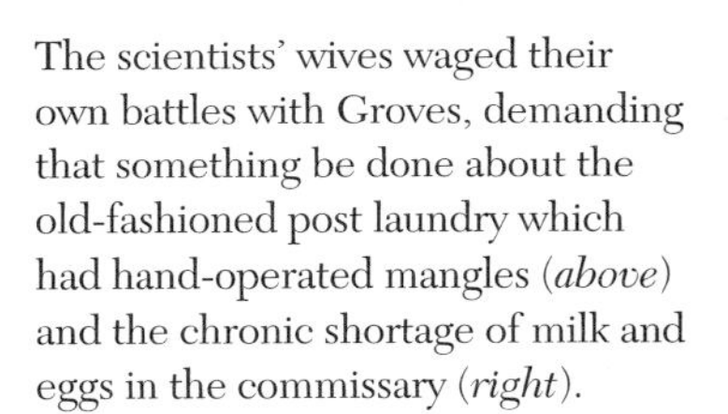

The scientists' wives waged their own battles with Groves, demanding that something be done about the old-fashioned post laundry which had hand-operated mangles (*above*) and the chronic shortage of milk and eggs in the commissary (*right*).

Scientists and soldiers lined up to buy newspapers, magazines, and cigarettes at the PX (*below*) but because of wartime shortages and uncertain deliveries, often had to do without.

22.

23.

The pressure to build the bomb worked on everyone's nerves, and drinking and dancing were popular ways to let off steam. A fair amount of carousing went on at the PX (*above*), which had a jukebox and a dance floor.

24.

Impromptu office parties like this SED bash in the electronics division were often fueled by Tech Area 200-proof liquor.

25.

Oppenheimer mandated that Sundays were a day of rest, and the scientists streamed down to the meadows for picnics. Dorothy McKibbin (*left*), taking a well-deserved nap. Robert Serber (*below*), one of Oppie's protégés, tried to plant rumors that Los Alamos was really a rocket factory.

26.

27.

28.

Serber's wife, Charlotte (*above*), was one of a handful of strong women Oppie counted on to help with the smooth operation of the laboratory, along with his assistant, Priscilla Greene (*right*), shown with Truchas, the English bulldog that Dorothy found for her.

Hiking was a popular pastime. After Dorothy took over the old Frijoles Canyon Lodge at Bandelier National Monument and turned it into a weekend retreat, the scientists and their families flocked to the picturesque spot to escape the drab confines of their mountain hideout. Dorothy recalled that "some rode their horses over, some came to fish, and most came to eat because the food was excellent."

30.

Shirley Barnett, one of Oppie's assistants, and her husband, Henry, Los Alamos's much in demand pediatrician. He had his hands full coping with the exploding birth rate at the laboratory.

31.

32.

Skiing was an obsession with some of the more intrepid scientists, including Enrico Fermi (*above, far left*) and Hans Bethe (*above, second from left*), who were undaunted by the steep slopes around Los Alamos. Niels Bohr (*left*) put younger men to shame on Sawyer's Hill.

33.

Oppenheimer (*left*) and Captain Peer de Silva, head of security at Los Alamos, scouting in spring 1944 for what became known as Trinity site, where the atomic bomb was to be first tested. Six months earlier, de Silva had sent Groves a memo saying he suspected Oppenheimer of playing a key part in attempts to pass "highly secret information" to the Soviet Union.

34.

(Above) The huge orange fireball and mushroom cloud of the Trinity test on July 16, 1945. When Conant first saw the burst of white light, he thought the world was coming to an end. *(Right)* Oppenheimer and Groves next to the charred remains of the steel tower at ground zero.

35.

36.

(*Above*) At the war's end, Oppenheimer received the Army/Navy E Award (for excellence) on behalf of the laboratory.

37.

Edward Teller congratulated Oppenheimer on receiving the Fermi Award on December 2, 1963. The long overdue recognition marked Oppenheimer's return from disgrace, but many of those present felt that Teller should not have attended the ceremony after he had questioned whether Oppie could be trusted with the nation's atomic secrets during the Gray Board hearings.

38.

Dorothy continued to run her small outpost at 109 East Palace and to serve as "the gatekeeper" to Los Alamos until 1963. When she retired after twenty years, the office was closed. On the wall behind her are pictures of two of her heroes, I. I. Rabi (*left*) and Jim Conant (*right*). Time did not diminish her feelings for Oppie, and she remained his most devoted friend and ally to the end.

39.

A battered hero: After the Gray Board found him to be a security risk and stripped him of his clearance, Oppenheimer was humbled but not destroyed. He continued to teach and write and remained a dignified if poignant figure.

Oppenheimer had first met Bohr at a particularly low point in his graduate studies and always credited the Danish laureate's straightforward manner and simple kindness with helping him to stay the course and resolve "to try to learn the trade of being a theoretical physicist." Dorothy could see that the sturdy, gray-haired theoretician's fatherly presence, with his slow, soft-spoken English, once again had a rallying effect. "He was the *master* scientist," she said. "He was older than Oppie . . . and he was a person who could get everyone all together." He was also modest and unassuming, and Oppie, who was amused by his code name, took to calling him Uncle Nick, and before long that was how he was affectionately known to everyone. "He was a genuine guru," said Priscilla Greene. "He was practically the only person Oppie unreservedly admired."

On one of Bohr's first evenings at Los Alamos, the Oppenheimers had a big dinner party for him. Most of the senior scientists and their wives were invited, along with the Agnews, who were rather more junior, but lived about seventy-five yards from Oppie's house so were often included. It started with Oppie's dry martinis, which had their usual salutary effect, and everyone had a wonderful time. Bohr entertained everyone with the story of his arrival in the United States: how he and his son were hustled from the boat to the hotel in New York by a squad of FBI men, who hurried them through the lobby and straight up to their room with all their luggage, and how everyone collapsed in exhaustion and relief that the Danish physicist's identity had not been discovered along the way. It was then that one of them noticed that written on the side of his suitcase in large black letters was NIELS BOHR.

At the end of the evening, Harold Agnew went to get his overcoat, found that it was gone, and went home without it. The next morning, he stopped by the Oppenheimers' house to see if it had turned up. There was an unclaimed coat, but it was too small to be his. Oppie suggested he wait around and maybe his would turn up. "In a little while, along comes Bohr," recalled Agnew. "His coat was dragging in the snow, his hands hidden by the long sleeves. 'I think this isn't my coat,' he says, 'because there are keys in the pocket, and I have no keys.' "

Bohr was not used to being on his own, without his devoted wife,

Margrethe, to look after him, and often showed up in comically mismatched clothes. Once he came to work in the Tech Area with a rope tied around his waist to hold up his pants because he could not find a belt. There were many anecdotes about Bohr's absentmindedness, though perhaps the best-known concerned an idea for a fruitful approach to an atomic bomb that had come to him just after the British scientists had first made him aware of their progress. Bohr had sent a cryptic telegram to England stating, in part, "AND TELL MAUD RAY KENT." The British scientists, including Peierls and Frisch, were convinced that the urgent message was a code, or even an anagram, warning them of something, and they tried to crack it, coming up with strange misspelt solutions like "Radium taken," which could have been pointing to the Nazis. When Bohr came to Los Alamos, one of the first things he was asked was the meaning of his mysterious cryptogram. It turned out that Maud Ray had been his children's governess and she had lived in Kent, but as to the germ of an idea the line was meant to convey, that had slipped his mind.

Everyone on the mesa tried to take care of Bohr and took turns wining and dining him at night. Oppie and Kitty often took him to dinner at Miss Warner's, where he struck up a friendship with the austere proprietress of the teahouse. "They recognized each other's qualities immediately," her goddaughter, Harriet "Peter" Miller, later recalled. "She always felt she had something to learn from him." After the war, Bohr wrote of their special rapport, noting: "[Edith] had an intuitive understanding which was a bond between us."

Bohr and his son Aage, a brilliant physicist in his own right who would earn his own Nobel, stayed for several weeks. Oppenheimer spent a lot of time with them and could always be seen walking with them between the Tech Area and Fuller Lodge. Both Bohrs were great hikers, and almost every afternoon they would go off on a long strenuous walk, sometimes accompanied by one or two other young physicists, while they would discuss a particularly thorny physics problem. Dorothy loved the way "Uncle Nick" looked in the silly, broad-brimmed straw hat he wore against the strong New Mexico sun, with his beloved pipe firmly planted in the corner of his mouth, always in need of lighting. Bohr also loved to ski, and she would never forget the sight of the legendary Dane expertly negotiating

Sawyer's Hill, the ski slope created by the Ranch School boys, which was only a short walk from the West Gate. He was not the least bit discouraged by the absence of a chair lift, and long after younger colleagues had called it quits, he could be seen energetically poling toward the top.

The Bohrs, like everyone on the Hill, were greatly heartened by the radio reports that the war was turning in their favor. At times the news was confusing, but Otto Frisch recalled that Bohr always listened closely and joked, "We must hear all the rumors before they are denied." Bohr had not come to Los Alamos to contribute at a technical level so much as to serve as a sounding board and benefactor. He had come to America as a wise man, a scientist-statesman, to alert Roosevelt and his advisors to the perils of hoarding atomic "secrets."

At the time, the dangers of the arms race he warned of sounded to the politicians like a distant threat compared with the far more pressing problem of winning the war. But his cautionary tone had a profound effect on Oppenheimer, as well as many other of the Los Alamos scientists, who were too caught up in their work to dwell on the ramifications of it all at the time but would later return to Bohr's whispered admonishments and advice. He never had a single word of criticism for Oppenheimer or his upstart operation, and his faith and confidence that their work was progressing well left every one of the mesa scientists feeling decidedly more encouraged.

Bohr could not stay long, and on January 17, 1944, left for Washington and then London, where he was scheduled to meet with Churchill to relay his concerns about the postwar implications of the bomb and the frightening possibility of the proliferation of nuclear weapons. He promised to try to return as soon as possible, and Oppenheimer, writing to Groves on the day of the Danish physicist's departure, made it clear that they would be most grateful for his continued collaboration:

> By word and deed Dr. Baker has done everything he could to support this project and indicate that he is sympathetic not only with its purposes and general method of procedure, but with the policies and achievements of the project's overall direction. I should like to make it quite clear that the effect of his presence

on the morale of those with whom he came into contact was always positive and always helpful, and I see every reason to anticipate that this will be true in the future.

For most of that first year at Los Alamos, before any significant amounts of nuclear material could be delivered, the main effort had gone into attempts to get as accurate an estimate as possible, as early as possible, of the critical mass of uranium 235. All of the experimental groups—Bob Wilson and the cyclotron, John Manley and the Cockcroft-Walton, and Johnny Williams on the Van de Graaff—were busy measuring cross-sections and other parameters, and then the Theoretical Division, led by Hans Bethe, would apply mathematical methods to the problem. The challenge was how to make fissionable material—either uranium 235 or plutonium 239—release its energy efficiently, at precisely the right time, and in a metal casing that could be carried by a B-29 bomber. It was apparent that there would also be fairly strict requirements for the size, shape, and weight of the actual gun assembly, and Ed McMillan's group was preoccupied with firing guns and testing assemblies down in an unoccupied canyon. By late1943, Bob Wilson's group had confirmed that an intense uranium fission reaction could take place before the bomb blew itself apart. A uranium bomb, using the gun method, was feasible and designing the assembly would be a relatively straightforward job, more engineering than physics.

But the gun method was an unwieldy design and extremely wasteful of the scarce uranium 235. As a backup, the implosion weapon based on plutonium was beginning to look better all the time. Seth Neddermeyer had been off on the sidelines exploring the difficulties inherent in implosion, the approach he had suggested early on. He had been pursuing implosion with Oppenheimer's tacit, if not exactly enthusiastic, approval, but now it was becoming too important to ignore. Since plutonium was thought to have a very fast fission rate, Neddermeyer was experimenting with all different kinds of explosive materials to see which assembly would be quickest.

At the same time, Groves and the Review Committee decided Los Alamos would be responsible for developing methods to purify and fabricate the plutonium into metal for all of the Pu-239 produced at the nuclear pile at Hanford, in Washington, and this would be done on site at Los Alamos, necessitating still more laboratory buildings, facilities, and staff. "That added a considerable extra effort to the Los Alamos activity," recalled John Manley. "It was a logical decision though, because the material was never very abundant, and if you did one experiment and a following test had to have a different shape, the material would have to be reworked. It made a lot of sense to do it right at Los Alamos."

The expanding work at the laboratory meant that more specialists were needed, and Oppenheimer rededicated himself to recruiting fresh talent for Los Alamos. Oppenheimer had already brought in David Hawkins, a philosopher from the University of California, to take over much of the responsibilities of post administration and allow him to focus on the work of the laboratory. Nineteen-year-old Frederic de Hoffman was part of a team of four physicists from Harvard selected to join the project. Arriving at Los Alamos just after Christmas, he recalled that a rather burly WAC sergeant was at the station at Lamy when the train arrived and hastily brought them to Dorothy McKibbin, who "quickly made a young Viennese feel at home in New Mexico."

There was a steady stream of new faces at 109 East Palace, among them the Polish mathematician Stanislaw Ulam, along with his attractive French wife, Françoise, who was two months pregnant. The Ulams, like the others before them, had been instructed only that they were headed for an undisclosed Manhattan Project site in the Southwest, near Santa Fe. As recent émigrés to the United States, they were totally unfamiliar with the region and, deciding to do a little advance research before their departure date, checked out a WPA guidebook on New Mexico from the college library. At the time, Stan Ulam was teaching at the University of Wisconsin at Madison, and there on the back leaf of the book, where borrowers signed out the volume, were the names of all the previous members of the physics department who had mysteriously disappeared to do hush-hush war work. The Ulams said nothing about their discovery until they arrived at Los Alamos, where they greeted their former colleagues without surprise, explaining that even the tight security surrounding "Project Y" had its loopholes.

One familiar face returning to Los Alamos was the Harvard chemist George Kistiakowsky, a strikingly handsome and charismatic forty-three-year-old Russian who in his previous visits to the post had been one of the most popular unattached men. When he finally arrived at 109 after the long dusty journey by train, which he had made once too often, he announced dramatically, "I am old, I am tired, and I am disgusted!" Dorothy had to laugh. She had come to know Kisty well from his travels to and fro, and had a weakness for his courtly Old World manners and droll charm. He always stopped by her office for a gossip, knowing her to be well versed in the latest news from the Hill. Unlike most of the scientists, he was also an excellent horseman. Dorothy had heard that he was a former Cossack and thought he cut a smart figure in the saddle. "He was very dashing," she recalled. "He had a tremendous personality. He was tall, and skinny, and had blond hair and blue eyes. He had been with the White Russian Army and had escaped with his life."

The leading civilian explosives expert at the time, Kistiakowsky was needed to head the division that would be taking up the experimental challenge of developing the implosion device for the plutonium weapon. He had been commuting back and forth from Harvard as a consultant and had been reluctant to commit to working on the bomb because he did not think it would be ready in time to make a difference in the war; it took considerable persuading by both Oppenheimer and Conant to convince him to move to Los Alamos full-time. When he finally relented, Oppenheimer saw to it that he got plum accommodations; he had the small stone powerhouse, which had housed the Ranch School's old generator, made over into a snug, one-room apartment. Oppie even sold him, for a nominal sum, one of his favorite saddle horses, Crisis, as the project kept him so busy he rarely found time to ride.

Kistiakowsky immediately began bringing in experienced detonation and explosives specialists, and set about doing experiments to figure out the best way to achieve implosion. Among them was the Harvard physicist Kenneth Bainbridge, who like Luis Alvarez had spent the last two years at the MIT Radiation Laboratory working on radar. Bainbridge would head a new group investigating full-scale high-explosive assemblies. The idea was to take a metallic pit, in the center of which was the plutonium fissionable material, and surround it with high explosives. In theory, the plutonium would then be imploded by the detonation of the explosive shell around it. This approach differed from that of firing the two halves of the bomb at each other in a gun barrel, as was sufficient for uranium. As Kistiakowsky recalled the challenge: "Our job was to induce the pit and plutonium to be compressed in an orderly fashion under the extreme pressure of a detonation wave, many millions of pounds per square inch, into something very much smaller than it normally was, whereupon it would become supercritical. A nuclear reaction would then spread and a big bang follow."

The experimental work progressed in the early part of 1944, using small samples of plutonium Oppenheimer had gotten from Chicago, and then later with small samples that began arriving from Hanford and Oak Ridge. John von Neumann had done mathematical work to show that the spherical shape of the charges could affect the bomb's explosive yield. James Tuck, one of the new British physicists, had invented a promising new explosive lens that could shape the shock wave, and the mesa scientists set about manufacturing more for their experiments. Von Neumann also brought the first computers to Los Alamos, and the machines, courtesy of IBM, were promptly put to work on calculations of implosion. At the same time, Parsons was pushing ahead on the ordnance development, including bomb hardware, arming and wiring mechanisms, and fusing devices. With the assistance of the army air forces, Parsons' group developed two bomb models: "Thin Man," named after President Roosevelt, was based on the gun design, while "Fat Man," in honor of Winston Churchill, was based on an implosion prototype. In March 1944 they began testing the models with B-29s.

That same month, Hans Bethe assigned Peierls and a small team of physicists to tackle the daunting calculations of implosion. Bethe had worked with the talented and unassuming German physicist in the past and knew him to be very capable, plus he brought with him the new techniques and perspective of the British mission. After a careful review of the British project's approach and results, Oppenheimer reported to Groves that the British method of calculating the blast wave, while "more laborious and less perspicuous than the methods used here," might be useful for studying the problem of the hydrodynamics of implosion. He concluded that they were "planning to attack the implosion problem along these lines with the highest possible urgency."

Peierls was favorably impressed by Oppenheimer's leadership and the efficiency and flexibility of the laboratory structure, enhanced by the weekly colloquiums held in the closely guarded theater. He was particularly pleased to see that no favoritism was exhibited to members of the American team over the British and that work was guided "by the necessity to get the best possible answer in the shortest possible time." Peierls could not help noticing the tension between Oppenheimer and Teller, and their obviously complex relationship. Teller was supposed to head the "hydrodynamics group" to look into the problems of detonation waves and shock waves, but as he insisted on devoting his time to his pet project, the hydrogen bomb, Bethe had asked Peierls to lead it instead. Teller complained loudly and to anyone who would listen about the lack of support being expended on developing "the Super." Finally, Oppenheimer agreed to let him carry on with a small group, even though it was clear that his work would not be contributing to their war work.

While Peierls had great admiration for Teller, "whose thoughts [were] often leaps and bounds ahead of his plodding colleagues," he later reflected in his memoir that the Hungarian's particular combination of ego and insecurity made him exceedingly difficult to manage. "Perhaps the reason was that Teller was anxious to earn Oppenheimer's respect, not only as a physicist—which he undoubtedly had—but as a person." Oppenheimer openly alluded to the tension after he gave a party in honor of Lord Cherwell at Los Alamos and, by some oversight, failed to invite Peierls, the deputy head of the British group. After discovering the mistake, Oppenheimer went to Peierls to apologize. "This is terrible," he told him, "but there is an element of comfort in the situation: it might have happened with Edward Teller."

TWELVE

Baby Boom

EVEN BEFORE the first explosions began echoing in the canyons around Los Alamos as the scientists tested their theory of implosion, ominous rumblings of change were disturbing the peaceful mesa. The pioneers, the small, tight-knit group of scientists who had first come to the site, now found there were many faces at dinner in Fuller Lodge they did not recognize. As more and more people arrived to work on the project in the spring of 1944, the housing crunch once again became acute. Families that were "expecting" signed up on the waiting list for larger apartments. The list was posted on a wall, so everyone knew who was pregnant, and for how long, and therefore had priority. Each week the list seemed to grow longer. Not long after Luis Alvarez and his family came to Los Alamos, his wife, Geraldine, became pregnant, and they signed up for a larger apartment. It caused a brief scandal when at the next Town Council meeting, the wife of a well-known European physicist complained that she did not want a Mexican family moving in next door. She was coolly informed that the Alvarez in question was the eminent Berkeley physicist, and the tall blond son of the famous Mayo Clinic physician Walter Alvarez.

Things got so bad that when Dorothy would call up to Los Alamos to say that she was sending up a new family for one of the vacant units, she would be informed by the Housing Office that someone had already claimed the property that she had reserved for them. Dorothy would then be forced to engage in elaborate delaying tactics down on her end while the Housing Office desperately scrambled to find them new quarters. She would take the scientists and their worn-out families on the lengthiest orientation tours of Santa Fe she could devise and treat them to their first enchiladas at the knee-high tables at La Fonda, while frequently excusing herself to phone up to the Hill until she was given the all clear. At one point, things got so bad there was even a rumor going around that the WAC in charge of counseling tried to dissuade one physicist from getting married rather than have to find new accommodations for him and his bride.

The town's services, originally designed with a small community of a hundred or so scientists in mind, were rapidly becoming insufficient for the population that had grown to more than 3,500 and counting. There were signs that the water supply was dangerously low, reminding them that they were hostage in a dry land. Army bulletins containing instructions on water conservation were distributed door-to-door by uniformed soldiers. Leisurely ablutions were to be replaced by a "good citizen's shower" lasting only a minute or two. The *Bulletin* issued a rash of edicts along the lines of "*REPORT LEAKY FAUCETS IMMEDIATELY!*" Toilet flushing was to be kept to a minimum, gardens were not to be watered, automobiles were not to be washed, and so on. The shortage was so great it became the main topic of conversation among the mothers wheeling their young around the mesa. Water had to be boiled, and typhoid shots were ordered for everyone. When long string worms starting turning up at the bottom of teakettles and drinking glasses, Doc Barnett would shrug and tell them of the "chow mein" that flowed from the faucets of San Francisco and remind them of the S. J. Perelman story about "the Filipino houseboy who arrived via the tap." It got so bad Elsie McMillan recalled that after disinfecting her baby's bottles in a pan of water, she would do the dishes in the same water, and then reuse the gray liquid to scrub the kitchen floor.

There were signs of overcrowding everywhere. Government trailers and Nissen huts sprung up on empty meadows, turning them overnight into urban slums. New arrivals sometimes waited weeks to get their plumbing connected. The back porches behind the apartment buildings were jammed with stinking garbage cans. Rows of parked cars sat hub deep in the spring mud. The lines of laundry flapping in the wind added to the Dogpatch look. Electricity was intermittent, and it was regularly turned off when a major experiment was in the offing. Food shortages were also a recurrent problem. Some anxious mothers even suspected Groves of instructing the Commissary not to carry sufficient inventories of milk and believed the chronic scarcity of diapers and formula to be part of a vast conspiracy to discourage them from expanding their families.

The mesa population's exploding growth was a constant source of jokes. Eighty babies were born in the first year, and by Doc Barnett's best estimates, the coming year would far exceed that figure. Another ten babies were born in April, and in the month after, and he had every reason to believe the birthrate would continue to be roughly the same. Most of the couples were young and healthy, and since the vast majority of infants he had delivered were first children, more were likely to follow. At one point, there were so many babies being born that nearly half the hospital was being used as a nursery. A June 22 memorandum on the Los Alamos baby boom from Colonel Stafford Warren, an officer in the Medical Corps, to General Groves, highlighted the problem:

> Item 3.d. Approximately one-fifth of the married women are now in some stage of pregnancy. (The birth rate over the nation elsewhere is decreasing.)
>
> Item 3.e. Approximately one-sixth of the population are children, one-third of whom are under two years of age.

When confronted by the escalating birthrate, and the corresponding rise in the cost of hospital services, Groves all but ordered Oppenheimer to do something about it. The laboratory staff was reproducing at such a rate that the situation was straining the post's resources and threatening to turn his top-secret garrison into a glorified pediatric ward. Oppenheimer objected that population control was not one of his duties as scientific director, not that it would do any good. What he probably did not mention, however, was that he was the last person who could take a stand on the issue as his own wife was with child.

Word of Groves' complaint quickly spread around the post grapevine and became fodder for all sorts of gleeful limericks and jingles, the politest of which went in part:

The General's in a stew
He trusted you and you
He thought you'd be scientific
Instead you're just prolific
And what is he to do?

After that, poking fun at the general's attempts to police their love lives became a favorite mesa sport. When a scientific journal published a report stating that there was statistical evidence linking male infertility to long immersion in hot water, several scientists suggested sending a copy of the article to Groves in a bid to have bathtubs installed in their homes. They would all have gladly risked a night of sterility in exchange for a good soak.

Groves, like any absentee landlord, did not see the humor in the "constant and unanticipated increase in the population." The town was overloaded, and the crowding aggravated the existing tensions between the military and civilian personnel. The scientists were not the only ones with wives and children on the post: officers could also bring their families, as well as the few civilians who manned key facilities, such as the fire station, steam plant, and lodge. Noncommissioned officers had to make do with visiting their wives in Albuquerque. More dormitories were built for married civilian couples, but married GIs were ineligible. This meant that the many soldiers and WACs who had fallen in love and married while at Los Alamos had to live apart, until the regulations were eventually changed by the end of the war and apartments were set aside where they could keep house for two-week periods. Lieutenant Colonel Ashbridge, who administered the post, was under constant fire from both groups, and suffered a mild heart attack. The Los Alamos inmates had defeated two COs, and would doubtless try for a third. "It was obvious," Groves wrote, "that until most of the innumerable petty annoyances could be removed, the frictions would continue."

While Groves did not like coddling the civilian population, or wasting an extra nickel on the temporary wartime town, he recognized that they had to build more satisfactory housing and that to better the morale of the community, some improvements had to be made for all the families. To alleviate the housing problem, he approved the construction of Morganville, and in early 1944 a swarm of tiny duplexes were erected at the north end of the alfalfa field near Fuller Lodge. They were built for the machinists coming in from Dallas, but the union rep took one look at the poorly built houses and refused to let the workers live there, demanding apartments instead. The Sundt and Morgan outfits were succeeded by the Robert McKee Construction Company as the primary contractor, and Groves ordered one hundred new prefabricated houses to be built in regimental formation in another field. In no time, seven city blocks of the boxy, flat-roofed McKeeville houses were erected. While they looked to be just as small and drab as the Morgan houses, they had the advantage of being single-family dwellings and thus immediately took precedence in the residential hierarchy. As a result, Morganville became known rather ignominiously as the neighborhood of last resort, the project's very own Siberia.

When fourteen families received memos from Oppenheimer explaining that they had lost out in the housing lottery and would have to give up their apartments to make room for larger families who merited the space, a minor revolution brewed. The notice to Eric and Eleanor Jette read in part:

> There is a very grave housing shortage at the present time, particularly in the larger size apartment, and it will be some months before new housing can be built. . . . When you were assigned your present apartment you were told that such an assignment was temporary and that you might be required to move to a smaller apartment at some time in the future. This time has unfortunately arrived. . . .

The Jettes did not have to check with Dorothy to know what it meant. They were headed for Morganville. Eleanor Jette declared bitterly that the only place she was moving was "back to Croton-on-Hudson." Their plight was the talk of the Tech Area. Everyone was in an uproar. The "Fighting Fourteen," as they became known, called a meeting and threatened to leave. A young draftsman quit and promptly got a better-paying job in Santa Fe. More threatened to follow suit. They drafted a lengthy petition and presented it before the army representatives, complete with impassioned oratory and questionable logic, at the Town Council the following week. Foreign scientists jumped up and down brandishing the Constitution and Bill of Rights. The threat of imminent insurrection was more than the administration could stomach, and the moving order was nullified. "Nobody will move unless they want to move," a subsequent notice pleaded.

In the months that followed, the scientists exacted a few other hard-won concessions. The original hospital, which was not much more than a shack, could no longer accommodate the soaring birthrate and was replaced by a new, larger project hospital. A full-time dentist was hired, so that if they had an aching tooth they no longer needed to go to Dorothy begging for relief. She had somehow always managed to get them a last-minute appointment with Dr. Lord, who had two sons in the service and was "cleared" to treat Hill people. But there was nothing she could do to spare them the long ride down the rough roads, each bump sending them into paroxysms of agony as they cursed the general anew. A dry-cleaning concession was established to solve the ongoing problem of lost clothes, which rankled the scientists and their wives. They had grown tired of sending their few good suits down to the cleaners in Santa Fe only to have them disappear for weeks on end, and in some cases never to be seen again. Another post laundry was also added, as the existing service was overwhelmed by soiled diapers. By popular request, an empty field east of the lawn in front of Fuller Lodge was turned into a victory garden, and plots were made available to those who wanted to grow their own vegetables. A barren pasture just west of the site was turned into a nine-hole golf course, but as it shared the acreage with grazing horses who were apt to gallop onto the greens, and the layout was so rambling a guide was essential on the first few outings, it was not quite the country-club amenity it sounded.

Some forms of progress were lamented. The original Trading Post, housed in one of the quaint old log cabins built by the Ranch School, had grown too small to serve their growing numbers, and the army announced plans to build a new one. The civilians protested the wanton destruction, but their concern for preserving the local architecture fell on deaf ears. They woke up one morning to find that the rustic little cabin had been leveled. The new PX, similar in construction to so many other recent additions to the site, was nothing to make John Gaw Meem proud. But it boasted a soda fountain, a jukebox, and a large dance floor, and quickly became popular with the soldiers and single girls.

By now, everyone at Los Alamos was used to making do. Emilio Segrè repaired Laura Fermi's broken washing machine, and Dick Feynman kept the Tech Area adding machines and computers running. Ed McMillan, Johnny Williams, and Bob Wilson kept an eye on Los Alamos's four old diesel-run generators, probably acquired from some defunct mining company, so the town had a more or less reliable source of power. They had learned to be tolerant of what Jane Wilson called their "Barnum & Bailey" world, a way of life that was backward, inefficient, and, more often than not, absurd. "If the laundry wasn't taking clothes one week, you were a little less fastidious," recalled Charlotte Serber. "If there was no gasoline in the local tank, you walked to work. If the PX had no cigarettes, you rolled your own." If the faucets deposited more algae than liquid, you gave up bathing and waited until you could take a day off and go to Dorothy's for a shower and shave. In the meantime, everyone from world-renowned scientists to buck privates grew beards and went around in rumpled khakis and looked more than a little scruffy. Women took to wearing overalls and pulling their hair back with bandannas. They joked among themselves that if the post were suddenly opened to the public, people would take one look at all the blue jeans and lumberjack shirts and assume the top-secret facility was being run by a bunch of cowboys. Elinor Pulitzer Hempelmann, who somehow always managed to look lovely, and clung stubbornly to the hat-and-glove requirements of good society, turned to Shirley Barnett one day and bemoaned the state of things, concluding in disgust, "I'm just so sick of looking at all the ugly people in the Commissary!"

That April, Kitty threw Oppenheimer a big fortieth birthday party. A group of his old Berkeley friends got together and made a spoof yearbook of his tenure at Los Alamos. It was organized like a high school album, complete with "campus" highlights and illustrated with "class pictures" consisting of poorly mimeographed copies of their pass photos. Most of these portraits were worse than mug shots, and some of the physicists resembled the pictures in the local museum of New Mexico's old hard-luck prospectors, with their unkempt hair and flowing beards. Everyone agreed it was a fitting way to commemorate their first year.

By late spring, the available housing lagged so far behind the number of potential occupants that Dorothy noted the situation was "again and yet and still more critical." She was called to a meeting on the Hill and informed that the project would be taking over the Frijoles Canyon Lodge in Bandelier National Monument, and they needed her to get it up and running in five days. They also asked if she would stay on and manage it for the next six weeks until the new dorms were completed.

Only twenty miles from Los Alamos, Bandelier National Monument, a fifty-square-mile park that bordered their mesa, had been closed to the public shortly after the government claimed the site, and the empty lodge was available for interim housing for the new laboratory staff. It was an ideal location, as it was much closer than the ranches in the Espanola Valley and an easier and less conspicuous commute than the long trek back down to Santa Fe. "I was asked if I would put on my Army cap instead of a scientific hat and go over there," recalled Dorothy. "We were told by personnel that there were eighty people coming that summer who had no place to sleep."

Before she even had a chance to think it over, Oppenheimer stuck his face in the door and said with beguiling simplicity, "Dorothy, I wish you'd do this." She started to object, stammering that someone would need to man the Santa Fe office, adding half-heartedly, "I've never run a hotel before, I don't know how. . . ." But nothing she said mattered. They both knew she would do as he asked.

The isolated mountain inn had been owned and operated for eighteen years by Evelyn Cecil Frey, a formidable old-time rancher. Back in 1925, the government had granted the newlyweds George and Evelyn Frey a ninety-nine-year lease on concessions, and they had worked tirelessly to develop the remote guesthouse, planting fruit trees and gardens, and providing hot meals and cabins for archaeologists and travelers who came down the steep trails by burro. After the Forest Service built a road to Frijoles in 1933, the canyon became easily accessible to tourists, and Mrs. Frey, who took over after her husband left her, was forced to stand by as the old stone lodge was torn down and a new, larger facility was constructed, along with an administration building, museum, pool, and extensive parking area. The new lodge proved extremely popular, but in June 1943, Mrs. Frey found herself displaced by the Park Service once again when they offered the lodge to the Los Alamos project. The secret installation had put a stop to the tourist traffic, and with no other source of income, she had little choice but to cooperate. The army took it over and used it as makeshift quarters for technical experts and consultants from the universities, private industry, and branches of the armed forces.

When Mrs. Frey was briefly allowed back on the premises to inventory the lodge's contents before turning it over to the army, she had been scandalized by all the damage. Between January and March 1944, a hundred men from the Robert McKee Construction Company who were working at the Los Alamos site had used the abandoned lodge and had left it in a terrible state of disarray. Mrs. Frey had just managed to stop their bulldozers from rolling over a parking area that was thick with bushes of purple and yellow asters and tall trees and was an integral part of the lodge's backwoods charm. The cabins the workers had occupied were filthy. The walls were coated with sooty fat spatters from the fireplaces, which the men had used to cook their dinners. "They were pretty rough," recalled Dorothy. "They just nearly destroyed the kitchen, the stove was so greasy, and they found dead rodents in the flour bin. The whole place had to be absolutely scraped clean." When Mrs. Frey heard Dorothy was coming in to prepare for another wave of recruits, she let her have it. "She didn't care a bit for the words 'Los Alamos' and I didn't know this," said Dorothy, who bore the brunt of her fury until she could persuade the old lady that this time the lodge was in better hands. "They just sent me into these battles like a lamb. I just waltzed in there and the worse it is, the cooler I get, so she and I became great friends."

In early June, Dorothy left an assistant in charge at 109 East Palace, and she and Kevin moved up to Bandelier for the summer so she could fulfill her new innkeeping duties. With a crew of girls from San Ildefonso and plenty of help from the army, Dorothy went straight to work setting the place to rights. In five days, the main hotel was habitable, and in just under two weeks, they had cleaned up the lodge and surrounding cabins and had eighty clean beds freshly made up and waiting. The army helped out by bringing the camp into the twentieth century, installing a generator "with enough power to light Chicago."

In typical Los Alamos fashion, the estimates for the influx of new personnel had been way off, and the eighty new recruits never showed up. With only half that number of guests, and occupancy running considerably lower than what they had prepared for, Dorothy decided to turn Frijoles into a weekend resort for the scientists for the rest of the season. She knew they were going a little stir crazy on their mountain hideout and would welcome the change of scenery. She organized buffet lunches and dinners on Saturdays and Sundays, and the grateful scientists and their families flocked down to the picturesque spot for some badly needed R & R. After months of working six-day weeks, pure recreation was a luxury they had almost forgotten. Groves' penny-pinching left no room for extras, and with little variety to their days, the weeks flattened into dull months. By comparison, Frijoles seemed like a desert oasis, and Dorothy's lavish spreads, and the many small ways she catered to their comfort, all made for a sweet escape. Those with automobiles took advantage of her hospitality and made it a frequent stop. Those with horses, led by Oppenheimer, came by way of the old Indian trails, which snaked down and through the canyons. "Some rode their horses over," recalled Dorothy. "Some came to fish, and most came to eat because the food was excellent."

The "Frijoles caper," as she always referred to it, turned out to be an idyllic interlude for Dorothy and her son. The army had provided her with a 1941 Ford station wagon to do her daily run up to Los Alamos, and Dorothy, who had gunned her aging jalopy up and down the bumpy road too many times to count, insisted the government foot the bill for a spare tire and the tools to change it. Being so close by meant that both she and Kevin could spend much more time at the post, and they solidified new friendships on the Hill. Since the start of the project, Dorothy had been working around the clock, and her time with Kevin had been cut short. A few of Dorothy's old friends even worried to her about whether she was giving the boy enough attention. But at Frijoles, they had adjoining cabins, and spent long, leisurely days together, and rediscovered a favorite pastime, spending many nights sleeping outside under the stars.

Thirteen-year-old Kevin was enchanted by Tyuonyi and the Ceremonial caves, and by the solitude and mystery of the canyon. The park was situated in the Jemez range, at the site of the historic remains of a compact city-state built by the Anasazi, who thrived in their immense, terraced, tenement caves for hundreds of years before suddenly abandoning the settlement for reasons that remain unclear. Dorothy and Kevin immersed themselves in the history of the plateau and the earlier peoples, who had left behind the ruined stone palaces to testify to the former greatness of their vanished race. "It was this ancient, haunted place, and I had it all to myself," said Kevin, who spent days exploring on his own and years later would return as chief ranger of Bandelier National Park. "It was an amazing time," he said, and then added softly, "Amazing and unforgettable."

All of them fell under the spell of their magnificent surroundings. The spring columbines were in full bloom, and the Apache plumes and gentians were bursting with color. They shook off the khaki-colored dust of their army settlement, where only a few tough clumps of grass still grew around the old school buildings, and learned to appreciate again the pink cliffs and lavender vistas of the Sangre de Cristos, and the exhilarating sunshine and mountain air. For Dorothy and the scientists, worn down from months of labor, one of the great pleasures was wandering along the shady paths by the cool river, where they encountered porcupines, marmots, badgers, herds of deer, and beautiful birds of all kinds. It was hard to remember on those slow Sundays that the world was in turmoil, that Allied troops were being slaughtered on the beach in Normandy, and that the first V-1s were dropping on London. Even the fighting in the Philippines, and news reports of the Japanese defenders' fierce resistance, seemed unreal. The protected valley with its sense of isolation made the war seem as distant and mute as a falling star. They felt as if they were on an adventure in a wild and romantic land, and it was all too easy to blot out the project and its fearful purpose.

Dorothy could not help feeling once again that they were intruding on sacred ground, treading on the ashes of an older, wiser civilization. She knew they were not the first invaders and that throughout the centuries the deep caves had sheltered all manner of marauding tribes, renegades, and bandits. The physicists were the latest in a long line of outlaws. Restless and homeless as the herds of deer, they, too, were nomads. But they were more respectful than most of the strangers who had besieged the canyon, and she hoped they would be forgiven for trespassing. Over those fleeting weeks, the warmth and spirituality of Frijoles infected them all and eased their worries and lightened their hearts. They felt safe and close to nature, and that renewed their faith in an orderly world:

> Again the wooded valley and the rushing stream knew the presence of visitors who walked miles over their trails. One afternoon a staff car coming through the woods hit a wild turkey and killed it, in a great swirl of flying feathers, breaking the windshield, and astonishing the bird and driver equally. At twilight every night in the canyon hundreds of bats flew out of a cave high in the cliffs, and the watchers sat below close to the frieze which had once decorated the wall of a house, and thought of the people who had lived in the valley long before.

THIRTEEN

Summer Lightning

BY AUGUST, Dorothy was back at her desk at 109 East Palace. True to her word, she had returned Frijoles Lodge to Mrs. Frey in a much better state than she had found it in. The Housing Office staff were busier than ever. People were arriving faster than they could find lodging for them, and they had to "work like dogs" to get them a place to hang their hats. The town was full to bursting. "They were streaming in," Dorothy wrote, "all eager and bright, apart from their fatigue from traveling, with the assurance that their furniture would arrive by van immediately. I always responded sourly that I would faint if it did."

It was hard not to be curious about why so many new hands were suddenly needed "up yonder." She tried not to ask questions—much. But it was impossible not to notice the tension on the faces of the project leaders coming through the office. High-ranking consultants were hurrying up to conferences on the Hill. James Conant, Groves' primary science advisor, was becoming such a regular visitor to Los Alamos he was commonly known as "Uncle Jim." Sir James and Lady Chadwick of the British mission moved from Washington with their seventeen-year-old twin daughters and crushed many families' hopes by snagging one of the last of the old ranch houses.

Security had been tightened again, and it was now standard practice to send a second car manned by G-2 agents to follow behind VIPs on their way up to the site, presumably for their protection or, more likely in Dorothy's opinion, in case of a breakdown. The pace of the work at the laboratory had picked up noticeably. The scientists were going at it day and night with no regard for the clock. At Frijoles, she often heard as many as a half dozen explosions in an afternoon. Even in Santa Fe, she could hear the distant rumble echoing in the hills well into the night, like summer lightning. Using the familiar code, Dorothy would comment in passing on all the thunder. But it was thunder that brought no rain, only billowing smoke from the mountains and the sense of building pressure. "We felt the suspense and excitement of being connected with the unknown," she wrote. "We realized the importance, the desperate intensity of the work."

Dorothy noticed a change in Oppenheimer as well. In earlier times, he had been able to slip away now and then for a quiet dinner. "He would come to the house and he would make the best dry martinis," she said. "And he would cook the steak and fresh asparagus, which was a ritual menu we'd have." These days, however, Oppenheimer was driving himself harder than ever and rarely left the post, and then only on business. Since wartime shortages had made cigarettes scarce, he had taken up smoking a pipe, and the way he puffed on it fiendishly day and night revealed more about the extreme anxiety that tormented him than anything he could have said.

He had always been rail thin, but over the summer he had dropped nearly ten pounds and was alarmingly gaunt. At times, he looked almost frail, as if the punishing workload might be too much for him. He seemed to have lost the transcendent calm that was the hallmark of his first year in the desert, a cool, reasoned detachment that had made him such a source of wisdom and inspiration to so many of the young scientists. All the uncertainty of the past few months, along with the project's helter-skelter expansion, had put them behind schedule. Oppie was testy and unsociable, and he left people with the impression that he was downcast. The whole laboratory was affected by his mood, and frustration and pessimism emanated from its doors like hot, dusty swirls of desert wind.

What Dorothy did not know was that the work at the laboratory had reached an impasse, and Oppenheimer was indeed at his lowest point since the beginning of the project. Ernest Lawrence, who was in charge of the enormous calutrons at Oak Ridge, had reported to Groves that the production was proceeding, but the yield was not what they had expected, and it was likely that they would be able to send only enough weapons-grade uranium 235 for one bomb. At the same time, it was becoming clear that the rate of production for plutonium was better, and its availability meant that more of the implosion bombs could be ready at an earlier date. But when the first reactor-produced plutonium had begun to arrive in Los Alamos from the plant in Hanford, the samples had turned out to be impure. Segrè's experiments confirmed the samples were poisoned by the isotope of plutonium 240, which was produced in a secondary reaction. The problem with Pu-240 was that it had a very high spontaneous rate of fission, too high to make the gun-assembly method practical. The danger of predetonation was too great, and it was probable that the bullet and target would melt before coming together. To purify the plutonium would require separating the isotopes, which would involve the construction of huge plants on a scale with those at Oak Ridge. That was prohibitive in both cost and effort.

Segrè's discovery of the high spontaneous fission rate of reactor-bred plutonium would turn out to be the most fateful single event since their arrival on the mesa. On July 17, Oppenheimer broke the news that they would have to abandon the plutonium-gun method. Work on the Thin Man was to be discontinued. There would be no shortcut to a second bomb. They would have to find a new technique for assembling the plutonium weapon. Oppie's announcement came as a "big jolt," recalled Manley.

> This terrible shock, and an inescapable one, was that the gun assembly method could not be used for plutonium. Of the two fissile explosives, U-235 and plutonium, we finally had to conclude that a gun just would not assemble plutonium fast enough. Another isotope besides the one that we wanted was also produced at the time in the piles. One could have separated out those bad plutonium isotopes from the good ones, but that would have meant duplicating everything that had been done for uranium

> isotope separation—all those big plants—and there was just no time to do that. The choice was to junk the whole discovery of the chain reaction that produced plutonium, and all of the investment in time and effort of the Hanford plant, unless somebody could come up with a way of assembling the plutonium material into a weapon that would explode.

For Groves, and the members of the Military Policy Committee who helped set the direction of the project, this meant the implosion method had become a priority. Neddermeyer's independent study was now to become a full-blown division with all the necessary support. The sizable quantity of plutonium promised by the Hanford reactors meant there would be enough for a field test of the plutonium weapon. This was crucial, as the new assembly method was so subtle and complicated that it would have to be thoroughly tested before the implosion bomb could be considered combat-ready. The success of the implosion bomb, Fat Man, depended on its design and the quality of the explosives, both of which Los Alamos would be responsible for developing.

For Oppenheimer, this turn of events was an unmitigated nightmare. In July 1944, implosion looked like anything but a sure thing. Yet the urgency of getting the implosion method to work would necessitate both expanding and reorganizing the laboratory to meet this massive new challenge. Rabi recalled sitting in a Los Alamos meeting room while the senior scientists debated the problem. "The question was asked: Should the laboratory be extended? The big problem was: Where was the enemy in the field of work?" They reviewed everything they knew of the Nazi bomb effort since the announcement of uranium fission in 1939, as well as the history of their own development, and tried to figure out if the Germans had better judgment or had made the same mistakes. "We finally arrived at the conclusion that they could be exactly up to us, or perhaps further," said Rabi. "We felt very solemn. One didn't know what the enemy had. One didn't want to lose a single day, a single week. And certainly, a month would be a calamity."

The shake-up of the laboratory would require the creation of new divisions and, far trickier from an administrative point of view, the appointment of new division heads. Here Neddermeyer's personality, which up to now Oppenheimer had managed to treat with forbearance, presented a major problem. Neddermeyer was extremely difficult. He liked to proceed slowly and methodically, and was constantly in conflict with his boss, Captain Deke Parsons, a navy officer accustomed to running a large and efficient military ordnance operation. When Kistiakowsky arrived, Oppenheimer was relieved to be able to find someone so qualified to assign as Parsons' deputy to run interference between the two scientists on the implosion project. But Kistiakowsky had quickly tired of refereeing Neddermeyer and Parsons, who were always "at each other's throats" and shot off a memo asking to be released from the project. An additional complication was that Jim Tuck, a member of the British mission, had been experimenting with explosive lenses and had found a way to bend the explosion wave going through the explosive. As Kistiakowsky put it, "If you have two explosives with different detonation velocities, and you put them together in the right way, you can shape the wave, and instead of having it expand, make it converge." Theoretically, it should be possible to line the core with explosive lenses and, by detonating them simultaneously, produce a tremendous symmetrical shock wave. It was the problem that Neddermeyer had been struggling for months to solve, but for which he had been unable to come up with an acceptable answer.

Oppenheimer's predicament was whether to leave Neddermeyer in place or risk alienating an extremely valuable member of the team, who had spearheaded implosion from the very beginning, by handing the new expanded implosion program over to someone else. During the last few weeks of July and the beginning of August, he agonized over what to do. "I remember Oppie's even calling me in and asking what I thought of Seth Neddermeyer," said Priscilla Greene. "He was hesitant about putting him in charge. [Neddermeyer] was sort of cranky, and didn't communicate well, and could be stubborn about his ideas." What followed was "a great period of tension" during which Oppie became more high-handed and abrupt than usual, and reverted to some of the bad habits he had been known for before the war. He relied increasingly on a close circle of intimate colleagues, shut others out, and summarily ended arguments with sardonic or wounding remarks. "But when he tried that on the scientists at Los Alamos," Greene noted, "Robert didn't always get away with putting them down."

Whether Oppenheimer was trying to goad Neddermeyer into action, or was simply at the end of his rope, the result was the same. Some of their fights were particularly bitter. Oppie's mishandling of Neddermeyer made him another enemy and alienated a number of the physicist's close colleagues. "From my point of view, he was an intellectual snob," Neddermeyer said after the war. "He could cut you cold and humiliate you right down to the ground. On the other hand, I could irritate him." But even Neddermeyer acknowledged his deliberate pace was partly to blame. "He became terribly, terribly impatient with me in the spring of 1944," he said. "I think he felt very badly because I seemed not to push things for war research but acted as though it was just a normal research situation."

In the end, Oppenheimer felt compelled to choose Kistiakowsky as head of the new X (Explosives) Division, the man he thought would best lead a crash effort on the daunting problem of producing an explosive lens for the implosion weapon. There was so much work to do on implosion that the whole project would essentially be transformed. Nearly every important scientist, except Teller, who was still single-mindedly focused on the Super, would be affected. Oppenheimer asked Alvarez to be Kistiakowsky's chief aide, and Neddermeyer went back to working in a corner, off by himself. Bob Bacher was put in charge of the new G (Gadget) Division, which focused on weapons physics and the plutonium sphere. Parsons continued to head the Ordnance Division, concentrating on bomb construction and delivery. Parsons was also appointed associate director of the laboratory, along with Fermi, who would be moving to Los Alamos permanently that fall to lend a hand. Double-tracking the two projects meant that more manpower would be required to supplement the existing staff, and Oppenheimer arranged for hundreds more SEDs—personnel from the military's Special Engineer Detachment—to come to the project to help with the implosion work and bomb construction.

Everyone at the laboratory took solace in the news that Enrico Fermi had finally been persuaded to come to Los Alamos full-time. Many of the physicists regarded Fermi as the soul of wisdom—he was sometimes referred to as "the pope" because of the weight of his pronouncements—and believed that his even temperament and common sense would help provide the support and balance that was badly needed at this stage. For his part, Oppenheimer hoped that Fermi could help mediate some of the running disputes between the different division heads. Fermi was unique even among the array of talents at Los Alamos in that he excelled at both experimental and theoretical physics, and could go toe-to-toe with anyone on the site.

Los Alamos was thick with experts and know-it-alls, with more arriving every day, so that Oppenheimer spent an inordinate amount of time settling disputes and preventing work from being disrupted. There was Teller's ongoing dissatisfaction with the project and refusal to do what he regarded as mundane work on the implosion bomb. Fermi liked Teller, and the two got along well, and Oppenheimer could only hope the Italian Nobel laureate would be able to convince Teller to stay and undertake some useful theoretical investigations. By now, he was no longer under any illusion that Teller could be won over to his point of view, but he was shrewd enough to realize that it was better to keep the troublemaker inside the tent, where he could at least serve as a useful critic in group debate. If Oppenheimer was fed up with Teller, he usually managed to hide his antagonism, though the physicist Charles Critchfield recalled that occasionally after talking about the progress of the lab, Oppenheimer would mutter a little prayer: "May the Lord preserve us from the enemy without and the Hungarians within."

Oppenheimer also had his hands full dealing with escalating tension between Parsons and Kistiakowsky, who clashed over everything from who knew more about explosive technology to scheduling access to the S-Site, the new testing facility located at the site of an old sawmill near Anchor Ranch. The problem between Parsons and Kistiakowsky could be boiled down to a difference in style, between that of the classic can-do military approach to developing detonation devices and that of theoretical and experimental physicists struggling to perfect a radical assembly design for implosion. Parsons had insisted on bringing in a veteran navy ordnance officer, who had proceeded to build an explosives casting plant that Kistiakowsky regarded as "a monstrosity." Kistiakowsky, who could be just as imperious, demanded that a completely new plant be built according to his group's specifications.

> Parsons argued that I was wanting to do an undoable job. He believed there was something else that should be done, a kind of smoothing out of the difficulties with plutonium, minimizing them rather than really overcoming them. Perhaps I am biased but I felt the way Oppenheimer handled this difference, his grasp of the technical problems, was really most impressive.
>
> He called a big meeting of all the group heads, and there he sprang on Parsons the fact that I had plans for completely redesigning the explosives establishment. Parsons was furious—he felt I had by-passed him and that was outrageous. I can understand perfectly how he felt but I was a civilian, so was Oppie, and I didn't have to go through him.

After much debate, Oppenheimer decided the S-Site would be expanded to accommodate Kistiakowsky's new explosives plant. Parsons never forgave Kistiakowsky, and the Russian noted that from then on they were never "on good terms." But Oppenheimer may have seen the rivalry between the two group leaders as productive, and from a political point of view, it may have been more convenient to have Parsons believe Kistiakowsky had bypassed him, rather than think the laboratory director had subverted his authority.

As adept as Oppenheimer was at reconciling divergent viewpoints, it was a measure of his exasperation with the more fractious members of his staff that he told the Administrative Board on August 3 that "certain" parties could be working harder. Groves had ordered a fifty-four-hour workweek for all military personnel at Los Alamos and canceled all leaves, and while Oppenheimer resisted imposing the same draconian measures, he instituted a number of organizational changes. Those "certain" few who were perceived as dragging their feet were transferred to different divisions, and several scientists were appointed to help hear complaints and expedite the work. He also announced that the Tech Area siren would sound at more frequent intervals to induce people to get to work on time and return from their lunch break in a more orderly fashion. When Fermi arrived that August to head up the F Division, which included theoretical and nuclear physics, his relaxed and unpretentious style, which was in its own way every bit as dominant as Oppie's, did much to help defuse the summer's simmering tensions and restore peace on the mesa.

By the end of the summer, Oppenheimer wrote Groves that he thought he had managed to bring the project leaders in line and restore harmony. He enclosed a report from Parsons, "with the general intent and spirit of which I am in full sympathy," but eloquently and resolutely defended the conduct of his physicists on several counts:

> I believe that Captain Parsons somewhat misjudges the temper of the responsible members of the laboratory. It is true that there are a few people here whose interests are exclusively "scientific" in the sense that they will abandon any problem that appears to be soluble. I believe that these men are now in appropriate positions in the organization. For the most part the men actually responsible for the prosecution of the work have proven records of carrying developments through the scientific and into the engineering stage. For the most part these men regard their work here not as a scientific adventure, but as a responsible mission which will have failed if it is let drop at the laboratory phase. I therefore do not expect to have to take heroic measures to insure something which I know to be the common desire of the overwhelming majority of our personnel.

It was hardly surprising that tempers ran high in the implosion divisions. The scientists at Los Alamos were working with explosives; and pressed for time and good alternatives, they often took enormous chances. "There was no time to build barricades, so we just worked," recalled Kistiakowsky. They made up new rules regarding safety as they went along, but manufacturing high-explosive castings was tough and dangerous work. At Los Alamos, they handled explosives by the ton, never forgetting that only one gram going off at the wrong time would "finish off the whole hand." One of Kistiakowsky's main concerns about the S-Site was that explosives had to be trucked all the way across the mesa, and right through the heart of the project. "With the whole theoretical division sitting in offices on one side of the road, Oppenheimer's office (and mine) on the other, and with hundreds of wild WACs and GIs driving trucks and jeeps there," he explained, "a truck loaded with five tons of high explosive (H.E.) going off there would have wiped out 90 percent of the brains in those temporary buildings." During the peak period of production, they were trucking up to five tons of high explosives up the Hill monthly over some of the worst roads he had ever seen.

Phil Morrison, one of Oppenheimer's Berkeley grads who had been working for Fermi in Chicago as a neutron engineer, was recruited to help with the difficult implosion work and assigned to the G Division. "We were working with critical assemblies, and it was clearly very dangerous work," said Morrison. "It was better than the front line, but not by much. I was very aware of the risks, but very motivated by the war . . . the urgency of the war. We believed the Germans could be ahead. We believed Heisenberg could do this, and that they had plenty of good people on their side."

Oppenheimer's driving desire to make headway on the implosion bomb spurred everyone on, and differences were put aside in the name of progress. He never spared himself in work, and as a consequence inspired an equally dedicated effort from his staff. He faced tremendous pressure from above. Groves insisted that two bombs were essential for victory in the war. From a strategic point of view, Groves believed they could not risk dropping one with no backup in reserve. General George Marshall thought that figure might be closer to nine. Forced to scratch the Thin Man, Groves presented a revised timetable for weapons delivery to Marshall on August 7, 1944, two months after the Normandy invasion. He promised the delivery of a small uranium gun bomb—nicknamed "Little Boy," Thin Man's lighter, smaller brother—by August 1945, with the addition of one or two more by the end of the year. If the experimental work went smoothly, small implosion bombs would be ready by the second quarter of 1945. There was no longer any escaping the fact that they were building instruments of war that would soon be used against another country. While the scientists had all been focused on Germany, Groves made it clear that if Hitler surrendered before the bombs were ready, Japan would become the target. As if underlining his own resolve, Oppenheimer wrote Groves, "The laboratory is operating under a directive to produce weapons; this directive has been and will be rigorously adhered to."

Looking ahead to the next twelve months, Oppenheimer must have thought Groves' estimates wildly optimistic. Given the time scale, and the slow rate of production of U-235, there was no point in even considering a test for the gun assembly. There was no way there would be enough U-235 to make that a possibility. They would simply have to trust that the gadget would work. Plutonium's rate of production promised to be better, and given the fact that the implosion device was so novel, Oppenheimer felt they had to consider making a field test. They could expect enough plutonium to be available by the following summer to allow for a test detonation of a single implosion bomb. But Groves was leery of wasting the little fissile material they had. If the test was a bust, and there was no nuclear explosion, he stated flatly that he could not afford to lose all that plutonium in the chemical blast. "Oppenheimer and I were pleading with Groves that there had to be a test because the whole scheme was so uncertain," recalled Kistiakowsky:

> General Groves was very sensitive about what would happen to him after the war and whenever he didn't like something, he'd say, "Think of me standing before a U.S. Senate committee after the war when it asks me: 'General Groves, why did you spend this million or *that* million of dollars?' "

To cover their desperation, Kistiakowsky proposed building a large prophylactic, a 200-ton ellipsoidal steel tank with twelve-inch-thick walls, dubbed "Jumbo." By testing the bomb inside the steel vessel, they hoped that if the weapon misfired, everything would remain contained, rather than being scattered over the countryside. Then, after the tank had cooled, it would be possible to salvage the plutonium for yet another test. If by some chance the weapon worked, the tank would be vaporized, but their job would be done. Oppenheimer had first written to Groves about a "sphere for proof firing" in March 1944 and, trying to sound confident about implosion, had stated "the probability that the reaction would not shatter the container is extremely small." It is a measure of Groves' own uncertainty that he went for the idea and gave the order for Jumbo to be built.

While the question of whether or not there should be field tests continued to be fiercely debated by the physicists on the Hill, Oppenheimer had to turn his attention to the problem of *where* such a test could be safely conducted. Beginning in early May, he organized a search team consisting of himself, Kenneth Bainbridge, who was in charge of developing the test site, Peer de Silva, and army major W. A. Stevens. They scoured maps of the area and visited several spots nearby where the land was relatively flat and isolated from any population centers, but still close enough to Los Alamos to facilitate the transportation of men and material. On their first scouting trip, they got caught in an unexpected snowstorm, and one of their trucks got stuck in the deep drifts. They had to dig it out and take a more circuitous route to San Luis and Estrella. They spent the next day exploring Star Lake, the center of an old coal region, and followed unmapped ranch trails past abandoned ranches too dry for farming and wind-beaten desert land. The front-wheel bearings of one of the vehicles finally gave out, and they had to leave it there and head back to Los Alamos. Bainbridge thought Oppenheimer was reluctant to see an end to their rugged three-day adventure, as he would not have time to participate in the subsequent site-hunting expeditions. "Oppenheimer had to return to more important duties, and could not again enjoy a trip into the open country which he loved."

On a following trip, they borrowed a small, seven-seater C-45 and flew low over the parched rural countryside, inspecting hundreds of miles of the western United States. By August, after all the other possibilities had been eliminated on the grounds of inaccessibility, they were left with a ninety-mile stretch of sandy desert in central New Mexico known as the Jornada del Muerto, located some three hundred miles south of Los Alamos. It was a desolate, forbidding area, battered by high winds, devoid of water, and with temperatures reaching well over a hundred degrees in July and August. Not surprisingly, the early Spanish settlers dreaded crossing the empty, unyielding region, endowing it with a name that roughly translates as "Journey of Death." The grim pioneer history added to the romantic aura of the site, and Oppenheimer, who was spending sleepless nights reading John Donne's *Holy Sonnets*, was inspired to give it the code name "Trinity" after some lines from Sonnet XIV:

> *Batter my heart, three-personed God; for you*
> *As yet but knock, breathe, shine, and seek to mend;*
> *That I may rise and stand, o'erthrow me, and bend*
> *Your force, to break, blow, burn, and make me new.*

The down-to-earth Bainbridge was less than enamored of the evocative name Oppenheimer had selected. Thinking more in terms of practicality than history, he dashed off a memo to Oppenheimer several months later requesting something more utilitarian:

> I would greatly appreciate it if the Trinity Project could be designated Project T. At present, there are too many different designations. Muncy's [business] office calls it A; Mitchell's [procurement] office calls it project T but ships things to S-45. . . . By actual usage, people are talking of Project T, our passes are stamped T and I would like to see the project, for simplicity, called Project T.

As summer waned, the debate over to test or not to test became an increasingly hot issue at Los Alamos. Even so, work on the site was in full swing. Most of the Jornada was already in government hands and, since the First World War, had been used as a gunnery and bombing range. Groves arranged to acquire the northwest corner of the bombing range not far from Socorro, a small settlement at the northern end that no doubt took its name from the fact that it had offered the only "succor" or sustenance to the old Spanish wagon trains that ran into trouble in the desert. There were a few ranches scattered around the region, and their owners had been compensated by the government for their loss of housing and income. Not all of them had left yet, and their cattle and sheep still occasionally wandered past the NO TRESPASSING signs posted by the army. The government bribed the remaining ranchers to leave, though a few needed prodding, and the MPs fired holes into their water tanks to discourage both the two-legged and four-legged inhabitants from drifting back. There were a few small townships thirty to forty miles away, but they had become accustomed to hearing the sounds of distant explosions over the years. Other than that, the area was literally barren—empty. The army leased the McDonald brothers' ranches and planned to renovate them for their use. Everything else would have to be trucked in and built from scratch. The roads were extremely primitive and would have to be improved to withstand the upcoming traffic, not to mention the transport of Jumbo from a stop along the El Paso–Albuquerque railroad line. The plans drawn up for the base camp called for housing for 160 military and civilian personnel. Trinity was going to be a tremendous field operation, and Groves ordered a hundred men from the Army Corps of Engineers to help with the construction.

The demands of the implosion work and Trinity site meant that once again Los Alamos was in desperate need of more manpower. Since late 1943, the military had begun bringing in SEDs, who were assisting in the endless work in the Tech Area. Now they were arriving by the busload. Most were students with some semblance of technical training—in chemistry, physics, or engineering—and served as laboratory grunts, doing everything and anything the scientists needed. In some cases, they were also promising young physics Ph.D.'s who were about to be drafted and whose professors thought they could probably serve their country better in a laboratory than in battle. In the beginning, they were crammed into barracks at the edge of town, as though the army was slightly ashamed of them and wanted to keep them as far away as possible from respectable folk.

The arrival of hundreds of very young, very raw male draftees altered the atmosphere on the mesa, lowered the average age of an already immature population, and proved to be another test of the community's creativity and adaptability. To Dorothy, they looked like a bunch of scrawny college boys, and she took many of them under her wing. They were thankful not to be in the line of fire on foreign soil, and their youthful high spirits showed. One of the very first to arrive was Bill Hudgins, who had been a few years ahead of Kevin in school and whom she had often seen at the home of her close friend Eleanor Gregg, a cousin of John Gaw Meem's. Hudgins had been enrolled in the engineering program at the University of New Mexico when he heard they were looking for qualified staff on a nearby army project. He was only eighteen, and green as could be when he walked into Dorothy's office at 109 East Palace and inquired about "a job with the government project." Panicked that there had been a breach in security, Dorothy nearly jumped down his throat. She interrogated him for more than an hour about what he had heard about the project, where he had heard it, whom he had heard it from, and how he came to have her office address. Only after she was satisfied that there had been no real harm done by the rumors did she relax and recover her usual equilibrium.

Afterward, feeling a little guilty about her rough treatment of the lad, she walked him through the application process and even attached a letter of recommendation. After he was accepted, she made a point of watching over him. "She mothered me whenever she could," said Hudgins, laughing. "She definitely looked out for me. Whenever I got into any kind of trouble, I could just call her and she would straighten it out. She was definitely a good person to know."

The SEDs were a breed apart from the post's army soldiers, and their brief five weeks of basic training did little to whip them into tip-top military shape. Whenever Groves was visiting with some high brass from Washington, there would be a formal military review on the baseball field in front of the Big House, and the whole post was treated to the laughable spectacle of watching the youngsters try to parade. "It was pathetic," said Hudgins, who worked as a technician in the Chemistry Division. "Here would be these people straggling in after pulling three straight shifts in the laboratory, just desperate for some sleep, and they would have to get up at dawn and line up for inspection. There were always a half dozen that couldn't keep time to anything to save their lives and were tripping over their feet. You have to understand, they had come from all over. Some were straight off the farm or from small family businesses. This was all new to them." After one visit, a general reportedly declared the detachment "a disgrace to the army."

Hudgins complained to Dorothy about the exhausting hours and tough drills, and although she commiserated, it was out of her hands. Dorothy, like most of the laboratory staff, was patriotic without being particularly pro-army, and failed to see the need for all the spit-and-polish drill on their remote base. Like Oppenheimer, she also had a strong sense that they were all in this together, and the disparities between the civilian and military privileges troubled her. Peer de Silva, the post military intelligence officer, was also the SEDs commander, and he arbitrarily refused to allow the young married men to get their wives jobs at Los Alamos, so they too could be quartered on the site. Dorothy felt bad that they were unfairly separated from their families, and underpaid and unappreciated to boot. "She stuck up for us," said Hudgins. "She was very informal and unpretentious, not a flag waver. She thought the army was giving us an awfully hard time."

Some of the senior scientists, who had barely escaped being put in uniform themselves, took pity on the SEDs and intervened on their behalf. Kistiakowsky, who had hundreds of SEDs working in his division, became their leading champion. He complained that the young enlisted personnel only rated what army regulations stipulated as "the minimum comforts: minimum housing, minimum recreation, minimum food facilities. And this means," Kistiakowsky continued, "40 square feet per man in the barracks, including part of the recreation area. Try to recreate yourself in that area." Their commanding officer was old school and, in Kistiakowsky's opinion, treated the junior technical personnel like pariahs. "Since he was not told, as many other military weren't (nor the machinists, of course) what the purpose of Los Alamos was, he loudly described all of us as draft dodgers who were just escaping Army service and having fun here. He insisted the SEDs be awakened by reveille and be mustered daily to do calisthenics and keep the barracks in order and even wear caps and salute officers on the streets." While this was traditional in the army, the SEDs were not regular army and were barred from going on to Officer Candidate School, which made the rigorous discipline all the more humiliating and unnecessary.

Kistiakowsky took up their cause with Oppenheimer, who as always worried about morale and argued with Groves to no avail. When Groves was next in Los Alamos, Kistiakowsky asked for permission to speak to him. Groves told him he could ride in his car with him as far as Albuquerque, and for the whole two-and-a-half-hour trip they argued about the SED problem. Kistiakowsky, getting nowhere with Groves, resorted to his "ultimate weapon," going so far as to threaten to quit if the hard-working youngsters did not receive better treatment. "You know, General," the imposing White Russian said, squarely facing Groves, "I didn't ask to come here and what's more you can't keep me here. I am too old to be drafted," he added, "and I'll leave." According to Kistiakowsky, Groves responded with "grunts and violent attacks on me for transgressing my authority as a civilian, and meddling in army affairs." But shortly thereafter, the SEDs got a new CO, Major T. O. Palmer, and the reveille and calisthenics were quietly dropped. Kistiakowsky, who apart from that dispute got along well with Groves, suspected the general tolerated his interference because he saw him "as more manly than the effete physicists" because of his explosives work: "I was to him a kind of kindred spirit."

For those who came to Los Alamos that second summer, the chaos of the first few months—when everything was half-finished and everyone was put to work setting up camp—had been replaced by an overgrown, relatively organized, and highly hierarchical society. As a consequence, those who had been with the project from the early days could not help taking a certain guilty pleasure in the misadventures that still beset newcomers. When Laura Fermi arrived at Lamy, Dorothy had arranged for her and her children to be met by an official car. In August 1944 Lamy was still the same small, unimpressive whistle-stop it was when the project began, a desolate depot with little more than a bar and a few squat adobe huts. Laura Fermi and her two children descended in the rain and looked around for their transport to the laboratory's front office on East Palace Avenue.

As the platform emptied, a uniformed WAC approached her and using the project code, inquired anxiously, "Are you Mrs. Farmer?" Unaware of her husband's alias, Laura Fermi shook her head no. The WAC then queried the remaining female passengers and, unable to locate the

VIP's wife, was about to depart when Laura Fermi stopped her and identified herself. She explained that she was going to "Site Y" and could the WAC by any chance take them there. The driver apologized, but maintained her orders were to pick up a "Mrs. Farmer." As Dorothy later related the sorry tale, "Out of her brilliant persuasiveness, coupled with extreme fatigue and some desperation, Mrs. Fermi, quietly insisting she was not as important as the Mrs. Farmer, nevertheless prevailed on the driver to let her and her children ride in the empty car."

Among the other late arrivals that August was Klaus Fuchs, a German member of the British mission. Fuchs had been doing research on gaseous diffusion for the Manhattan Project at the Kellex Corporation in New York and was being transferred to Los Alamos at Peierls' request to work on implosion. He was a very quiet young man, and like many of the émigré scientists, had been through a difficult time. After he sought asylum in England as a destitute refugee from Hitler, he was briefly detained as an enemy alien, then interned for six months at a prison camp on the Isle of Man, and finally banished to Canada. When leading German physicists such as Peierls and Frisch joined the British effort, Fuchs, who had by then been granted political asylum, was recruited to help with the war work. Physicists were in desperately short supply in wartime England, and British intelligence made allowances for his past Communist activity and cleared him to work on Tube Alloys. Given the urgency of the bomb project, when it was agreed the British Technical Mission would be joining the American effort, clearances for all their scientists were put through by the State Department and were never vetted by the FBI.

At Los Alamos, Fuchs quickly earned a reputation as extremely bright and hard-working. He was assigned to an important position in the Theoretical Division, and he attended the weekly colloquiums and important planning discussions. Bethe, who had known him since the 1930s, called him "one of the most valuable men in my division." But, even for a lab rat, Fuchs was so pale and withdrawn that Segrè's wife, Elfriede, nicknamed him "Il Poverino," Italian for "poor soul," and it suited him so well that a number of people used the nickname when referring to him. Fuchs was extremely shy and serious, and while painfully polite, he had almost nothing to say for himself and usually replied in monosyllables. He had grown close to the Peierlses early in the war while a lodger in their house in Birmingham, England, and as they treated him as a member of the family, he was quickly included on Sunday outings, group picnics, and hikes. He was assigned the room adjacent to Dick Feynman in the Big House, which was a dormitory for single scientists. Although the two men could not have been more different, they soon became friends. Fuchs had his own car, a dilapidated blue Buick, and often drove Feynman to Albuquerque to see his wife, Arline, who did not have long to live. Realizing that her condition was precarious, Fuchs told Feynman he could borrow the car anytime he needed it, so that in case of an emergency he could get to the hospital quickly. It was the sort of generous gesture that endeared him to many in the tight-knit community.

"We all liked him," recalled Dorothy, who thought he was "very gentle and sweet," and looked rather like an owl with his large sad eyes framed by round tortoiseshell glasses. He was so solicitous of their children that he became a popular babysitter and often took care of the Tellers' son, Paul. "He had quite a reputation for being so good and kind," said Dorothy, who on more than one occasion asked him to keep an eye on Kevin, who was becoming a rambunctious adolescent. As far as she was concerned, Fuchs was a likable young man, and she invited the lonely-looking bachelor to dinners and Saturday night gatherings. It turned out he was quite a good dancer, and he became a pet of many of the Hill wives. "In Los Alamos, we all trusted him and saw him frequently," recalled Laura Fermi. "We had frequent parties, and Fuchs came often. He seemed to enjoy himself, played 'murder' or charades with the others, and said only a few words. We all thought him pleasant and knew nothing about him."

Five years later, when the news broke of Fuchs' arrest, Groves' former assistant Anne Wilson happened to be meeting Oppenheimer for an early breakfast at the Oyster Bar in Grand Central in New York City. She would never forget that morning of February 4, 1950, as Oppie's face crumpled as he took in the headlines splashed across the front page of

The New York Times. "The papers had just come out, and there it was. Klaus Fuchs was a Soviet spy," she said. "Robert was in shock. He just stared at the story shaking his head. He could not believe the man had done this."

Fuchs was the last person anyone at Los Alamos would have pegged as a security risk. Feynman even went so far as to joke with him once about which of the two of them would be the "most likely candidate as a suspect for possible espionage." There was no question; they both agreed that it was Feynman, the Tech Area's speediest code breaker and nimble-fingered safecracker. Dorothy never had the slightest doubt about Fuchs' integrity until his treachery was made public. By the time agents from the British Secret Service Bureau, MI5, confronted him and Fuchs confessed, he had returned to England and was working at Harwell, the center of Britain's nuclear weapons research.

"When this was discovered, British intelligence came over here," recalled Dorothy, who was beside herself when she heard of his arrest. "The only thing that they could trace on the Continent that got to Russia could have come from only one place in the world, and that was Los Alamos. . . . They narrowed it down to the information and the way it was dispatched." It turned out that the private Fuchs was a resolute Communist and had been passing classified technical information to the Soviets through an American courier named Harry Gold for almost three years before being transferred to the Manhattan Project's New Mexico site. Gold, a biochemistry technician at Philadelphia's General Hospital, had been doing industrial espionage for the Soviets since the early 1930s and was told to contact Fuchs in March 1944, while Fuchs was still in New York. The whole time he was at Los Alamos, Fuchs had been stealing atomic secrets, driving off with them in his battered car to Santa Fe, and handing them off to Gold at the Castillo Street Bridge. G-2, which was obsessively investigating Oppenheimer, never noticed a thing.

Suspicion did not fall on Fuchs until four years after the war, when British code breakers, in the process of cracking the Soviet cipher system, intercepted a startling document—a report on the Manhattan Project authored by Fuchs while he was at Los Alamos and sent from New

York to Moscow. This set off alarms, and Fuchs came under investigation. As soon as he realized the British were on to him, Fuchs blurted out a full confession. "[They] just tapped him on the back and he made no resistance whatever, and went with them," said Dorothy. She found the idea that he had given the Russians documents about the implosion bomb deeply disturbing. She had formed an impression of him as a harmless creature and had failed utterly to recognize his deception. He was the saboteur that all the security precautions, clearances, and passes had been designed to catch, and instead he had been operating right under their very noses all during the war, and even afterward.

In retrospect, Dorothy had to admit Fuchs was the ideal double agent. As Bethe put it, "If he was a spy, he played his part beautifully." Fuchs was so quiet and unassuming, he faded into the background. When Dorothy and several of her friends on the Hill later compared notes about him, they were amazed at how little Fuchs had ever revealed about his background or politics. No one was more traumatized by his arrest than Rudolf Peierls, who had been his mentor. When Fuchs confessed, he told investigators that he had had no trouble keeping his friendships and ideals separate: "It had been possible for me in one half of my mind to be friends with people, to be close friends, and at the same time to deceive and to endanger them." Peierls went to see him in Brixton Prison, and during a long conversation, Fuchs admitted everything and "expressed regret." He told Peierls, "You must remember what I went through under the Nazis."

Fuchs' revolutionary rage had been sparked by the death of his mother, who had been driven to suicide by Nazi persecution, and stoked by the arrest of his father, who had been imprisoned in a concentration camp. He became a fervent Communist long before he escaped to England, only to find himself distrusted and mistreated by the British government until they realized his skills as a physicist might make him useful to the empire. His passion to serve the Soviet cause became the one pure thing left in his life, and his idealism led him to betray all of his friends and colleagues. In his mind, aiding the Russians was not morally wrong. As Laura Fermi observed, "Fuchs' intelligence was far above the average; his judgment below, probably distorted by the circumstances of his life."

Fuchs pleaded guilty and stated in court that he hoped his admission would help atone for his wrongdoing. He was sentenced to fourteen years, the maximum under British law. But his bizarre, abbreviated trial, which took all of ninety minutes and included scant evidence beyond his confessions and the testimony of the MI5 agent, left Dorothy feeling frustrated. The British understandably wanted to limit the damage and prevent the exposure of more official secrets, but a satisfying account of Fuchs' motives never emerged. He had let them all down, done irreparable harm, and now they would never really know why.

Later, most of the Los Alamos scientists concluded that the documents Fuchs leaked probably shortened the Soviet atom bomb development by one to two years. "He was an extremely clever guy," said Joseph Hirschfelder. "He set out to discover what was going on in the Manhattan Project. He very systematically went to each of the laboratories in the Manhattan Project, got himself placed in a key position, and transmitted his information. For example, at Los Alamos, he ended up as the editor of the twenty-five-volume secret *Los Alamos Encyclopedia,* which summarizes all of the research which was carried out."

Unbeknownst to him, Fuchs was not the only security leak at Los Alamos. Gold was running another operative inside the classified laboratory, a relative grunt by comparison, but his information may have helped confirm what the Soviets were learning about the bomb project from Fuchs. David Greenglass was a young SED who worked as a technician in the machine shop, and while he had only a fraction of Fuchs' knowledge and access, he supplied whatever information he could, including a list of the scientists, the layout of the project and technical buildings, and a very crude sketch of the explosive lens molds used in the plutonium bomb. Greenglass was married to Ruth Rosenberg and had been recruited by his brother-in-law, Julius. Ruth, Julius, and Julius' wife, Ethel, were members of a Soviet espionage ring in New York. Greenglass was arrested in June 1950, six months after Fuchs, and received a thirty-year sentence. His case might only have been a footnote had it not blown the lid off the far greater treachery of the Rosenbergs and triggered the events that would lead to the most infamous espionage trial of the century. The Rosenbergs were executed in 1953.

On August 24, 1944, the news that Paris had been liberated swept the mesa late in the evening and sparked an impromptu party. The French 2nd Armored Division had fought their way into the heart of the city, overcome a last stand by the Germans, and cleared the way for General Charles de Gaulle's return. The European scientists, many of whom had spent the previous night glued to the radio, were jubilant. It was an intensely emotional moment for those with family members who were still in France when the country was divided in two and Marshal Henri Philippe Pétain headed the Vichy government in the south.

Françoise Ulam was nursing her baby daughter, who had been born in their secret city a month earlier, when a sharp rapping brought her to the window and a none-too-sober Rabi delivered the happy news. As word spread through the apartment complexes, people gathered outside to shake hands, hug, and cheer. Rabi and Victor Weisskopf apparently decided the event wasn't being properly celebrated and began marching around the residential area bellowing "La Marseillaise" and inviting everyone to join their parade. They proceeded to serenade Françoise, the only French citizen on the Hill, with the strains of her proud anthem, which Rabi played on his pocket comb. Even in the midst of all the excitement, Françoise felt a pang of guilt that she and her husband and baby were safely ensconced on their small mesa while war was raging across the world. She had no idea about the fate of her family, left behind in the embattled cities of her homeland, and no way to find out if they had survived the occupation. Communication with relatives trapped in Hitler's Europe had long since ceased to be possible. There was nothing to do but wait, and trust that friends or relatives would find some way to get word to her. It was not until a year later that she would finally receive the news that her mother had been sent to a concentration camp and had not survived.

Françoise was not alone in receiving painful reports from home as the Allies penetrated German-occupied zones. Not long after the first troops had entered Rome in June, Oppenheimer had called Emilio Segrè into his office and gravely informed him that military intelligence had learned that although his father, a well-to-do Jewish paper manufacturer, was safe, his mother had been captured by the Nazis in the fall of 1943. Her fate was unknown. Segrè was too shocked to respond at first, and unsure he had made himself understood, Oppenheimer was forced to repeat the news to him several times.

In the days that followed the liberation of Paris, elation soon gave way to the familiar driving anxiety. Hitler had been dealt a major defeat, and for a brief moment, it had seemed that after years of devastation and struggle, Europe was finally rising up and shaking off the chains of German occupation. In the days that followed, Toulon, Marseilles, and Brussels were liberated in rapid succession, and the battle for France seemed almost over. Allied forces were poised to advance into Germany. But then, as so many times before when the scientists had allowed themselves to believe the end was in sight, progress stalled. American and British paratroopers dropped behind enemy lines at Arnhem were killed and captured in staggering numbers as the German reinforcements surrounded them. Their grief at the numbers of men who lay dead on the bridges over the Rhine was compounded by the losses on the beaches of Guam. In the last bittersweet days of summer, their spirits rose and sank with every new radio dispatch, and the slender hope that what they were working on at Los Alamos might not be needed after all flickered and went out.

FOURTEEN

A Bad Case of the Jitters

AN EARLY SEPTEMBER frost turned the aspens lining the road to Los Alamos brilliant yellow overnight. The summer holiday was over, and it was with a feeling of relief that Dorothy packed Kevin off to school. He and some teenage pals had recently gone joyriding around the Plaza and had been hosing down bystanders with large bottles of seltzer before being apprehended by two MPs. Dorothy had been called down to the local jail and had had to do some fast talking to smooth things over. It was Fiesta time in Santa Fe again. The scientists piled into cars and went to see the traditional Corn Dances in San Ildefonso, and the triumphant return of the conquistadors, a colorful pageant that was reenacted every year in the Plaza. The entire British mission, almost to a man, begged for that Saturday off so they could join in the ritual festivities. For three days, everyone in town donned costumes and their best silver and turquoise jewelry, including Dorothy, who raised eyebrows by coming to work at 109 in long, brightly patterned Navajo skirts and white fiesta blouses. Since the scientists were supposed to remain apart from the locals, they refrained from dressing up, but, as she pointed out, they looked more conspicuous than ever on the crowded streets among the Indians, señores, señoritas, singers, and dancing clowns.

For a few days, the Plaza teemed with carnival-booth hawkers, performers of every kind and quality, and hundreds upon hundreds of camera-carrying tourists. Then, thankfully, it was over and order was restored. The weeks flew by so quickly that it was only by these seasonal markers that Dorothy, looking back over the jumble of events, could fix them in her memory. There was never a moment to think about what they were doing, the real purpose of the research in the Tech Area, or the ramifications of the powerful "gadget" that was buzzed about on the grapevine. She barely had time enough at the end of each week to catch her breath, do the Sunday wash, and get ready for business on Monday.

Despite the feverish pace, there was one isolated moment that would always remain etched in her mind. It was on a serenely blue New Mexico afternoon that she got a call from the director's office reporting that a Japanese fire balloon had been sighted. Her heart had skipped a beat. Any threat to the laboratory was taken seriously by security, which was always on the lookout for intruders. The idea that they would be under attack by Japanese balloons seemed incredible, but in her confusion and alarm, she did not know what to think.

She remembered noticing that it was five o'clock and, glancing out the window, seeing the sun was about to set. The sky was clear and would soon be studded with stars. She thought of the horrifying possibility of an enemy presence floating over the valley under the cover of night. The news had been transmitted to the Albuquerque Army Air Base, and a fighter plane had been dispatched on a search-and-destroy mission. In the meantime, the caller from the Hill persisted, wanting to know if Dorothy would leave the office and look around "to see if there was anything in the sky that might be a Japanese balloon":

> The object was situated so many degrees from the sun, etc., and they had been observing it from the site and would like to know how it appeared from Santa Fe. I bustled out and scanned the skies from the Plaza and then drove to the top of old Fort Marcy and looked and looked. There was a certain apprehension and fear that crept around the mind and heart in contemplating such a possibility, with full knowledge of the danger to the Project. I

> could not see the object in question, but I did see little puffs of cloud, very frail and tenuous, which formed and reformed like vapor, and each one I imagined to be a small parachute with missile attached.

Dorothy stood on Fort Marcy Hill searching the horizon in the deepening twilight. When it grew dark, she could make out the small cluster of lights of Los Alamos miles away. She thought of her friends who lived on the site and wondered if they were safe. She shivered, less from the cold than the dread of what might be passing silently overhead.

What American military intelligence knew, but did not reveal until close to the end of the war, was that a Japanese balloon bomb offensive against the U.S. mainland had begun in the fall of 1944. The first reported launch took place on November 3, when a navy vessel spotted a balloon floating off the coast of California. Another was spotted outside the town of Thermopolis, Wyoming, and exploded in the sky. A few days later, two woodchoppers found the remains of a large Japanese balloon in the forest near Kalispell, Montana. In the months to come, balloon debris was found in Arizona, Alaska, Iowa, and Nebraska. All that fall and winter, fighter planes scrambled to intercept and shoot down balloons, and the exploded remains were taken to military labs to be examined. By spring, there were several reports per day, and exploded and unexploded balloon remains were turning up everywhere from Alaska and Washington to Colorado, Mexico, and Texas. In January 1945, the government, as part of its policy of wartime censorship, instructed the news media to withhold all stories about the Japanese "inter-continental free-flight balloons" to avoid panicking the civilian population, and to prevent the enemy from learning about the success rate of its attacks. Three months later, a confidential note to editors and broadcasters stated:

> Cooperation from the press and radio under this request has been excellent despite the fact that Japanese free balloons are reaching the United States, Canada, and Mexico in increasing numbers. . . . There is no question that your refusal to publish or broadcast information about these balloons has baffled the Japa-

nese, annoyed and hindered them, and has been an important contribution to security.

The Japanese called the rubberized silk and paper balloons "Fu-Go," and they were launched from sites along the east coast of Honshu. Filled with hydrogen, they would quickly climb to a cruising altitude of 16,000 feet, where the high-speed wind currents would carry them across the Pacific to North America. Each balloon was anchored to an aluminum ring that was wired with thirty-two "blow-out plugs," each supporting a sandbag. The plugs were designed to fire whenever the balloon dipped below a predetermined altitude, cutting free two bags of ballast and thus propelling the lightened Fu-Go back up into the jet stream. The last ballast bag was a weapon. If the balloons survived the 6,200 mile trip across the ocean, and everything went according to plan, the flash bomb would detonate over U.S. territory, causing damage and at the same time incinerating all evidence of the mysterious weapon. As part of their ambitious war plan, the Japanese sent an estimated 9,000 Fu-Go firebombs raining down on North America as a reprisal for the bold "Doolittle Raid" in April 1942, when sixteen American B-25s under the command of Lieutenant Colonel James Doolittle attacked Tokyo, and the subsequent escalation of the strategic bombing of Japan. Seven hundred of the lethal devices were launched in November 1944, and almost twice as many the next month, with an increasing number sent drifting over North America in the early months of 1945.

Designed as a terror weapon, the random attacks by the Fu-Gos were supposed to frighten civilians and ignite forest fires. The latter scenario was troubling, particularly in the dry summer months, when forest fires could endanger large portions of the Pacific Northwest. U.S. military intelligence was also afraid the balloons might be used as anti-aircraft devices and, ultimately, in biological warfare. On the West Coast, the military organized the secret Firefly mission, at one point numbering three thousand soldiers, to put out fires started by Fu-Go bomb balloons. The closest the bombs came to wrecking disaster was on March 10, when a balloon descended in the vicinity of the Manhattan Project's Hanford, Washington, production site. The parachute and cables caught on an electric line that fed power to the building containing the plutonium reactor, temporarily shutting it down.

In the end, because of mechanical malfunctions and the combined obstacles of distance and weather, only about a thousand Fu-Gos ever reached America. The Japanese ultimately considered the balloon barrage a failure, and it was halted after five months. Though it never became the propaganda weapon they had hoped for, the Fu-Gos succeeded in causing the only wartime fatalities within the continental United States. In May 1945, a balloon bomb in Oregon killed six members of a church picnic, five of them children, when it went off as they tried to drag it clear of the woods. Two weeks later, the War Department lifted the secrecy order, and the government issued a statement warning the public not to tamper with the hazardous balloons.

The Los Alamos project leaders had been informed that a number of Japanese fire balloons had landed in the Southwest. The laboratory staff knew little about the airborne threat but they were keenly aware that Site Y represented a vital target. With the approaching test, "people at Los Alamos were naturally in a state of some tension," Oppenheimer admitted in a letter to Eleanor Roosevelt written in 1950.* He added that the fact that they all immediately jumped to the conclusion they were under attack showed that "even a group of scientists is not proof against the errors of suggestion and hysteria":

> Almost the whole project was out of doors staring at a bright object in the sky through glasses, binoculars, and whatever else they could find; and nearby Kirtland field reported to us that they had no interceptors which had enabled them to come within range of the object.

Dorothy's account confirmed that Oppie and company had a bad case of the jitters. "The entire staff," she noted, "from the Director down, was in a dither that day." Work came to a standstill. Some of the scientists

*Oppenheimer's letter states the incident took place "very shortly before the test," but several other sources indicate it was at least several months prior to Trinity.

were convinced they could see some sort of suspicious "gondola" hanging from the balloon. Not everyone was convinced that what they were looking at was a balloon, and a few cooler heads suggested the mysterious orb might be a new star. "The scientists spent the afternoon craning their necks and evolving fantastic theories about the phenomenon," she reported. "The personnel director, an astronomer by profession, was called on in his dual capacity to settle the argument in order to get the staff back to work. Since he would not make a flat statement, the speculation continued."

At noon the following day, the same luminescent object again appeared in the sky. "Only then did all the experts agree that it was nothing more or less than the planet Venus, rarely seen in broad daylight," wrote Dorothy. "But it did give many of us a few hours of rather quiet terror."

With the brisk fall air, a back-to-school feeling permeated the mesa. The transformation of the laboratory had been completed over the summer and had cleared the way for a fresh start. The work had been messy and painful at times, particularly since none of the scientists were used to subjugating their interests and instincts to a large, cumbersome framework. But a number of chronic problems had been fixed in the process, and those individuals who could not adjust, or were not team players, had been threatened with removal or transferred to other divisions. A lot of anger and frustration had been vented, and the great majority, who had come to Los Alamos in the first place because of a strong sense of patriotic duty, were ready to put aside their differences and plunge back into the fray. Most of the physicists seemed reinvigorated, and even the Tech Area administration, which had never really settled into any kind of rhythm but lurched from one crisis to the next, seemed to function better. The new work hours on implosion were more grueling, but the divisions and groups were operating better, and overall there was less griping. As work on implosion progressed, Oppenheimer devoted himself to coordinating the different parts of the bomb project and making preparations for the test detonation of the plutonium device, which had been set for July 1945.

Stan Ulam, who had watched Oppenheimer overhaul the lab from top to bottom, marveled at what he called the "American talent for cooperation" and how it contrasted with what he had experienced in Europe:

> People here were willing to assume minor roles for the sake of contributing to a common enterprise. This spirit of teamwork must have been characteristic of life in the nineteenth century and was what made great industrial empires possible. One of its humorous side effects in Los Alamos was a fascination with organizational charts. At meetings, theoretical talks were interesting enough, but whenever an organizational chart was displayed, I could feel the whole audience come to life with pleasure at seeing something concrete and definite ("Who is responsible to whom," etc.).

As equally striking to Ulam as the cooperation was the conviviality, not only among the physicists, both theoretical and experimental, who differed greatly in temperament, but also among mathematicians, chemists, and engineers. "People visited each other constantly at all hours after work," he wrote. "They considered not only the main problem—the construction of the atomic bomb and related physical questions about phenomena that would attend the explosion—the strictly project work—but also general questions about the nature of physics, the future of physics, the impact of nuclear experiments on technology of the future, and contrastingly its influence on the future development of theory." Beyond this, there were wide-ranging discussions of the philosophy of science, and of course the world situation, from daily progress on the war fronts to the prospects of victory in the months to come. "The intellectual quality of so many interesting persons and their being constantly together was unique. In the entire history of science there had never been anything even remotely approaching such a concentration."

Luis Alvarez, who had come to Los Alamos after stints at both the Rad Lab in Cambridge and the Met Lab (later called the Argonne National Laboratory) near Chicago, was amazed at how effective Oppenheimer was at the helm of Los Alamos:

> Remembering the unworldly and longhaired prewar Robert, I was surprised to see the extent to which he had developed into an excellent laboratory director and a marvelous leader of men. His haircut was almost as short as a military officer's; he ran an organization of thousands, including some of the best theoretical and experimental physicists and engineers in the world. The laboratory's fantastic morale could be traced directly to the personal quality of Robert's guidance.

No one recognized better how difficult it was to achieve such solidarity and concentration of effort than James Conant, who had been traveling to the site every three weeks since August and was becoming a close advisor to Oppenheimer during the last, difficult stage of the project. The two tall, thin men, who shared a love of the mountains and rugged hiking trails, could often be seen heading into the hills at a rapid stride, deep in conversation. Conant was a man who had distinguished himself far beyond his profession as a chemist; he had served as chairman of the S-1 executive committee, was a member of Groves' Military Policy Committee, and was one of the elite few with the background and knowledge of the new weapon's development who could act as an intermediary between the White House and the scientists on the Hill. He had a firm grasp of the political and military problems facing them in the months ahead, and according to Manley, Oppenheimer came to view Conant as a statesmanlike figure, "a very wise, elderly person who in a normal sense of events he would have liked to have had as a godfather."

By October, their relationship had reached the point where Conant often tried to cheer Oppie on, and perhaps understood that he was one of the very few outsiders who could provide the kind of encouragement and moral support Oppenheimer needed at this critical juncture. "Just a line to tell you again how satisfactory I think everything is going at Y," Conant wrote after one of his inspection visits. "In all seriousness, you are to be congratulated on the progress made and the organization as it now stands. I enjoyed my trip immensely, and I am particularly grateful to you and your wife for your hospitality."

Oppenheimer was restored to his old self again, if more nervous and distracted than ever. He paced constantly. From behind her typewriter, Shirley Barnett watched him do rings around the large conference table in his office, trailing plumes of gray smoke from his pipe. With so much doubt and uncertainty hanging over the project, everyone was showing signs of strain. On his last fly-by visit, Groves, who was always instituting more security strictures, berated Oppenheimer for wearing his trademark porkpie hat. "He said it made him too recognizable when he left the site, and that anybody could pick him out in an instant," recalled Barnett. "Well, Oppie didn't take kindly to that sort of thing." The next time Groves came to visit he was prepared. He had an elaborate Indian headdress of eagle feathers that someone had given him as a gift, which he kept hanging on a wall. "It was enormous," said Barnett. "So he put it on, and greeted the general by saying, 'Is this better, sir?' " Oppenheimer was so pleased with his prank, he told everyone. While people laughed at the story, what they remarked on later was that despite all the battles of the past few months, their director and general still enjoyed a remarkably good relationship. "Both of them had a sense of humor," added Barnett, "and they had a real mutual respect, which went a long way."

It was always a wonder to Dorothy that with so much work to do the scientists still found time to stir up trouble. That autumn, the talk on the mesa turned to the coming presidential election, and people became agitated about their right to vote. Roosevelt stirred deep emotions in the scientific populace, and they were worried that because of the usual last-minute problems, they might find themselves disenfranchised. They had surrendered their freedom and privacy in order to serve their country, but those who were citizens were not about to give up their vote. People raised their voices at town meetings, and the issue became the focus of much excited chatter on the vine. Dorothy learned that before the war Los Alamos had been entitled to its own ballot box. But the box had disappeared along with many of the school's more civilized features, and it seemed unlikely the army would be in a hurry to put it back anytime soon.

In the beginning, the post administration assured the scientists that they would be able to vote. The first group of citizens eager to register volunteered to make the seventy-five-mile trip to the county seat, Bernalillo, which at the thirty-five-mile-per-hour wartime speed limit promised to take the better part of a day. "Ten of us in two groups went, and all of us were Democrats," recalled Priscilla Greene. The only problem was that when the local sheriff caught wind of the expedition, he worried that hundreds of liberal voters would swing the county and possibly lose him his job in the New Mexico Republican administration. "Los Alamos was part of Sandoval County, which at that point had a one hundred percent Republican majority," explained Priscilla, "and they got an injunction [against letting the Hill people register]." The army promised to look into contesting the injunction, but kept putting them off with various excuses, and finally decided against doing so because the scientists' names would have to go on the voters' rolls if they registered, and that was out of the question. All lists of names were restricted and had to remain classified information. The senior scientists had been told all along that if they wanted to vote, they would have to vote by absentee ballot, so the few who had made arrangements to do so were the only civilian members of the Los Alamos community who voted that year. There was nothing the rest could do but chalk it up to the stupidity and madness of life under military rule.

Nevertheless, there were many victory bashes after FDR's reelection, not that they needed much of an excuse to celebrate. Despite the accelerated work schedule, their enthusiasm for parties was undimmed. Saturday night gatherings, whether large and loud or small and intimate, usually lasted until dawn. The walls were too thin, their private lives too exposed, and the liquor too strong for it to be any other way. It was impossible to seriously misbehave, but for all their awareness of being monitored, they were, as Bob Bacher's wife, Jean, put it, "peculiarly uninhibited and completely unrelaxed." They went to see *Our Hearts Were Young and Gay* in the post theater and felt incongruously happy and safe in their secretive little utopia, where no one was old, or sick, or handicapped, and children could roam freely without fear of strangers or crime. They jammed incredible numbers of people into their little GI apartments for cocktails, and couples danced in the hallways and out on the porch. For all their worldly sophistication, even the most serious scientists engaged in adolescent binges and frat-house hijinks. Pranks, practical jokes, and ridiculous propositions were usually the climax of a long evening of drinking. During a particularly rowdy affair at the Bainbridges' house on Bathtub Row, a pretty young wife wagered five dollars that no one would dare take advantage of the enviable porcelain fixture—one of only six on the mesa—to "take a bath then and there." Naturally, a young man came forward to take up her challenge. "When he discovered the door to the bathroom had no lock, he borrowed a pair of Ken's trunks, drew a tub full of hot water and prepared to soak in comfort," recalled Bernice Brode. "But of course, everyone crowded in as witnesses, offering to scrub his back and wash behind his ears. When a couple of not-entirely-sober friends tried to climb in too, the water started to run into the hall, and Peg put a stop to the fun."

After another Saturday night bash, a number of guests deliberately hung around until the early hours of the morning to arrange several dozen empty liquor bottles beside the front door of the stone house where George Kistiakowsky lived. They placed a large sign over the bottles, which read, "Milkman, only two quarts today, please." Apparently, Kistiakowsky, who was not religious, could be counted on to sleep in while churchgoing families passed by his house on the way to and from Sunday services. "George was quite angry about the prank, and none confessed to it," recalled Brode. "However, the next Monday morning, the Bainbridge family could not get out either door, as huge piles of logs had been banked against them. Ken had to climb out a window to go to work."

Planning parties and events became a staple of mealtime conversation, and the preparations usually focused on rounding up enough alcohol to guarantee a hardy punch. Dorothy would look on in amusement as they piled cases of cheap rum in her office and passed up lunch at La Fonda to scurry around town laying in supplies for that weekend's debauchery. Personal inspiration, private deals, and devious wangling accounted for much of the alcohol that found its way up to the Hill. When word spread that Sam Allison was coming from Chicago and would be driving his car, people rushed to put in orders for their favorite drink. On the last leg of his trip, Allison was hit by a rock slide on a steep patch of highway between Taos and Santa Fe and knocked over the edge and into the river. He survived unscathed and called the laboratory, which dispatched a jeepload of GIs to rescue him. The army boys succeeded in hauling out his car, which had somehow managed to plunge forty feet down into the Rio Grande and land upright on its wheels, its trunk still intact and packed to the brim with booze. When he finally made it up to the Hill, he found his colleagues were far more concerned about the possibility of broken bottles than broken bones.

They had become adept at making their own fun. They found any excuse to dress up, flocking to dances in Fuller Lodge in black tie—the women in floor-length gowns and carefully hoarded, prewar nylons. Costume parties drew the biggest crowds, and at one particularly wild affair, the normally retiring Klaus Fuchs led a conga line through the post Commissary before passing out cold behind the bar. By now there was a club or mesa organization for everything, and all staged magic shows, variety shows, and quiz shows, the latter requiring the theorists to display their erudition in more plebian realms. The high point came when Teller was called on to answer this trivia question: "The General wants to know whether the whistle blows at 7:00 or 7:30 in the morning." This was greeted with hoots of laughter, as everyone was well aware that Teller disdained the Tech Area siren, slept late, and enjoyed playing his piano at all hours of the night. But he received a round of applause when he gamely stood and faced the crowd.

The biggest hit of the season was the Little Theater Group's presentation of *Arsenic and Old Lace,* which played to a packed house, in part because nearly everyone on the post had been approached about auditioning for a role or providing props or assistance of some kind. The climax came at the end of the last act, when the dead bodies were brought up from the cellar, and the audience was delighted to see Oppenheimer, his face heavily dusted with flour, playing one of the stiffs, carried in on a board and laid out on the floor. After his scene-stealing appearance, a dozen of the leading men on the Hill joined Oppie on stage as corpses—including Deke Parsons, Bob Bacher, Cyril Smith, Otto

Frisch, and Harold Agnew—each to wild applause. Afterward, everyone congratulated Oppie on his rigid performance, and Cyril Smith joked that, for his part, it was "the most restful occupation he had on the mesa."

Winter came suddenly that year, and they woke up one morning to find a blizzard had dumped ten inches of snow on their mountain stronghold. The children whooped for joy as they headed out the West Gate for Sawyer's Hill. Entertainment was in short supply, and skiing was a favorite pastime. They were surrounded on all sides by wonderful ski country, but while Hans Bethe could go up or down any kind of mountain in skiis no matter how steep or thickly studded with trees, the lack of a clear run made it difficult for the less intrepid. George Kistiakowsky, an expert skier and their resident demolition expert, decided it was high time to do something about the situation. He teamed up with Hugh Bradner and Seth Neddermeyer, and together they scrounged some Primacord and Composition C—a puttylike plastic explosive—from their experiments and set about creating a long extension to Sawyer's Hill. They set up half necklaces around the trees, which when detonated, "cuts as if you had a chain saw," recalled Kistiakowsky, "and it's faster. A little noisier, though." After a dozen jaw-rattling explosions, they managed to clear one medium-steep hill, though, according to Bradner, "in most cases the trees failed to fall the way we wanted them to." They then liberated a length of rope from one of the laboratories and, using a snow tractor borrowed from the army, managed to rig up a reasonable facsimile of a rope tow. Such frivolous use of supplies was officially discouraged, but as on so many other occasions, they knew they could count on Oppie to look the other way.

As their second Christmas on the Hill approached, women began descending on Santa Fe in a fruitless hunt for goods that had long ago disappeared off the shelves. Dorothy was deluged with requests for skates and skis, baby clothes and anything that might pass as a maternity dress. The rapid population growth on the Hill had strained the resources of the little town, and the shops were out of everything. Because of wartime rationing, the amount of goods the local stores were allowed to order was based on the tiny population of Santa Fe alone, and not on the needs of a burgeoning mountaintop city that for all intents and purposes did not exist. There were no new boots to be had, no cribs, no electric ovens, no tires for the car—new, used, or even patched and tread-worn. None. The town was tapped out. There was nothing to do but try to order things from a store back in civilization and have it shipped at considerable expense, or wait and hope the war would soon be over and they would be allowed to leave.

People haunted Dorothy's office at 109 East Palace in hopes of snagging the first coveted copies of the Sears Roebuck catalog. When they returned to the post, they would run down the street yelling, and everyone in earshot would gather around for a look. Since nearly all personal shopping had to be done by mail, they would pore over the pages, nearly swooning at the wealth of merchandise on display. For some mysterious reason the Montgomery Ward catalog was freely available, while others were so much in demand at Los Alamos that one company dispatched a stern note: "Sir: We don't know what you are doing to our catalogs. We have sent more than 100 catalogs to this address and will send no more." The treasured volumes, thick as telephone books, were jealously guarded and passed along from family to family until they were dog-eared and torn. On one occasion the Tech Area operator, by request, sent a pointed message over the laboratory loudspeaker system: "Attention, please. Will the person who took the Sears Roebuck catalog from Harold Agnew's room please return it immediately. Repeat—immediately!"

Despite Groves' objections, Los Alamos's baby boom continued unabated. One after another, the scientists' wives announced they were pregnant. Oppenheimer's assistant, Priscilla Greene, was expecting, as were Elsie McMillan, Beverly Agnew, and Rose Bethe. Phyllis Fisher, the wife of the young physicist Leon Fisher, was reassured to learn that a whole group of her friends would be "waddling around the mesa together," as though these real-life babies-to-be would somehow counterbalance the "awesome baby" that the project would give birth to in a few months. "It seemed to me that we were all striving to maintain the fiction that somehow life could be normal in that very abnormal setting," she observed, adding, "Perhaps that was a vain hope."

On December 7, 1944, the birth of Katherine Oppenheimer was announced, and the whole town shared in their director's pride and delight.

Everyone was in the habit of dropping by the hospital to see the newborns, and friends often stood on packing crates outside the maternity ward's window for a peek and a chat. So many people wanted to see the boss's baby daughter that the hospital staff was finally forced to hang a hand-lettered sign that read OPPENHEIMER over her crib, and visitors' hours were suspended to make way for the steady stream of well-wishers who filed by for days. Undeterred by the bad weather, Dorothy drove up with chains on her tires just to see the new addition to Box 1663. "Little Toni," as she was called, had been born into such great secrecy that when Dorothy filled out her birth certificate, she could only list the benign rural post office box in Sandoval County as the home address. She brought Kevin with her; and as it had been his birthday the day before, she had arranged a special treat. Dorothy had worked out a deal with the sergeant who was the head of the post motor pool "to borrow" one of the heavy, four-wheel-drive army vehicles for the fifteen-year-old to take for a spin. "The snow must have been two feet deep," recalled Kevin. "I remember driving up and down the main street, and I thought I had died and gone to heaven."

It was a madly busy holiday season. There was something in the air—no one was in the mood for quiet reflection. People threw themselves into the festivities with abandon, singing, drinking, and dancing until dawn. There were endless rounds of parties, and vats of heavily spiked eggnog were consumed. A group of families banded together and organized a big New Year's Eve bash at Fuller Lodge, and at the stroke of midnight, they rang in 1945 with the old Ranch School bell used to call the boys to meals. It was such a memorable evening that they all agreed it should become a mesa tradition—that is, if they were all still captive there in twelve months' time. Even as they joined hands and broke into a rousing rendition of "Auld Lang Syne," they heard the quaintly nostalgic words as they never had before, and sang their hearts out, not knowing what the coming year would bring.

FIFTEEN

Playing with Fire

THERE WERE MANY lighthearted moments that winter, but while relieving the boredom of people's regimented lives, those good times did nothing to dispel the deepening chill that had settled over the community. The construction of the Trinity site had been completed, and as the laboratory moved into the final production and testing phase, a growing sense of foreboding tugged at everyone's thoughts. In February, Groves swept in for a conference, and they came to the decision that the time had come to "freeze" the implosion program and focus their efforts on one of the several designs they had been considering. Oppenheimer had set July 20, 1945, as the target date for having both bombs ready. He drew up a schedule and methodically drove the divisions forward, checking off stages as they were completed. The men toiled at a frenzied pace to prepare for the test shot. Even at parties they never seemed to unwind, but gathered in tight knots and discussed complex equations in low, urgent voices.

The fissionable materials—plutonium and uranium 235—were now coming off the production lines in large quantities, and the physicists were faced with the long, arduous task of converting them into metal, casting them, and finally delivering the explosive devices to be used in the bombs. At the Omega site, an operations lab set up in a canyon well removed from the crowded mesa, Don Kerst and Fermi had just about completed work on what was called the "Water Boiler," code for a small nuclear reactor, and were ready to begin preliminary tests to measure the critical mass of the uranium 235 bomb. As the critical assembly tests progressed, Fermi grew increasingly nervous and kept finding excuses to take his crew on long hikes in the mountains.

All of the work at the Omega site focused on preparing experiments that could help reveal the performance of the bomb. But as Segrè recalled, a nuclear explosion was such a completely new and complex event—with mechanical, thermal, optical, chemical, and nuclear aspects—that in some cases they did not even know the order of magnitude of the quantities to be measured, and consequently needed equipment that could cover a vast range. The physicists were dealing with a host of unknowns, under extremely unusual conditions, and, as Segrè noted, this "worried everyone":

> There was obviously plenty to measure; the energy released was the overall central parameter, which could be inferred in many different, independent ways. Each measurement had its particular difficulties, but one was common to all of them: the experiment could not be repeated. If something failed, there was no second chance.

During this period, Otto Frisch proposed an extremely bold experiment that would simulate bomb conditions and go, as he put it, "as near as we could possibly go towards starting an atomic explosion without actually being blown up," a risky procedure that Feynman likened, with false levity, to "tickling the tail of a sleeping dragon." Nonetheless, Frisch managed to convince Oppenheimer and the senior physicists that with the proper precautions, the "Dragon Experiment" could be conducted with complete safety.

Once his experiment was given the green light, the experimental setup was completed in a few weeks, with Frisch as group leader. His idea consisted of using some of the U-235 to assemble an explosive device, but leaving a sizable hole, so that when the missing core of enriched uranium hydride was dropped into the ring, it would cause the material to go supercritical for a fraction of a second. The core would immediately be pulled back and could subsequently be reinserted to enable more measurements. This would allow them, through repeated efforts, to learn exactly how much uranium would be needed for Little Boy.

Everything went as planned. "When the core was dropped through the hole," recalled Frisch, "we got a large burst of neutrons and a temperature rise of several degrees in that very short split second during which the chain reaction proceeded as a sort of stifled explosion." There was always the danger of a runaway reaction. They were all aware that the assembly could become critical in no time, and that even a minor mistake could result in death. They devised a reliable system of safety checks and made a strict rule that no one could work alone, so that someone would always be monitoring the equipment and careless accidents could be avoided.

Despite all his precautions, however, Frisch nearly made a fatal mistake while working on an unusual assembly he dubbed "Lady Godiva." He and his assistant were standing by the neutron-counting equipment when they both saw the red signal lamps start blinking faster and faster. His assistant, a young graduate student, panicked and pulled the plug on the meter. Frisch yelled out, "Do put the meter back, I am about to go critical." Out of the corner of his eye, he saw the signal lamps had stopped flickering and were now glowing red. Thinking quickly, he removed some of the blocks of uranium compound he had just added and the lights began flickering again as the meter slowed. "It was clear to me what had happened," he recalled. "By leaning forward I had reflected some neutrons back into Lady Godiva and thus caused her to become critical." He had not felt anything, but after carefully completing the experiment, he checked the radioactivity counter and found that he had received a rather large dose, though within the lab's permissible parameters. Had he hesitated for even two seconds before removing the fissionable material, the dose would have been lethal.

Even after his close call, Frisch continued to work with critical masses, maintaining that the danger was largely psychological: "Assembling a mass of uranium-235 was something we completely understood, and as long as we hadn't reached the critical amount—when the chain reaction began to grow spontaneously—the assembly was completely harmless." Overeagerness and haste were the real threats, and a bright young physicist who pushed too hard for results would pay the price. They were all working under enormous pressure. The uranium had to be returned shortly to be turned into metal and assembled into a real atom bomb, so they worked at a frantic pace to complete their experiments, putting in seventeen-hour days and snatching a few hours sleep from dawn till mid-morning.

Progress was being made in other areas. James Conant visited Los Alamos again to obtain a firsthand report on the "Christy bomb," project slang for the ingenious proposal made by Bob Christy, one of Oppie's former students, to simplify the design of the plutonium bomb. His modification promised to save the physicists an enormous amount of work. After the meeting, Oppenheimer and the others left, leaving only Conant and Teller in the room, sitting together in contemplative silence for a few moments. Then Conant muttered, more to himself than anyone else, "This is the first time I really thought it would work." Teller stared at him. "That was the first indication I had of how little confidence those in the highest scientific quarters had in our work," he recalled. "I was slightly shocked."

There were still too many unknowns and uncertainties, and Oppenheimer decided they needed to have some sort of dry run. They had to have at least a working idea of the conditions they could expect at ground zero, to check their equipment and correct for any weaknesses in their plans. The idea was to explode one hundred tons of conventional explosives at the Trinity site and then perform all the same measurements they would later make on the atomic bomb. Although TNT could only approximate the atom bomb's effect—for example, it did not emit neutrons or gamma rays—it would at least give them a chance to study the blast effects of a huge explosion and begin to calibrate their instruments for the final shot. This was the best they could do without wasting any of their valuable nuclear explosives.

Bainbridge was asked to hastily organize the dress rehearsal. By March, Project Trinity was formalized, and Oppenheimer tapped Bainbridge, a three-year veteran of the MIT Rad Lab, to be director of the test program. Oppie appointed Johnny Williams as deputy director to help oversee the construction crews and make sure the scientific facilities and shelters conformed to the project's needs and were completed on time. Dozens of physicists were called away from their divisions and assigned to work on experiments designed to obtain measurements on the blast, heat, and radiation effects of a nuclear explosion. He formed the Cowpuncher Committee to "ride herd" on the implosion program and make sure they met their deadline. As Bainbridge observed, "The great push to solve the problems of the implosion method had meant that only a small amount of staff time, shop time, and money could be spent in preparation for a test." Every decision they made had to be considered on the basis of the time scale for completion. There would be no delay of the implosion test.

As the secrecy intensified, the already high-strung character of the mesa altered perceptibly. The military personnel, feeling more useless and beside the point than ever, were increasingly irritable and impatient. The scientists, dashing between the mesa and the remote test site, were harassed and exhausted. Wives who had been close friends became guarded about their husbands' new responsibilities and whereabouts, and no longer felt free to confide their mutual worries. For the first time since they had arrived at Los Alamos, the wives could not rely on one another to help defuse the tension, and their fears and anxieties hung heavily in the air. "The wives couldn't talk to each other," said Dorothy, who had by then gleaned the purpose of "the gadget," but was in no position to enlighten anyone else. "Each didn't know how much the other knew about what their husbands were doing."

The laboratory work continued around the clock, and the men pulled double shifts and returned to their experiments at all hours. The lights burned in the Tech Area all night, and the intermittent power outages that had plagued the community from the beginning became more frequent, the lights often blinking off and on during the busy dinnertime hours. A supplementary power line was run from the post to Albuquerque to help alleviate the problem. Ironically, it resulted in one of the worst security breaches by inadvertently carrying back Los Alamos's closed-circuit radio broadcast to civilization. Dorothy, who often brought morsels of news from town, reported that several friends from Albuquerque who had been visiting her were "simply agog" at what they were picking up on this mysterious station. "They can't imagine where the broadcasts come from or why none of the entertainers have last names," Dorothy told them. "Children's stories are read by Betty, newscasts compiled by Bob, and Mozart's piano Sonatas played by Otto have them guessing."

They all laughed because it was too serious not to. "We laughed all the time, and at everything, or else we would have lost our minds," said Marguerite Schreiber, who was married to the physicist Raemer Schreiber, "because really, the atmosphere was icy. There was very little conversation between husbands and wives. It was a very cold, lonely and difficult time. We simply put one foot in front of the other, minded our p's and q's, and tried to get through it. We didn't look ahead because we didn't know how long it would last." Marge Bradner never asked her husband what was going on, even when he went off to the distant operation sites for days at a time. "I had no idea what my husband was doing," she said. "I didn't know, and I didn't speculate because I didn't want to know."

Even for those who knew about the "gadget," there were still too many unknowns not to lie in bed awake worrying into the night. Would the bomb work? Would it blow apart New Mexico and them with it? Or the world? Would it finally end the war? Would everything be all right? For many of the refugee scientists, who had already been through terrible times, the future was fraught with terrifying questions that haunted their sleep. What had happened to their families? Would they be able to find them after the war? Where would they go? When they left Los Alamos, they would be without savings, homes, jobs, or even countries to call their own anymore. Many of them had left everything behind when they fled Europe, and they knew it was doubtful that their property or possessions could be reclaimed. When one distraught wife discovered she was pregnant and told her husband, after a long, anguished discussion, they decided they could not afford a baby at that time. Moreover, they did not see how they could bring a child into such an uncertain world. The unhappy woman went to Dorothy with their problem.

Dorothy did not presume to judge them, nor did she try to convince the woman to change her mind. She simply promised to help. After making discreet inquiries in town, she located a reputable woman doctor in Santa Fe, who in turn provided the name of another physician in the neighboring town of Espanola who was not frequented by the scientific community. The clandestine arrangements were necessary not simply because of the desire to avoid scandal, but because abortions were illegal at the time.

The mounting tension manifested itself in different ways all over town and exacerbated some long-standing problems on the overcrowded mesa. The spread of Quonset huts and trailers had grown to slumlike proportions, and the sanitary conditions were beyond belief. A group of Hill wives made a study of the Spanish American quarter and reported to the army administration that the lack of latrines and adequate facilities could lead to the spread of disease. A new influx of rowdy machinists was also creating havoc. Rumor had it they had been offered inducement wages to leave their families and work on the isolated site, but Dorothy, relaying complaints from local Santa Feans, was concerned that their hell-raising behavior was "giving the Hill a terrible reputation." For certain, it was keeping security busy both on the post and in town.

Groves had brought in yet another new commanding officer, Lieutenant Colonel Gerald Tyler, to try to get a grip on the situation before it spiraled out of control. Tyler, much to everyone's amazement, took it upon himself to actually try to improve the quality of life on the mesa for both the civilians and the troops. He risked the general's wrath by asking the higher authorities in Washington for more money and oversaw the construction of a new cafeteria, additional barracks for the WACs, and a recreation hall for the soldiers. The food at the new cafeteria was positively gourmet compared to the mystery meat and gray vegetables that congealed on their trays in the mess hall, and even the most work-obsessed physicists began turning up for steak night once a week.

To everyone's relief, the army finally modified the regulations restricting their movements. By late 1944, even their hard-driving general recognized that the isolation and stress were becoming too much for the mesa's inhabitants. They were jumpy and restless. The scientists and their families had been cooped up for too long and were badly in need of a break, and Groves determined that "the improvement in morale would outweigh the increased security risks." Short, one-week vacations were encouraged, as long as the usual precautions were observed. The men needed to get away from the constant pressure of hurrying to finish the bomb, hurrying to end the horror and killing once and for all. They bottled it all up inside, recalled Elsie McMillan, and going to a party or two or taking a Sunday off was not enough. "We were tired," she wrote. "We were deathly tired. We had parties, yes, once in a while, and I've never drunk so much as there at the few parties, because you had to let off steam, you had to let off this feeling eating your soul, oh God are we doing right?"

Some couples took advantage of the new rules to take a winter sojourn to Denver or Colorado Springs, while others made a quick trip home to see ailing parents who were full of questions and recriminations and could not understand why visits had been barred for so long. Oppenheimer was far too busy to get away, but Kitty began escaping the confines of the post as often as possible. "She would go off on a shopping trip for days to Albuquerque or even to the West Coast and leave the children in the hands of a maid," recalled Jackie Oppenheimer, who had recently moved to the Hill with Frank, who had been working at the Oak Ridge plant before Groves transferred him to Los Alamos. Assigned to Bainbridge's safety crew, Frank was often away at the Trinity site, and the two sisters-in-law were thrown together. The close proximity did not improve their relationship, however, and Jackie was appalled by Kitty's behavior:

> When we went up to Los Alamos, Kitty made a dead set at me. It was known that we didn't get on together and she seemed determined that we should be seen together. On one occasion she asked me to cocktails—this was four o'clock in the afternoon. When I arrived, there was Kitty and just four or five other women—drinking companions—and we just sat there with very little conversation—drinking. It was awful and I never went again.

Many of the young Hill wives found Kitty disconcerting and kept their distance. She in turn avoided most of their clubs and societies. Kitty had "no friends at Los Alamos," according to Priscilla Greene, and the few she did have were generally men. She was increasingly rude and impossible, and the mesa buzzed with stories about her erratic behavior. "She was really rotten to Jackie," said Shirley Barnett. "She made a point of it, and it did not go unnoticed."

Kitty had brought her own troubles to the mesa, and she reacted to the stress of that bleak winter by becoming even more melancholy and withdrawn. Barnett, who sometimes went to visit, remembered thinking the atmosphere of the bungalow was infused with gloom. After the birth of her second child, Kitty seemed to go into a depression, often going days without leaving the house. "She was often ill, and took various drugs to quiet her nerves," recalled Barnett, who spent many hours listening to her while she chain-smoked cigarettes. "She spent a lot of afternoons [lying] on the couch with the curtains closed, 'suffering from the vapors.' Really, I think it was nothing more than she had overindulged. I don't know what Oppie thought. He wasn't in great shape himself at that point, and he had so much to do he couldn't fret too much about it."

It is not clear why Oppie wanted Frank and Jackie at Los Alamos, except perhaps that having his brother by his side was a comfort. They were close, and Frank was someone he could talk to and consult about things that were weighing on his mind. For Frank, however, being at Los Alamos was fraught with problems. For one thing, he was painfully aware of how wretched Kitty was to his wife. While he himself was warmly received on the Hill, people were forever commenting on his striking resemblance to his older brother and their shared mannerisms—they could often be seen in the Tech Area talking and walking in circles, and rubbing their palms together in identical fashion—and the comparisons could be trying. Frank tried to fit in, but he was a late arrival in the mesa's tribal society. "I think it wasn't easy being Oppie's little brother at Los Alamos," observed Barnett. "He had chosen to follow in his footsteps, but he ended up being so overshadowed. I think Frank had a hard time dealing with the fact that his brother had become such an important man."

Groves had repeatedly brought Frank to Los Alamos in hopes that he would have a soothing effect on Oppenheimer, who, as usual, looked at death's door. Oppie was sleeping only a few hours a night, his weight had dropped to 114 pounds, and he appeared to be living on nervous energy.

Groves was genuinely concerned that his poor body might not be able to take much more abuse and wanted to surround him with people who would help sustain him during the nerve-wracking run-up to the test. "Frank turned into one of the prime worriers of all time, that was how he tried to keep Robert well," recalled Rabi, then director of the Radiation Laboratory at MIT, whom Groves brought in a few weeks later for much the same reason. "As for me, I was invited just because Robert liked me. He'd asked me to join his project as experimental physics director and I'd refused because of my radar work. Now he was under tremendous strain and I was supposed to watch him and look after him."

Dorothy saw little of Oppenheimer during that period. He was always on the move, running back and forth between the Tech Area and the top-secret desert location to the south. But she was always available at a moment's notice if he needed her. When a bad case of chicken pox confined him to his bed, she fussed over him as she would over a sick child, berating his staff and shooing people out of the room so he could get some rest. "She was absolutely devoted to him," said Barnett. "She was someone he could talk to, someone he knew he could trust." For Oppenheimer, her warm embrace must have been wonderfully reassuring in an agonizing time. "She loved him," said Marge Schreiber. "She tried to take care of him. She mothered him, and he needed it, God knows. He was carrying such an awful burden. He needed all the help he could get, and she was there, and she was up to it."

To complicate matters, his faithful assistant, Priscilla Greene, was getting bigger and bigger by the week and wanted to stop working. Oppenheimer kept putting her off. No matter how many people she suggested as her replacement, he found a reason to disqualify them. But the office was busier than ever with the test preparations, and neither Shirley nor Dorothy could cope with all the responsibility. Finding Oppie a new secretary became so imperative even Groves got involved. He trolled the War Department for names and came back with a list of suitable candidates, all of whom Oppie vetoed. Exasperated, Groves demanded, "Do you have ideas?" Oppenheimer replied, "Yes. I'd like to have Anne Wilson come here," referring to the lively, doe-eyed admiral's daughter who was a member of the general's small staff. Surprised, Groves called Wilson into his office and asked her if she wanted to go. When she immediately replied, "Yes," he sighed and said, "You deserve each other."

Before Wilson left for Los Alamos, they had a little send-off for her at the War Department office, complete with a cake in honor of her twenty-first birthday. Everyone gave her packs of cigarettes as gifts because they were so hard to come by. At one point, Groves took her aside and warned her that the scientists on the Hill might not welcome her with open arms. "They'll just think I'm sending in one of my spies," he told her, and then he proceeded to make it clear that nothing could be further from the case. By choosing the Los Alamos assignment she needed to understand that this part of her life was over and she was starting something completely new.

Not everyone in the War Department wanted her to make such a clean break. On one of her last days at work, John Lansdale sauntered over and asked casually, "How would you like to make an extra $100 a week?" When she asked what she would have to do for that kind of money, he replied, "Just send me a little three-page report once in a while and tell me how they are doing." Wilson was flabbergasted. "I was really outraged," she said. "I told him I couldn't believe he thought I would do such a thing." She understood perfectly that he was attempting to enlist her to spy on one of their own. The more she thought about it, the angrier she got. She had listened in on and transcribed hundreds of the general's telephone conversations with Oppenheimer and the other Manhattan Project leaders, and he had never once expressed disapproval of Los Alamos's director. "Groves always trusted Oppenheimer," she said. "He had picked him and he didn't second guess himself. He couldn't. There he was all by himself, running this huge project. He had complete confidence in himself and his judgment—otherwise he couldn't have done what he did."

When Wilson arrived at 109 East Palace, it was still very cold, and as she headed up the steep, winding road to the laboratory, it started to snow. As soon as she arrived, she was told at the gate to go straight to Oppenheimer's house on Bathtub Row, as he had organized a party so she could meet everybody. When Oppie opened the door, she was shocked at his appearance. He was still recovering from the chicken pox, and a week's growth of dark beard covered a mottled red face that was still too tender to shave. He had suffered from fevers of 104 degrees and looked emaciated. She remembered thinking, he was the skinniest man she had ever seen in her life and she wondered how he managed to keep chain-smoking cigarettes in his condition.

Oppenheimer made her one of his famous martinis and introduced her around. Before she knew it, the room was spinning. "Nobody told me you could not drink at that altitude, so I had two martinis, and the second one just plowed me under," she said. "I wondered why Robert kept pressing little snacks on me during the party. Everybody up there knew exactly what was going to happen. I think they enjoyed doing that to newcomers. It was very embarrassing, and Robert had to walk me back to my room in the nurse's quarters." A few weeks later, Kitty stopped in to see Wilson at the office and asked how she was settling in. Then Kitty let her have it. "She gave me a little chat to the effect of 'Lay off,' " said Wilson. "I was really astounded. I was still quite innocent, and though I thought Robert was marvelous, I had never thought about him *that way.*"

Week swiftly followed week, and there was an inescapable sense that they were hurtling to the finish line. The Big Three—FDR, Churchill, and Stalin—met at Yalta in February 1945 and plotted the fate of postwar Europe following the inevitable German surrender. But the fighting in the Pacific was bloody as Allied forces slugged their way through the heavily fortified islands and atolls, and a land invasion of Japan loomed as an awful prospect, particularly for those with family members in uniform. The amphibious assault on Iwo Jima had resulted in one of the worst battles of the war. By the time the eight-square-mile volcanic island was secured in mid-March, 6,821 Americans had been killed and another 20,000 wounded. The battle for Okinawa was even worse. Victory came at a terrible price, with nearly 50,000 casualties and as many as 12,000 dead. The fact that more than 110,000 Japanese had died trying to prevent the island's capture was an ominous sign of how fiercely the Japanese would fight to defend their mainland.

With the approach of spring came the howling dry winds and drought, and the fear of what a fire could do to their tinderbox of a town. Their uncontrollable furnaces kept their flimsy houses heated between

90 to 100 degrees no matter what they did, and everyone's backyard looked like a coal bin. Explosions from the canyons below shook the mesa several times a day, each blast rattling windows and knocking paintings off the walls. At those times, the sweet scent of pine, which permeated the mesa, mingled with the acrid smell of explosives, serving as a constant reminder of the laboratory's deadly purpose. Everyone was on the alert for a sudden blaze. Every time the siren wailed—and it wailed repeatedly that spring—people stopped in their tracks, paralyzed by fear. The extreme cold and occasional flurries did nothing for the water table, and there were rumors the community might run out of water. The dire warnings in the *Bulletin* returned, ordering people to conserve in order to avoid "drastic restrictions" and reminding them over and over again to carefully extinguish cigarettes, cigars, pipe ashes, and, most of all, campfires. "Fighting fires is dangerous work and requires manpower greatly needed for important Project work," the paper chided.

Then one evening their worst fears were realized. The siren howled, bringing them all running from their dinner tables. Eleanor Jette remembered looking out the window and seeing "the orange glow of flames reflected in the night sky":

> C-Shop, our main machine shop, was ablaze. It was just inside the Tech Area fence. Frantic MPs struggled to keep the fence and road clear of the townspeople and their children. The buildings inside the Tech Area had brick firewalls. The fire escape from the administration building opened onto its firewall and overlooked the C-Shop. The Commanding Officer, the Director of the Laboratory, the Tech Board and its alternates, which included Eric [Jette], watched the fire fighters from the fire escape. The flickering light of the flames illuminated their grim, set faces. The faces reflected fear for the work program and concern lest the fire spread and wipe out the entire town.

White-faced, the onlookers watched as the firefighters finally doused the flames, using vast quantities of their dwindling water. By the time the shop was reduced to smoldering ruins, there was only an hour's supply of water left in the storage reservoir. Someone in the crowd asked if it was sabotage. Another worried out loud what would have happened if it had been the D building, which was already "hot" from all the radiation. No one answered. It terrified them to think how much damage could have been done and how many months it could have delayed their work—delayed the end of the war. They were playing with fire, literally and figuratively, and the dangers were too great to contemplate.

When the announcement crackled over the Tech Area loudspeaker that President Roosevelt had died suddenly on April 12, the scientists rushed into the hallways. The news, Serber recalled, struck them "like a blow." Grief and shock drained the color from their faces. Everything seemed lost. Oppenheimer emerged from his office and spoke briefly to those assembled on the steps outside, his rousing, impromptu tribute rallying those bordering on despair. Word reached Dorothy later that day. She was on the Hill, caring for Kevin, who was in the post hospital with pneumonia. When she sunk heavily down on the bed to tell him the news, there were tears in her eyes.

Oppenheimer organized a memorial service for the following Sunday, and everyone went, crowding into the cold theater. A night's worth of snow had hushed the normally busy mesa and veiled the town in white. "It was no costume for mourning," Phil Morrison recalled, "but it seemed recognition of something we needed, a gesture of consolation." Oppie stood before the lowered American flag, looking pale and gaunt, and strangely naked without his beloved hat. Aware as always of the community's precarious morale, he spoke movingly and eloquently of their fallen leader in a voice so low many found themselves standing on their toes, straining to catch every word:

> When, three days ago, the world had word of the death of President Roosevelt, many wept who are unaccustomed to tears, many men and women, little enough accustomed to prayer, prayed to God. Many of us looked with deep trouble to the future; many of us felt less certain that our works would be to a good end; all of us were reminded of how precious a thing human greatness is.

> We have been living through years of great evil, and of great terror. Roosevelt has been our President, our Commander-in-Chief and, in an old and unperverted sense, our leader. All over the world men have looked to him for guidance, and have seen symbolized in him their hope that the evils of this time would not be repeated; that the terrible sacrifices which have been made, and those that still have to be made, would lead to a world more fit for human habitation. It is in such times of evil that men recognize their helplessness and their profound dependence. One is reminded of medieval days, when the death of a good and wise and just king plunged his country into despair, and mourning.
>
> In the Hindu scripture, in the Bhagavad Gita, it says, "Man is a creature whose substance is faith. What his faith is, he is." The faith of Roosevelt is one that is shared by millions of men and women in every country of the world. For this reason it is possible to maintain the hope, for this reason it is right that we should dedicate ourselves to the hope, that his good works will not have ended with his death.

Afterward, the scientists and their families stood there bowed and silent, too saddened to speak. It was a terrible loss, and a very personal one, for they had thought of themselves as working directly for the president and had pinned their hopes on him. Los Alamos had been Roosevelt's clandestine project, and his successor, Harry S. Truman, knew nothing of the bomb—or of their existence. It was terrifying to think that the government's new leader, a man about whom they knew so little, was only then being informed of the massive Manhattan Project, their secret city in the wilds of New Mexico, and the $2 billion that had been wagered on the development of an experimental new weapon without the consent of Congress, or the American people. High on their mountaintop, they felt more alone than ever.

SIXTEEN

A Dirty Trick

ON MAY 2, 1945, barely three weeks after FDR's death, Berlin surrendered. On May 4, Oppenheimer circulated a memo from Washington. The War Department sent its congratulations to the scientists, but cautioned them against thinking that this development meant their services were no longer required. There would be no break in the all-out drive to complete the bomb.

RESTRICTED

May 4, 1945

FROM: J. R. Oppenheimer
TO: All Project Employees

The following message is from the Honorable Robert P. Patterson, Under Secretary of War, to all project employees:

"I want to congratulate you on the vital war contribution you have made in developing this project. I thank you on behalf of the Army for the great work you have done here and the sacrifices many of you have made in coming to work here.

"The work you are doing is of tremendous importance and must go forward with all possible speed. At the same time, it must be kept secret from the enemy.

"The army cannot give you and your work the public recognition it deserves. Nevertheless, I want to tell you of the importance of the work, and that is why I am giving you this message.

"The importance of this project will not pass away with the collapse of Germany. We still have the war against Japan to win. The work you are doing must continue without interruption or delay, and it must continue to be a secret.

"We still have a hard task ahead. Every worker employed on this project is needed! Every man-hour of work will help smash Japan and bring our fighting boys home.

"You know the kind of war we are up against in the Pacific. Pearl Harbor—Bataan—Corregidor—Tarawa—Iwo Jima—and other bloody battles will never be forgotten.

"We have begun to repay the Japanese for their brutalities and their mass murders of helpless civilians and prisoners of war. We will not quit until they are completely crushed. You have an important part to play in their defeat. [There] must not be a let-up!"

R E S T R I C T E D

May 8 was V-E Day, but after so many tumultuous months it was something of a letdown at Los Alamos. In Italy, Mussolini had been shot and his body brought to Milan for public display. He was strung up in the main square in the same place where partisan bodies had been left as a warning to members of the resistance in 1944. Thousands crowded around to spit on and kick his corpse. The Los Alamos inhabitants read of Hitler's suicide in an underground bunker, and of Germany's surrender, while still working doggedly to set up the preliminary shot in the desert. In the days that followed, the papers were full of monstrous accounts of Hitler's butchery and of the six million Jewish men, women, and children who were systematically hunted down, murdered, gassed, tortured, and condemned to concentration camps as he tried to fulfill his twin ambitions: the establishment of a "master race" and the total domination of Europe. Safe in their sky-high fortress, the scientists could not help the sick feeling that they were too late. They had all been told, had absolutely believed, the atomic bomb would be needed to bring the Nazis to their knees. But that had happened without them. "For me, Hitler was the personification of evil and the primary justification for the atomic bomb work," recalled Segrè. "Now that the bomb could not be used against the Nazis, doubts arose. Those doubts, even if they do not appear in official reports, were discussed in many private conversations."

It was a dirty trick of war that the end of the fighting in Europe in no way slowed the pace of the work on the Hill, which had taken on a momentum of its own and was nearing its climax. The atomic bomb, once a hypothetical, had become a necessity, and all indications were that Japan was now the inevitable target. Events were moving so quickly that the scientists were powerless to turn back. Most of them alternated between feeling elated and deeply apprehensive. They could not keep their thoughts from turning to the life and work they had left behind, and all the family and friends that awaited them outside the wire in civilized society. At the same time, their relief that the war was coming to a close was tempered by the alarming estimates of casualties that would be incurred in the coming invasion of Japan. The worst was far from over in the Pacific, and the radio broadcast dire predictions of what lay in store in the island battles for the Philippines and Okinawa. They could not share their doubts and misgivings with their wives, but the scientists realized with growing certainty that before they were released from their semicaptivity on the mesa, the terrifyingly powerful weapon they had been working on for the past two years would be unleashed on the world.

Ironically, the full-dress rehearsal shot at Trinity, which had been planned for May 5, took place on May 7, 1945, the last day of the European war. Hundreds of crates containing TNT were gingerly stacked on the platform of a twenty-foot wooden tower. Kistiakowsky had procured the special, fast-acting explosives, and had had the foresight to have them packaged so they could withstand considerable manhandling. This proved fortuitous when several crates fell off one of the panel trucks transporting them from the depot and several others were knocked off the elevator while being loaded onto the platform. To simulate as closely as possible the radioactive effects of the implosion bomb, 1,000 tubes of dissolved reactor fuel from Hanford were interspersed in the stack of crates. When detonated, the 100 tons of TNT produced an orange fireball that could be seen for sixty miles. The physicists, who watched the dawn sky light up from 10,000 yards away from ground zero, took grim satisfaction in the success of their experiment. Their optical cameras and acoustical gauges had duly recorded one of the largest explosions in history. But the celebration at base camp that night was muted. They knew the blast they had just witnessed would be dwarfed by what was to come.

After the preliminary test, a new urgency took hold. The run-through had revealed many shortcomings in their equipment and turned up flaws in the test operations and organization of the test site. Trinity base camp, rushed into construction during the winter of 1944, had become a satellite city to Los Alamos, complete with scientific buildings, barracks, and an elaborate communications system requiring hundreds of miles of wire running along an extensive network of new roads. But the hasty construction showed, and the large group of physicists who had packed into GI sedans for the field test returned to Los Alamos full of complaints. The test site was understaffed, the shelters were dodgy, and the existing communications were woefully inadequate and had broken down repeatedly. Even worse was the state of the roads. The crumbling surface was so slippery that their vehicles careened all over and constantly got mired in pits of sand. The clouds of dust got into everything, and some of the physicists worried their delicate equipment would never survive the jarring 230-mile journey.

The living conditions at the base camp were almost intolerable. The desert was broiling hot, with temperatures reaching above 100 degrees Fahrenheit, and even though they stripped down to shorts and shoes, they had to retreat from the brutal midday sun. The hard water and alkaline dust clogged their nostrils and lungs, and dehydration and dysentery were rampant. Work began at 6:00 A.M. and often continued after dinner and well into the night, when it was cooler. Days that stretched for ten to eighteen hours were not unusual, but bone tired as they were, when they finally collapsed in their bunks, it was hard to sleep, especially when bedrolls had to be carefully combed for tarantulas and scorpions before being used. It did not help that on two separate nights B-29s from the Alamogordo Air Base mistakenly took their camp as a lighted target for their night exercises and nearly obliterated the site. The airmen apparently needed the practice because their one-hundred-pound bombs, carrying five-pound black powder flash units, fell wide, and all they succeeded in blowing up was the Trinity carpentry shop and stables. The barracks, where all the scientists and soldiers slept, survived unscathed. The truth was no one wanted to stay in the "dread Jornada" a moment longer than necessary.

Despite the hardships, Oppenheimer kept the Trinity crew to a tight schedule, running them through drills designed to weed out as many technical problems as possible before the actual test. Groves approved the installation of additional phone lines and a public address system for the shelters, and agreed to blacktop twenty-five miles of road at a cost of $125,000. Laboratory personnel shuttled back and forth between Los Alamos and Trinity in buses and cars. Truckloads of equipment took off for the remote southern site, convoys of two to ten trucks departing every evening after dark to avoid detection and the broiling desert sun. An entire detachment of MPs was transferred from Los Alamos to Trinity to guard the base camp during the ongoing preparations. Top army officials came through for inspection tours. In late May, Brigadier General Thomas F. Farrell, deputy military commander of the Manhattan Project, and Richard Tolman, one of Groves' chief scientific advisors, paid the test site a visit and, as Baindridge put it, treated him to "a friendly between-the-halves fight talk." Their agenda, probably at Groves' suggestion, was to see how much time Oppie was spending at Trinity and to see if his presence there could be curtailed. "Essentially, they ordered me to keep Robert Oppenheimer away from the tower and the bomb before the final test for his own safety, and not let him know I was trying to do it," recalled Bainbridge. "No way! The bomb was Robert's baby and he would and did follow every detail of its development."

The next day, Jumbo arrived at Trinity. Just as Groves had predicted, the huge steel bottle turned out to be a monumentally expensive headache from beginning to end. At first, the feasibility of designing the massive container had seemed doubtful. Then, even after the Los Alamos physicists and engineers had satisfied themselves that it was possible, virtually every steel company they consulted had expressed serious doubts. After much haggling back and forth, the Babcock and Wilcox Company was finally persuaded to take up the challenge, and Jumbo was commissioned at enormous expense.

The cost was further compounded by the fact that the 25-foot-long, 214-ton vessel had to be carefully transported over railroads in specially reinforced cars to New Mexico and then transferred to a custom-built, sixty-four-wheel flatbed trailer and towed by tractor thirty miles overland to the test site. But by then, more plutonium was being delivered, and the tremendous steel container was no longer in favor. Oppenheimer and his division leaders were feeling much more positive about implosion, and there was a substantial lobby opposed to using Jumbo at all, on the grounds that it would throw off all their measurements and possibly even create additional dangers. One concern was that if the heat from the explosion did not melt the steel casing, it might send pieces of jagged steel hurtling great distances. Jumbo was sidelined, and the $500,000 albatross was set up a half mile away from ground zero, where it remained unused.

All that spring, as they hurried to complete their preparations for the test, a steady stream of top advisors came to see Oppenheimer, and the Los Alamos scientists were keenly aware that the high-level deliberations concerned the use and consequences of the atomic bomb. Following the Yalta conference, there had been a debate about whether it would be better to encircle Japan and isolate it or to defeat it by direct attack. Both General Douglas MacArthur and Admiral Chester Nimitz voted for direct assault, and with their support, the Joint Chiefs of Staff approved the Kyushu invasion plan in April. The target date was set for November 1, 1945. "It was estimated that a force of 36 divisions—1,532,000 men in all—would be required for the final assault, and it was recognized that casualties would be heavy," Groves reported in his memoir. "In such a climate, no one who held a position of responsibility in the Manhattan Project could doubt that we were trying to perfect a weapon that, however repugnant to us as human beings, could nonetheless save untold numbers of American lives."*

In Washington, Secretary of War Henry Stimson had briefed the new president on the issues created by the development of the bomb, not just as a weapon in the United States' arsenal, but as a powerful new force in the world. But Stimson was deeply concerned about the international reaction that would result from the bomb's being tested and used without any advance notice. Niels Bohr had been warning the American government that diplomacy with the Soviet Union was vital and that any sudden use of the bomb would inevitably trigger an arms race. Conant and Vannevar Bush, the head of the OSRD, who were sympathetic to his views, had been lobbying for greater consideration of the issues at stake.

On May 2, Stimson received approval to set up the Interim Committee to advise Truman on the postwar implications of the bomb. It was chaired by Stimson, who, at seventy-seven, was in some ways the wisest and most respected member of the administration and was uniquely qualified to undertake such a grave responsibility. He had been secretary of war under two administrations and secretary of state under Herbert Hoover, and had served as a colonel in World War I. He had also overseen the bomb project from the very beginning, had backed the appointment of Groves, and was one of the few who was knowledgeable about both the military situation and the global ramifications of using atomic weapons. The committee included General George C. Marshall; Ralph A. Bard, undersecretary of the navy; William L. Clayton, assistant secretary of state; George Harrison, a special advisor to Stimson; the incoming

*The question of how heavy the sacrifice would be remains murky: Secretary of War Stimson asserted the November invasion would cost between a half million and a million Allied casualties, and at least as many Japanese lives. Army Chief of Staff Marshall reportedly put the figure closer to forty thousand. After the war, these casualty estimates became controversial and subject to a great deal of second guessing. Stimson was accused of inflating the figures by those who believed there was already ample evidence Japan had been pushed to the brink and could have been made to surrender without the bomb.

secretary of state, James Francis Byrnes; and three key scientific advisors—Karl Compton, Vannevar Bush, and James Conant.

Conant, reluctant to make bomb policy, asked to be excused from the Interim Committee. He was already aware, he reported to Stimson, of "a growing restlessness among scientists actively involved in the program." Many of the scientists who had joined the Manhattan Project, particularly those who had fled Nazi persecution, had done so because of their fear of a German head start on atomic weapons and because they viewed Hitler's regime as the supreme enemy. They did not have the same level of conviction about destroying the Japanese. He made no reference to what he felt sure was also their "lack of confidence in those of us who had been determining policy." Conant feared that as one of the chief administrators of the Manhattan Project, he would be held responsible for the dangers posed by the bomb in the postwar world. He suspected that in helping to craft the vast covert undertaking, and in preserving the extreme secrecy that covered every phase of the research that resulted in hundreds of scientists being kept in the dark, he would ultimately be the focus of considerable suspicion and hostility. He had "serious doubt," he wrote Stimson, that he and Bush were the proper men to serve on such a committee—"for we have been primarily distant administrators rather than active participants."

Conant and Bush had in fact given considerable thought to the international problems that would arise from the use of the atomic bomb and had spent the past year trying to convince Stimson and Roosevelt of the necessity of preparing plans for the postwar era. During an inspection trip to Los Alamos in the summer of 1944, Conant and Bush had taken advantage of the opportunity "to discuss at leisure and in complete privacy what the policy of the United States should be after the war was over." On the basis of that exchange and subsequent conversations, Bush had drafted a memorandum to Stimson in September 1944, which he asked Conant to sign. Conant recalled in his memoir:

> This was the first of several papers we sent to the Secretary, signed by both of us, in which we pointed out totally new and alarming situations which would result if no U.S. policy was developed before the war ended and the knowledge of the exis-

> tence of the bomb was made public. We advocated free exchange of scientific information with other nations, including Russia, under arrangements by which the staff of an international office would have unimpeded access to scientific laboratories, industrial plants and military establishments throughout the world.

Stimson knew Conant did not want to be drawn into the debate, but prevailed on him to serve on the Interim Committee. Conant finally agreed to do so, but only after insisting that some of the more "active participants" in the project also had a seat at the table. As a result, on May 14, at his urging, a four-man Scientific Panel, consisting of Oppenheimer, Fermi, Lawrence, and Met Lab head Arthur Compton, was appointed to provide advice. At the end of the month, Oppenheimer and Fermi traveled to Washington for the top-secret meetings, which were held over two days, on May 31 and June 1. During the first morning session, Stimson dictated the agenda and led a rather lofty discussion that centered on the future of the Manhattan Project, atomic energy, and the need to solidify America's role as the leader of this unprecedented endeavor, and its potential benefits for mankind. It was brave-new-world talk, and Stimson held the floor as he warmed to his theme, making it clear the administration already had a stake in the postwar advantages of the soon-to-be-tested weapon.

That afternoon, the committee took up the much thornier issue of how to use the bomb against Japan. The question of whether or not the bomb should be used, which, as Conant later observed, "was the most important matter on which an opinion was to be recorded," received scant consideration. The discussion dealt with the advisability of using it without warning as opposed to conducting a harmless demonstration of its destructive power and then delivering the Japanese an ultimatum. It was primarily a debate over tactics, and with his forceful presence, cool analytic mind, and extraordinary powers of elucidation, it was Oppenheimer's moment to shine. Perhaps he knew he had been brought in for the day to play the role of atomic expert, or perhaps by then he had reconciled himself to the position advocated by Stimson and Groves, that the bomb was built as a weapon of war and should be used at the earliest date to quickly end the fighting. Certainly, the vision of Fat Man's ferocity that he articulated for the committee bolstered their case: a single bomb, he told them, would wipe out a city in a flash, a "brilliant luminescence that would rise to the height of 10,000 to 20,000 feet."

During lunch, the scientists "threshed over" the alternatives to direct military use. Lawrence favored a prior demonstration, but Oppenheimer could not conceive of a technical demonstration that would be spectacular enough to actually induce the Japanese to surrender. He cut through the morass of objections to dropping the bomb and exposed the heart of the problem: any harmless demonstration would result in losing the overwhelming shock effect of surprise. All the security measures surrounding the project had been designed to preserve that element of surprise. The debate tested Oppenheimer's novice political skills, and eager to prove himself in this new arena, he did much of the talking for the panel. "He told us the uranium bomb couldn't be tested, because material was being supplied too slowly," recalled Ralph Bard. "He said the plutonium bomb might be a dud, and would have to be tested, but that even after that he couldn't guarantee the force of the explosion of the next one. He didn't say drop the bomb or don't drop it. He just tried to do his job, which was to give us the technical background."

The official minutes of the Interim Committee recorded the fateful outcome of the May 31 meeting: "After much discussion, the Secretary [Stimson] expressed the conclusion, on which there was general agreement, that we could not give the Japanese any warning; that we could not concentrate on a civilian area; but that we should make a profound psychological impression on as many of the inhabitants as possible. At the suggestion of Dr. Conant, the Secretary agreed that the most desirable target would be a vital war plant employing a large number of workers and closely surrounded by worker's houses."

Afterward, the members of the Scientific Panel agreed they would meet again in mid-June to finalize their recommendations to the Interim Committee. In the meantime, the foursome were told they could inform their laboratory personnel about the committee's work in dealing with the future control of atomic weapons, though without identifying any individual members, and should generally impress upon their colleagues that the government was taking an active role in developing policy. But on his return to Los Alamos, Oppenheimer found the scientists on the Hill obsessed with the outcome of their wartime labors. At the Met Lab, Arthur Compton contended with far greater skepticism and emotional turmoil. Leo Szilard, who had earlier convinced Einstein to alert Roosevelt to the necessity of starting an American bomb program, was once again trying to precipitate American policy and had been agitating the Chicago scientists to oppose combat use of the bomb on "moral grounds" and to strongly push for a public demonstration before killing a multitude of Japanese. Szilard's protest echoed some of Bohr's arguments, and it stirred up old doubts and anxieties and resulted in heated discussions about the need for more responsibility on the part of the scientists in making recommendations for the weapon. In an effort to quell the dissension, Compton dispatched a series of committees to study and report on the implications of the atomic bomb.

On June 21, the Scientific Panel met at Los Alamos and, as promised, took time to consider the divergent points of view. The most important of these was written by members of the Met Lab, including Szilard, and chaired by James Franck, Oppenheimer's old Göttingen professor. The Franck Report stated that a surprise atomic assault on Japan would destroy America's credibility and precipitate an arms race. It urged that a demonstration take place in the desert or on an uninhabited island, and thereby end the war without any further bloodshed. While Stimson and Groves used bureaucratic channels to effectively block Szilard's protest from reaching Truman's ears, Franck was too well liked and respected to ignore, and his document, signed by seven physicists, and carrying a cover letter by Karl Compton, could not be swept under the rug. Instead, they passed the political hot potato to Oppenheimer, who had distinguished himself as the leader of the scientific panel and was already committed to the position that Fat Man might fizzle and that a bloodless demonstration could not be risked.

At Los Alamos, Serber recounted, Oppenheimer presented the problems facing the Interim Committee, the plans for the fall invasion, and the fact that the medical units of the armed services had been told to prepare for half a million casualties. "Given this background, we had no doubts about the necessity of using the bomb," he recalled. "We spoke of it as a 'psychological weapon,' and were sure dropping a bomb on a Japanese city would end the war." Both Oppenheimer and Fermi argued that physicists "had no claim to special competence in solving political, social or military questions which are presented by the advent of atomic power," as they wrote in their final report. In the end, the panel came to the conclusion Stimson and Groves had been counting on: "We can propose no technical demonstration likely to bring an end to the war; we see no acceptable alternatives to direct military use."

There were, however, still a number of physicists at Los Alamos who were troubled by their consciences. Bob Wilson, always an instigator, decided to convene a meeting in his laboratory one evening in June to examine the "impact of the gadget" and the issues raised by Szilard's petition. Wilson had strong moral objections and felt that more discussion was warranted. Oppenheimer, who was consumed with preparations for the test, was not happy about the meeting and at first considered not attending, but when it became clear that a great many of the Tech Area personnel planned to go, he changed his mind. Most of the mesa physicists had long ago come to the conclusion that the bomb would be used, if only because Groves would not want his hugely expensive wartime effort to have been wasted. Priscilla Greene, who attended the meeting with her husband, Bob Duffield, recalled that many also believed the decision should be left to the president, even though they did not have much faith in him, and to experienced leaders like Stimson and Marshall, who understood the military situation. Phil Morrison joined the crowd that jammed the room, and while he sympathized with Wilson's instinct, he thought it was a dangerous attitude to take at the time. "I supported Oppie and opposed Wilson, because I knew it was inevitable," he said, "and I thought it was unwise of us to pretend to be owners of the bomb."

Oppenheimer stood silently to one side and let Wilson have his say. But at some point, he had clearly had enough. As laboratory director, his job was to see to it that nothing distracted his staff from completing their work on the bomb. While he personally may have shared many of Wilson's concerns about the use of the bomb, not to mention those of the Met Lab scientists, and had devoted many hours to the subject of postwar planning with Niels Bohr, he did not feel this was the time and place for such a discussion. There would be plenty of time to revive the topic when the war was over. He then delivered a brilliant impromptu speech spelling out his conviction that the weapon had to be made known to the world and that was the only way its potential destructiveness would ever be understood and ultimately controlled.

"Oppie totally dominated the meeting," said Phil Morrison. "He was the boss and had every right to do so. He succeeded in swaying people, but it wasn't so much to his point of view, as [it was] simply a question of what would happen." By the time Oppie was finished, Morrison added, there was no support for the petition: "We went away sheepishly."

By early July, Szilard, realizing he was running out of time, made a last attempt to enlist support for his cause by writing to Edward Teller at Los Alamos, begging him to circulate a petition to the president among his Tech Area colleagues. But Teller, who was still nursing his dream of the Super, and even bigger, more powerful bombs, lined up squarely with Oppenheimer: "The things we are working on are so terrible that no amount of protesting or fiddling with politics will save our souls," he wrote Szilard. "The accident that we worked out this dreadful thing should not give us the responsibility of having a voice in how it is to be used. This responsibility must in the end be shifted to the people as a whole and that can only be done by making the facts known." Later, Teller would complain that he had been perfectly willing to support Szilard, but Oppenhiemer talked him out of it, telling him "in a polite but convincing way that he thought it was improper for a scientist to use his prestige as a platform for political pronouncements." If Teller is to be believed, it would have been the first time he was inclined to do as Oppenheimer asked since arriving on the Hill.

As the Trinity test neared, Dorothy, in her small office in town, felt the tremendous jump in activity. Although she was never officially told of the date, the term "Trinity" was in the air, and she knew the decisive moment was near. Security had imposed super-secrecy measures, and G-2 was swarming all over town. Trucks, loaded with tons of batteries, cables, and instruments, barreled through town—straight through the main street of Santa Fe—without stopping on their way to the classified test site in the south. The telephone was boiling over, but the usual friendly banter with men in the Tech Area had been replaced by barked orders, and the voices "showed strain and tautness." A steady stream of project employees, from generals and GIs to technicians and engineers, shuttled back and forth between the Hill and the distant test site. As many as seventy new people a day were checking in, including high army brass and War Department officials. There were more hotel rooms to be arranged for and very few to be found. "I in my backroom felt the tension," she recalled. "I just felt it in my bones."

A *New York Times* reporter by the name of William L. Laurence caused a stir when he arrived with orders from Groves to be taken directly up to the site. At first, no one could believe it. As far as Dorothy knew, he was the first and only reporter ever invited up to the classified weapons installation. After a careful double-check of his credentials, he was cleared. But his brief tour of the laboratory was the talk of the Hill. A number of air force personnel started showing up, and the word was they were training at a secret air base in Wendover, Utah. It did not require much imagination to conclude they were part of the select crew who would be in charge of dropping the bomb on Japan and finally finishing the war. American troops in Europe were being redeployed, and several of Dorothy's friends on the Hill had received letters informing them their loved one's unit was on the move. She only hoped the super weapon the scientists were working on would be ready before they had to begin the assault on Japan. A few young soldiers, proudly sporting battle ribbons from Anzio, had arrived at Los Alamos, and she could only imagine what they had been through by how grateful they looked to be there.

In late May, an earnest-looking young man rushed into her office at the start of the lunch hour, just as the offices on the Hill were closing for an hour. He explained he was a lieutenant colonel in the air force, and had the papers to prove it, and insisted he was late for an important meeting "up there." Dorothy had not been notified of his arrival and explained that regulations dictated that she had to detain him until she could confirm his identity with the director's office. But he appeared very presentable and honest, and there was something in the urgency of his pleas that made her reconsider. "I didn't want to hold him up," she said. "He looked like a person for whom time was very valuable." She studied him carefully one more time and then decided, as she put it, "to shoot my whole future to the winds with one wild and unprecedented action." In a firm, unshaking hand, she wrote out a pass for Colonel Paul W. Tibbets. It was the one and only time she ever issued a pass without previous authorization. She did not know then that he had been assigned to pilot the *Enola Gay,* which would drop the first atomic bomb. She only hoped the tardy flier with the lovely smile would not get her into too much trouble.

Tibbets was on his way to a meeting of the Target Committee in Oppenheimer's office to discuss combat employment of the bomb. A number of the top laboratory scientists were present, including Bethe, Penney, von Neumann, and Wilson. Deke Parsons was also there, of course, as Los Alamos's only navy captain, and would be flying on the bombing mission. The main purpose of the meetings, first held on May 10 and 11 at Los Alamos, was to select the target cities and review the status of the conventional bombing of Japan. Of the initial seventeen targets selected by Stimson's War Department staff, the number was winnowed down, partly because some of the cities had already been destroyed by fire bombing. Priority was to be given to cities that were still untouched, so as to provide incontrovertible proof of the bomb's devastating power—and to provoke the maximum psychological effect—to induce Japan's prompt surrender. By the meeting on May 28 that Tibbets attended, the list had been shortened to Kyoto, Hiroshima, and Niigata.

The Target Committee members, including air corps officers and Manhattan Project consultants, went over the combat employment of the bomb, the physics of the explosion, and the proper burst height. Norman Ramsey explained that the bomb would probably explode with the force of 20,000 tons of TNT. "Even though it was still theory," Tibbets recalled, "I wanted to ask Oppenheimer how to get away from the bomb after we dropped it. I told him that when we dropped bombs in Europe and North Africa, we'd flown straight ahead after dropping them—which is also the trajectory of the bomb." Tibbets wanted to know exactly what he should do the moment Little Boy dropped out of the bomb bay. Oppenheimer's reply made working out the flight maneuvers sound like a simple mathematical problem. "You can't fly straight ahead because you'd be right over the top when it blows up and nobody would ever know you were there," the slender, blue-eyed physicist told him. "Turn either way 159 degrees. You will then be tangent to it. That way you will get your greatest distance in the shortest length of time from the point at which the bomb explodes."

In June, the tempo of activity at 109 East Palace jumped markedly. Rumor had it that the army was finally going to cut a new road to Los Alamos. Apparently, some generals were treated to a particularly bumpy ride down the switchbacks—rumor had it a disgruntled soldier stuck wood in the jeep's springs—and demanded something be done about it. But Dorothy could tell that wherever the convoys were going, the roads had to be even worse than the deeply rutted washboard connecting Santa Fe to the Hill. She might as well have been in the repair business, judging by the number of urgent requests for automotive parts that came across her desk. Drivers came in swearing under their breath, demanding special replacement parts that needed to be installed immediately in dust-caked jeeps outside. Or she would get a call telling her to get a new battery or tire to a broken-down truck outside of town. The warm weather was making the going rougher than usual. Spring had arrived with an unexpected vengeance that year, and the sun had baked the arroyos dry, the hardened creek beds cratered and barely passable in places.

Complicating matters, the stringent new security measures forbade all recreational trips to town. The truck drivers and Trinity staff were barred from making any pit stops in the little junctions, and Dana Mitchell, assistant director of the laboratory, had issued a stern travel advisory: "Under no condition, when you are south of Albuquerque, are you to disclose that you are in any way connected with Santa Fe. If you are stopped for any reason and you have to give out information, state that you are employed by the Engineers in Albuquerque. Under no circumstances are telephone calls or stops for gasoline to be made between Albuquerque and your destination." The confidential memo further instructed them to "stop for meals at Roys in Belen," though more than one parched driver ignored the regulations and pulled into Meira's bar and service station in the one-horse town of San Antonio to refuel.

The atmosphere on the Hill tightened with each passing week to the point where it was almost unbearable. Dry electric storms swept across the mesa and lit up the sky, but still no rain came. Rumors swirled. The test date, once scheduled for Independence Day, was postponed, and there was talk the deadline was now mid-July. Some of the scientists and medical staff were being inoculated for tropical diseases and would be leaving soon for the Pacific. This was supposed to be a secret, but some of their wives were wild with worry and could not help confiding in one another. People shook their heads over the fact that Jim Nolan and Henry Barnett, the post's obstetrician and pediatrician, were both being sent overseas, while Louis Hempelmann, their trained radiologist, was not. Apparently, Hempelmann was the only member of the medical staff who was not in the army, and he was fit to be tied at being excluded. Everyone's nerves were on edge. Sam Allison's wife, Helen, took to pestering Dorothy daily to check the local jewelers to see if they had completed the repair job on her husband's watch. The woman seemed positively frantic on the phone, and Dorothy, with all she had to do, could not imagine what could be so important about the old timepiece. She later heard Allison had just been ordered to Trinity to conduct the official countdown of the test and had hoped to have his own watch for good luck.

A meteorologist named Jack Hubbard had joined the staff and was heading a team that was closely monitoring the weather conditions for the days surrounding the test. Clear weather was vital to the experiment and to their ability to get accurate measurements from their instruments. Rain, either before or during the test, could damage the electrical circuits and operating equipment, interfering with the firing of the bomb and wreaking havoc on their instruments. There was also the potential problem of fallout. One of the main reasons they had decided to explode the bomb from the top of a one-hundred-foot tower was that a ground detonation, in addition to not really revealing what it could do as a weapon, would create a tremendous amount of fallout at a low elevation. There was a possibility that strong winds in the wrong direction could carry the poisonous radioactive cloud over inhabited areas, most notable Amarillo, which was about three hundred miles away. Very little was known about fallout, but they could not ignore the potential danger, and evacuation teams were being organized. Drawing on information from myriad sources, including the Army Air Forces weather stations at Alamogordo and Albuquerque, Hubbard finally pinpointed the middle of July as the best time to test the gadget.

On the last day of June, all the division leaders reported to Oppenheimer, and it was decided that July 16 was the earliest possible date they could be ready. Groves, however, was determined that Truman would be armed with knowledge of the test's outcome when he met with Stalin and Churchill at the Potsdam conference, which Stimson, dragging his feet all the way, had managed to delay until July 15. The first week in July, with the Potsdam deadline looming, Groves fixed the final test date for July 16. He was pushing up hard against Hubbard's long-range prediction for that weekend, which did not look promising, but that was a chance they would have to take. Oppenheimer instructed the Trinity team that orders from Washington were that as soon as the plutonium for the bomb was ready, the test would go forward. There must be no delays. As Bainbridge noted, "A successful test was a card which Truman had to have in his hand." Because of all the uncertainty, Groves took off for the Hanford site, taking Bush with him. They would meet up with Conant on the Pacific Coast. "This would enable us," Groves wrote, "to get to Alamogordo promptly if the date of the test was advanced."

In the days immediately preceding the test, dozens of high-ranking project consultants and Nobel laureates returned to Los Alamos: Richard Tolman, Ernest Lawrence, Isidor Rabi, Sir James Chadwick, and, making another appearance the day before the test, Bill Laurence, the sole member of the press assigned to document the event. Groves, warned not to invite too many observers, well exceeded his ration, and at the last minute Dorothy had to scramble to find sleeping quarters for an extra general. He had requested a room at La Fonda, but the hotel was overbooked. "Not only did he have to settle for a second choice billet at the De Vargas hotel," she recalled, "he had to share a double room with a sergeant." She told him what she told everybody, "You know, there's a war on!" Memos came down from the Hill daily with eight or nine new names to expect, and calls came from Washington saying, "We have the following coming in. . . ." There were so many people coming and going that a number of G-2 agents worked in the office helping her check papers and issue security passes.

One afternoon, Dana Mitchell, who had worked at 109 with Dorothy in the early days of the project, stopped by to make an important call. Dorothy politely stepped into the other room to give him some privacy, but he did not seem to notice. She overheard him tell someone in a voice that was louder and more strident than usual that they had "ambulances ready in Albuquerque" in case they were needed. Dorothy knew then that things were moving very fast. "Time, time, time. Speed, rush. Care, care, and worry," she wrote. "Anxiety and work. All hours of the day and night, not sleeping, not eating regularly, losing weight. Always tension, excitement, pride, rising in the great crescendo of the test at Trinity."

SEVENTEEN

Everything Was Different

IN JULY, THE RAINS CAME. At midday, bright blue skies would suddenly darken as black clouds amassed overhead, lightning streaked across the sky, and thunder cracked with frightening violence. Brief, soaking downpours followed. Everyone on the Hill had become obsessed with the unpredictable midsummer weather patterns. Physicists with no particular expertise in meteorology would scan the cloudless horizon with furrowed brows as if they could divine signs of trouble. Even men who were usually careful not to talk about their work told their wives to pray for a good forecast.

Doubt and pessimism blew into town with the thunderheads. There was a prevailing skepticism in the air, as if the scientists could not believe the witching hour had arrived and that the bomb's fearful power would prove all their experiments and calculations correct and finally put an end to their long endeavor. Instead, they distrusted their own handiwork and took refuge in the idea that the Trinity test's many uncertainties would probably result in a fizzle. Their lack of faith was never clearer than when some of the physicists organized an informal betting pool to see who could most accurately predict the explosive yield of the bomb. Their own blackboard estimate put the gadget's potential at roughly 20,000 tons, or

20 kilotons, of TNT, but the pool ran from zero to 45,000 tons. Rabi bet 18,000 tons. Bethe guessed 8,000 tons. Kistiakowsky thought the figure would be closer to 1,400. Oppenheimer conservatively settled on 300. But his was not the lowest bet: Johnny Williams figured on 200 pounds, and more than a few pessimists thought it would be zero. It was Teller, who had done the least direct work on the bomb and had the least at stake, who unhesitatingly went for the biggest bang—45,000 tons.

They were all so consumed with their own worries that the news that Feynman's young wife, Arline, had passed away put them all to shame. Richard had borrowed Fuchs' car and managed to get to Albuquerque in time to be at her side and say good-bye. Dorothy heard that he was in such a rush to get to the hospital that he got not one but two flat tires and ended up hitchhiking the last thirty miles. Distraught as he was, Feynman returned to his Tech Area office the next day and told people he intended to bury his sorrow in work.

"That last week in many ways dragged," recalled Elsie McMillan, and "in many ways it flew on wings. It was hard to behave normally; it was hard not to think; it was hard not to let off steam. We also found it hard not to overindulge in all natural activities of life." She had long ago guessed that the Trinity test was for an atomic bomb and now asked her husband, "in all innocence," what would happen. She needed to know. Not knowing was worse, and she was afraid she was transferring her mounting fears to her newborn son, as she rocked him to sleep in her arms.

Slowly, and with some difficulty, Ed McMillan told her what he thought she could expect. "There will be about fifty of us present, the key workers," he explained.

> We ourselves are not absolutely certain what will happen. In spite of calculations we are going into the unknown. We know that there are three possibilities: One, that we will be blown to bits if it is more powerful than we expect. If this happens you and the world will be immediately told. Two, it may be a complete dud. If this happens, when I return home I will tell you. Third, it may as we hope, be a success, we pray without loss of any lives. In this case, there will be a broadcast to the world with a plausible expla-

nation for the noise and the tremendous flash of light which will appear in the sky. Next week we will quietly and separately leave the mesa starting around 3:00 A.M., the cars to reconvene at the test site. In all probability the zero hour will be about 5:00 A.M. on the morning of the next day. If all goes well, I will be home sometime in the early evening of that day.

At the end of the day on Wednesday, July 11, Oppenheimer gave some final instructions to his secretary, Anne Wilson, tucked an extra carton of cigarettes under his arm, and took off for Trinity. "I thought I was queen for a day because he left me in charge of the whole place," she said. "I thought he was mad because I was all of 21. Everyone who was anyone was going to the test. They piled into buses and left in droves, and I would go out into the street in front of the Tech Area and wave goodbye to them."

On Thursday, July 12, the explosive casing for the test bomb was finished. For safety reasons, the nuclear and non-nuclear parts of the bomb would be moved separately and then assembled at the Trinity site. Just "to be whimsical," Kistiakowsky decided to transport the finished bomb assembly from Los Alamos to Trinity on Friday the thirteenth, hoping such bravado would reverse the date's traditional bad luck. They took off at ten minutes past midnight, a whole convoy of trucks, including the one carrying the gadget, a big spherical aluminum ball, carefully tethered in place and covered by a concealing tarp. Because of the number of scientists and soldiers in their party who were somewhat anxious all the shaking might cause the gadget to explode en route, Kistiakowsky jumped into the cab of the truck alone and took it for a quick spin over the rough roads. For security reasons, the bomb was escorted by an entire entourage of guards, with military police cars in front and back. Every time they came to a town, they blared their sirens and flashed their lights, reportedly in an effort to fend off any drunk drivers, but this defeated the whole point of their secret nighttime expedition by virtually announcing their presence to the sleepy inhabitants as they barreled through.

By contrast, Bob Bacher had arranged for the plutonium core of the bomb to have an extremely quiet trip to Trinity. His contingent left Los

Alamos at 3 P.M. that Thursday, winding their way down the mountain and through Espanola and Santa Fe and then on to Albuquerque unnoticed in an ordinary government sedan, a carload of MPs leading the way. Phil Morrison, who accompanied him, was also bringing the initiator. He recalled that they drove no more than thirty miles an hour and for the whole trip down were "apprehensive about an automobile crash or some catastrophe of the sort that might make it very difficult to run the test." They knew that only when the two hemispheres of plutonium were united could they achieve critical mass. Still, they took every conceivable precaution, packing the two halves of the nuclear core in specially designed cases, protecting them from the shock of impact, corrosion, overheating, overcooling—anything they could think of that might affect their precious cargo as it made its way across the desert and, when the time came, across the ocean to the Pacific.

Early on Friday, the two parts of the bomb were delivered to the temporary "clean room" set up in the abandoned McDonald ranch house, which had been vacuumed and whose windows had been sealed with black electrical tape against the pervasive dust. That morning, the final assembly phase began, with eight scientists outfitted in white surgical coats bending over a makeshift laboratory table where the plutonium pieces had been arranged on sanitized brown paper. The initiator, Hans Bethe's creation, was a spherical shell of beryllium containing polonium, which on implosion would mix and produce the triggering neutrons. Warm to the touch, it was very carefully placed inside the two halves of the plutonium core, the mating pieces of the globe fitting together like a particularly lethal puzzle. Oppenheimer hovered in the background next to Thomas Farrell, the boyish young brigadier general whom Groves had chosen to represent him at the assembly site because of his rule that both project leaders not be present in situations where there was "an element of danger." Waiting on the dirt road outside stood four jeeps, their motors idling, just in case the core accidentally went critical.

By afternoon, the core was mounted in a cylindrical plug of uranium and ready to go. The scientists gingerly placed it on a small litter and carried it out to the backseat of the same sedan they had driven to the desert. They eased the car into gear and rolled toward the one-hundred-foot tower. At ground zero, at the base of the tower, was a tentlike enclosure covering Kistiakowsky's partially assembled implosion device, a five-foot sphere complete save for the cylindrical plug containing the plutonium and the initiator. The plug was attached to a manually operated hoist and, while everyone held his breath, gently lowered into the bowel of the high-explosive shell. The first attempt to insert the plug failed, inducing a horrifying moment of panic. A few people stepped out of the tent to steady their nerves, while Oppenheimer, Kistiakowsky, and Bacher tried to calmly assess what had gone wrong. Bacher quickly surmised that the sweltering heat inside the farmhouse, combined with the car ride, had raised the temperature of the plutonium core and caused it to expand, whereas deep down in the shell it was still very cold. With prolonged contact, however, the temperatures would equalize, and the minutely calibrated plug would probably fit. A few minutes later, the plug finally slid into place. With that crisis averted, the explosive blocks were packed in without further complication, and the test device was buttoned up for the night.

Early Saturday morning at Trinity brought a singularly peculiar sight. The tent was removed, and the physicists prepared to raise the bomb assembly to the top of the tower using a huge winch, which Groves had procured for the astronomical price of $20,000. But the scientists, novices when it came to large-scale construction, had been so worried by the possibility that the cable would snap and the five-ton device would fall that they arranged to take some highly unusual precautions. Standing by all around the base of the tower were large army trucks containing hundreds of mattresses, the great mountains of striped cotton ticking visible above the metal siding. After they had hesitantly wound the device about fifteen feet in the air, a contingent of GIs rushed in and stacked nearly twelve feet of bedding on the ground directly underneath it. Just in case Fat Man fell, it would have something soft to land on. The gadget was then safely hoisted onto the tower, brought to the corrugated-steel shack on the top of the platform, where the detonators were inserted, and hooked up to the firing console.

With the finished bomb assembly poised on the tower, it looked like everything was going according to plan but the weather. A hot, sultry air mass had moved in, and the atmosphere thickened palpably. It began to drizzle lightly, promising the usual late afternoon thunderstorm. Watching the wind and lightning playing around the one-hundred-foot steel structure, it was hard not to imagine the tower as a giant lightning rod sticking out above the flat desert plain. The mood at base camp was strained. They were all keyed up in the extreme, and the waiting was making everything worse. While the instrument crews worked desperately to complete the last-minute checks on their vast array of equipment, senior scientists like Rabi, who had completed their part of the preparations, tried to relax by playing poker. Just when it seemed like the only threat to the test going forward as scheduled was the darkening sky, Oppenheimer received crushing news from Los Alamos. A coded telephone message reported that the dummy rig that was to be a dry run for Trinity, using an almost identical explosive assembly but without the plutonium core, had been tested in an empty canyon and had not worked properly. The experiment's failure virtually guaranteed the implosion method would not create a nuclear reaction. The bomb would be a dud.

It was a shocking setback. Oppenheimer called an emergency meeting, but it quickly deteriorated into a shouting match with Kistiakowsky identified as the chief villain. "After that a perfectly ghastly scene developed," recalled Kistiakowsky. "Oppenheimer, of course, who was responsible for the whole thing, being top dog at Los Alamos, was at the point of complete nervous exhaustion. You can't blame him. He was really emotional, essentially telling me I might be responsible for the total failure of the project and how terrible it was. And how could I have trusted all these young people who worked for me, and were probably incompetent, etc., etc." The fault finding continued, with Groves and Conant, who had by now arrived in Albuquerque, hauling Kistiakowsky over the carpet for what seemed like hours. "All of which was very unpleasant," added Kistiakowsky. "So I finally said, at one point, 'Look, Oppie, I bet you a month's salary of mine against ten dollars that my part of the bomb will work.' "

They spent a long, miserable night contemplating what the failure of the bomb at Trinity might mean. Bush recalled a gloomy dinner he shared with Oppenheimer in Albuquerque that night, after a meeting at the Hilton Hotel with a large number of VIPs and army generals that Groves had invited to observe the test. Oppenheimer had put on a brave face, but his growing despair was expressed in the stanza from the Bhagavad Gita that he quoted before turning in, though whether for inspiration or consolation, Bush could not be sure:

In battle, in forest, at the precipice in the mountains,
On the dark great sea, in the midst of javelins and arrows,
In sleep, in confusion, in the depths of shame,
The good deeds a man has done before defend him.

Early on Sunday, July 15, another phone call came, this one bringing a reprieve. Bethe had stayed up all night recalculating the results of the experiment and had discovered an error, rendering the dummy test results meaningless. Chances were good the gadget would work after all. Kistiakowsky was out of the doghouse and, as he drolly observed, "became again acceptable to local high society." Oppenheimer cheered up immediately and went off to find Hubbard and his meteorologists. By Saturday evening, Groves had arrived at base camp with Conant and Bush, and they had dinner with Oppie in the mess, followed by a general meeting covering such topics as how to avoid eye damage and the recommended evacuation routes in case fallout came their way. Conant took an interest in the betting pool on the size of the explosion, and privately figured on 4,400 tons but declined to sign up. "The atmosphere was a bit tense as might be expected," he noted, "but everyone felt confident the bomb would explode."

Groves was considerably less sanguine about the state of affairs. "The weather was distinctly unfavorable," he recalled in his memoir. "There was an air of excitement at the camp that I did not like, for this was a time when calm deliberation was most essential." Oppenheimer was "getting advice from all sides on what should and should not be done," and Groves did not like what he was hearing.

The test had originally been scheduled for 4:00 A.M. on July 16, when most of the surrounding population would be sound asleep and there would be the least number of witnesses. But the weather was interfering with their plans, and there was talk of a postponement. It had become increasingly misty and blustery, and it rained intermittently. Some of the scientists were afraid there might be "a reversal," that the winds would change direction and blow the radioactive debris over Trinity base camp and the outlying areas. Others expressed concern that the excessive moisture might have damaged the connections and increased the chances of short circuits, even a misfire. They were urging the test be postponed for at least twenty-four hours.

When he had heard enough from the doomsayers, Groves pulled Oppenheimer into an empty office where they could discuss the matter privately. As the night wore on, Groves had become increasingly anxious about security and was convinced they should carry out the test, even under less than ideal circumstances. He worried that "every hour of delay would increase the possibility of someone's attempting to sabotage the tests." While Oppenheimer and his top advisors had held up admirably, Groves also worried that someone might become unnerved. If they lost a key physicist, it could cripple the operation and affect the test. "The strain had been great on all our people, and it was impossible to predict just when someone might give under to it," he reasoned.

Groves was still annoyed with Fermi, who had remained completely cool amid all the bedlam, but that evening had unaccountably announced his intention of taking bets from his fellow scientists "on whether or not the bomb would ignite the atmosphere, and if so, whether it would merely destroy New Mexico or destroy the world." Bainbridge had also been furious when he heard talk of the atmosphere being detonated. That possibility had been raised by Teller back in Berkeley, rehashed at Los Alamos, and roundly shot down by Bethe. Bainbridge considered it "thoughtless bravado to bring up the subject as a table and barracks topic before soldiers unacquainted with nuclear physics and with the results of Bethe's studies."

At the end of their discussion, Groves and Oppenheimer were in agreement: there was no need to postpone the test for a day, but they might have to put it off for an hour or two. Groves had decided to put his faith in Hubbard's forecast, which called for a clear early morning with light winds. They would meet again at 1:00 A.M. and review the situation.

Groves elected to go to bed and urged Oppenheimer to get some rest, but noted later that he "did not accept my advice and remained awake, I imagine constantly worrying."

At 1:00 A.M. Groves got up and prepared to join Oppie in the forward barricade. Conant, who was quartered in the same tent as Groves, recalled that from 10:30 P.M. to 1:00 A.M. the wind blew very hard—the canvas tent flaps slapped loudly—and then it poured for an hour. The storm had kept him up, and he was amazed the general had managed to sleep right through it. Oppenheimer and Groves drove the three miles to the control station at South 10,000. To protect the scientists and their equipment, the control dugout was built of wood and reinforced concrete, and buried under huge mounds of earth, 10,000 yards, or 6.2 miles, from ground zero. General Farrell was waiting for them as well as Bainbridge, who supervised all the detailed arrangements for the test, and Kistiakowsky, who was among the scientists keeping watch over Fat Man during its last night cradled in the tower. Hubbard, the weather expert, was there, along with a handful of army officers and soldiers. The dugout was a beehive of activity, but there was far less confusion than before, and the atmosphere was now one of forced calm. The physicists had been rehearsing for this moment for months and were focused on checking and rechecking their instruments and radios. The main worry was still the rain, which had let up some but had not stopped. After a brief huddle, they decided to postpone the test—first for an hour, then later for another thirty minutes.

Every five or ten minutes, Groves and Oppenheimer left the dugout and went outside to see if the weather was showing any signs of improving. By 3:30 A.M., the sky was still heavily overcast, but a few stars were becoming visible. They decided they would be able to go forward. Farrell remembered the scene inside the control station as "dramatic beyond words":

> In and around the shelter were some twenty-odd people concerned with last-minute arrangements prior to firing the shot. . . . For some hectic two hours preceding the blast, General Groves stayed with the Director, walking with him and steadying his

> tense excitement. Every time the Director would be about to explode because of something untoward happening, General Groves would take him off and walk with him in the rain, counseling him and reassuring him that everything would be all right.

With daybreak only an hour away, the decision was made—the firing would occur at 5:30 A.M. Thirty minutes before zero hour, the five men guarding the bomb at the tower threw the last safety switches and hightailed it back to the shelter in their jeeps. In case of car trouble, they should have time to make it back on foot, but Groves comforted himself with the idea that "since Kistiakowsky was one of the five . . . they would find a safe position even in the event of a complete breakdown." With twenty minutes to go, Groves returned to base camp, the safest close observation point, leaving Farrell, Oppenheimer, and his men behind. Farrell, in his account for the War Department, reported that once the countdown began and the intervals shrank from minute to seconds, "the tension increased by leaps and bounds."

> Everyone in that room knew the awful potentialities of the thing that they thought was about to happen. The scientists felt that their figuring must be right and the bomb had to go off but there was in everyone's mind a strong measure of doubt. The feeling of many could be expressed by "Lord, I believe; help Thou mine unbelief." We were reaching into the unknown and we did not know what might come of it. It can be safely said that most of those present—Christian, Jew and atheist—were praying and praying harder than they had ever prayed before. If the shot were successful, it was a justification of the several years of intensive effort of tens of thousands of people—statesmen, scientists, engineers, manufacturers, soldiers, and many others in every walk of life.

Through the loudspeaker, they heard Allison counting the seconds—minus 45, minus 40, minus 30, minus 20, minus 10. As the last seconds ticked off, Oppenheimer's thin body tensed. Farrell kept his eyes on Oppenheimer, who had carried this "very heavy burden" for the past twenty-eight months. "He scarcely breathed," he remembered. "He held a post to steady himself. For the last few seconds, he stared directly ahead and then when the announcer shouted, 'Now!' and there came this tremendous burst of light followed shortly thereafter by the deep growling roar of the explosion, his face relaxed into an expression of tremendous relief."

Kistiakowsky, who was standing outside the barricade, was knocked flat by the blast. When he reached Oppenheimer, he slapped him on the back and blurted out the first thing that came to mind: "Oppie, you owe me ten dollars!" It was "a silly thing to say," he admitted later, but he was still smarting over the accusations that his explosives work might have been shoddy. Oppenheimer, still at a loss for words as he watched the huge orange ball of fire rise slowly above the plain, pulled out his wallet and stared numbly at it for a moment before shaking his head. It was empty.

Then Oppenheimer was surrounded by jubilant physicists, who were jumping up and down, pounding each other on the back, and shouting congratulations. Bainbridge went around personally thanking everyone on his team, and when he reached Oppie, he said, "Now we're all sons of bitches." Conscious that they were waiting for his reaction, and fully aware of the moment's historic proportions, Oppenheimer walked out of the shelter and stood on the sand. Gazing out at the twisting column of smoke, he solemnly quoted a line from sacred Hindu writings that his assistant, Priscilla Greene, had no doubt he had prepared in advance for the occasion. As Oppenheimer later famously painted the scene:

> A few people laughed, a few people cried. Most people were silent. I remembered the line from the Hindu scripture, the Bhagavad Gita: Vishnu is trying to persuade the Prince that he should do his duty and to impress him he takes on his multi-armored form and says, "I am become Death, the destroyer of worlds." I suppose we all thought that, one way or another.

Back at base camp, Groves, Conant, and Bush had viewed the explosion from a shallow trench. "It was agreed," Conant wrote in his diary of the day's events, "that because of the expected (or hoped!) bright flash and the ultra violet light (no ozone to absorb it) it would be advisable to lie flat and look away at the start, then look through the heavy dark glass." They lay there belly down, facing 180 degrees away from the spot on the tarpaulin. The last ten seconds, he whispered to Groves, seemed terribly long. Conant kept his eyes open, looking at the horizon opposite the spot:

> Then came a burst of white light that seemed to fill the sky and seemed to last for seconds. I had expected a relatively quick and bright flash. The enormity of the light and its length quite stunned me. My instantaneous reaction was that something had gone wrong and that the thermal nuclear transformation of the atmosphere, once discussed as a possibility and only jokingly referred to a few minutes earlier, had actually occurred.

Slightly blinded for a second, Conant turned on his back as quickly as possible, and watched the fireball through the welder's glass. "At this stage it looked like an enormous pyrotechnic display with great boiling of luminous vapors, some spots being brighter than others," he recalled. "Very shortly this began to fade and without thinking the glass was lowered and the scene viewed with the naked eye. The ball of gas was enlarging rapidly and turning into a mushroom. It was reddish purple, and against the early dawn very luminous." Then someone shouted to look out for the detonation wave, which hit them like a sharp gust of wind some forty seconds after zero. While impressive, Conant recalled that both the blast and the sound were less startling than he had expected because the shock of the first sensory image was still so dominant in his mind: "My first impression remains the most vivid, a cosmic phenomenon like an eclipse. The whole sky suddenly full of white light like the end of the world. Perhaps my impression was only premature on a time scale of years!"

Still sitting there on the ground, Conant, who was next to Groves, reached over to acknowledge their achievement with a silent handshake. Bush, who was on the general's other side, did the same. As he rose, Conant heard Groves mutter, "Well, there must be something in nucleonics after all." They could hear the continuous reverberation of the initial report, like a loud rifle shot, echoing back as the sound waves bounced off the surrounding hills. After about sixty seconds, as the dust cloud billowed up, Conant remembered that "the whole assembled group, including many MPs, gave out a spontaneous cheer." They had suddenly come to the thrilling realization that the force of the blast exceeded even their most optimistic expectations. Fat Man had been powerful far beyond their wildest hopes, and all the pent-up emotions of the past twenty-eight months were released as they broke into a gleeful celebration at base camp, carrying on, Eric Jette recalled, "like a bunch of college freshmen after a football victory."

A few minutes later, Oppenheimer arrived by jeep, followed by Farrell, who bounded over to Groves and declared, "The war is over." Groves' response was more tempered: "Yes, after we drop two bombs on Japan." Groves quietly congratulated his director and said, "I am proud of all of you." Oppenheimer replied with "a simple 'Thank you,' " and Groves, noting his restraint, observed, "We were both, I am sure, already thinking of the future and whether we could repeat our success soon and bring the war to an end."

By that time, the reports were flooding in. "The most exciting news was that the steel tower over 'Jumbo' 800 yards away had disappeared," Conant recalled. "This was unexpected and showed a very much more powerful effect than expected." Tolman, comparing the huge ball of fire, which had mushroomed to a height of over 10,000 feet before dissipating, to the 100-ton TNT shot, agreed that this was "entirely different." There was, he agreed, "no question they got a nuclear reaction." The best estimates of the blast measurement seemed to be between 10,000 and 15,000 tons, though Rabi maintained 18,000 would yet prove right. There was some fallout, but it was nothing dangerous, though Conant noted that those at the North 10,000 shelter had to evacuate in a hurry when their meter went off the scale. They later swore the cloud of smoke "seemed to chase them!" but that turned out to be a false alarm. Several men told stories of being knocked down by the detonation wave. The only serious injury was one man at base camp who made the mistake of looking directly at the explosion without protective glasses, which immediately resulted in a bad burn to his corneas. He had to be given morphine, though the prognosis was that he would not lose his sight.

The crowd of celebrating scientists at base camp swelled as people returned from the shelters, and Oppenheimer and Groves were mobbed by men pounding them on the back and showering them with congratulations. But not everyone was overcome with joy. When Conant bumped into Sam Allison, whose strong authoritative voice calling out "Zero" had been the last thing he had heard before the landscape lit up, Allison seemed almost distraught. "Oh, Mr. Conant," he said, his voice breaking as though it had only just occurred to him what they were doing in the desert. "They're going to take this thing and fry hundreds of Japanese!"

On the morning of Sunday, July 15, Dorothy got a call from a friend on the Hill saying that a few couples were going on a picnic and would she like to join them. She was happy to tag along, but when he told her what time to meet them, she expressed surprise. "But, goodness, it's the evening isn't it?" she asked puzzled. "Yes," he continued in the same light but insistent tone, but they were going "way out into the country," and she should be sure to tell her son that if it got too late, they might stay with some acquaintances that lived out that way.

It had taken her a minute, but Dorothy understood the meaning of his call. This was it. She dutifully repeated the story about the overnight trip to Kevin, but he did not swallow it as easily as she might have hoped. He was used to going on hiking trips with her and the Hill people—it was one of the things they always did together—and was hurt at being excluded. He demanded to know what was so special about this particular outing that he could not join them. But she could not tell him, and he was still sulking when she left. "This was a great secret, my going," she explained. "We all would have been shot at sunrise for anything like that, particularly the young scientists who asked me to join them."

She jumped in her friends' car, and they sped toward Albuquerque. They climbed to the top of Sandia Peak and got out, and looked south toward Trinity. There was nothing but black sky. The test was scheduled to take place around 4:00 A.M., but it was a bad night, and it showered sporadically while they waited. In the distance, Dorothy could see a plane dragging instruments and balloons designed to measure the humidity, temperature, and wind speed and direction. If the storm picked up, the observation planes would not be able to fly. She knew that meant the test might not proceed as planned. They did not have to tell her what a delay would mean. Everyone's emotions were already keyed up to a fever pitch. It started to pour again. She pictured Oppenheimer, already spent from too many twenty-four-hour days, now forced to do battle with the elements. Her heart went out to him, and she remembered thinking, "Poor Oppie, he's down there and it's raining."

It was a long wait till the first pale hint of dawn. Then, without warning, the whole sky lit up. The flash was so bright, it illuminated the northern sky from its apex to the horizon. She could not remember hearing any noise, but the shock was so great it left her feeling weak. "The feeling of awe that I had when that light hit us was remarkable," she said. "I don't think anyone has ever seen such an explosion. . . . The leaves of the green native trees were kind of shining with the gold. It was different. Everything was different. The world was changed."

EIGHTEEN

A Rain of Ruin

MOST OF THE SENIOR scientists not required at base camp took off late Sunday night to watch the first atomic shot from Compañia Hill, the officially designated observation point twenty miles northwest of ground zero. Many others staked out unofficial viewing sites of their own, or held all-night vigils with colleagues. Those wives who were aware of the test were told to stay tuned to the radio, and to look out their windows to the south for something that might look like the glow of the midday sun on the dark horizon. Unable to sleep, Elsie McMillan had fixed her husband a hearty breakfast at 2:30 A.M. Monday morning, and was sitting up and watching the clock when a light tap came on her door. It was her neighbor Lois Bradbury, whose husband, Norris, was "out there, too." They sat together throughout that long night, consuming gallons of coffee and talking. "We talked of many things, our men whom we loved so much, of the children, their futures, of the war with all its horrors. We kept the radio on softly," she recalled. "We dared not turn it off."

They were in the baby's room, staring out the window, when they saw it—a blinding light in the sky. The news on the radio broke the stunned silence in the room: "FLASH! The explosives dump at the Alamogordo Air Base has blown up. No lives are lost. The explosion is what caused the tremendous sound and the light in the sky. I repeat for the benefit of the many phone calls coming in: the explosive dump at the Alamogordo Air

Base has blown up. No lives are lost." They recognized the prepared press release. They knew from past experience that army intelligence would have prepared other fictitious stories blaming the explosion on a plane crash, an earthquake, or even a meteor, anything but the truth. "We looked at each other," recalled Elsie McMillan. "It was a success. Could we believe the announcement, 'No lives are lost'? They had not said no injuries. We had hours to wait to be absolutely sure. At least it was over!"

That Monday morning dawned clear and dazzlingly bright, and the overjoyed inhabitants of Los Alamos gathered in groups all over town to celebrate. "There were tears and laughter," Eleanor Jette recalled. "We beat each other on the back, our elation knew no bounds; the long months of loneliness and worry were almost over, the work was a success—the gadget worked!" There was such a great demand for newspapers that the PX ran out by noon and the remaining copies had to be passed around. Dorothy ran around town scooping up all the editions she could find and had them sent up.

The Santa Fe New Mexican reported only that "an ammunition magazine" had blown up and buried the story on the back page. The story in *The Albuquerque Tribune* was next to the weather, and they were all certain they saw security's handiwork in the brief report, which included the hilarious understatement: "That big flash to the southwest this morning wasn't sheet lightning." But the El Paso papers appeared to have escaped the long arm of G-2 and carried headlines about the explosion.* The *El Paso Herald-Post* even included an eyewitness account of a railroad engineer in Belen who said he had a "front row to the greatest fireworks show he had ever seen." He described seeing "a tremendous white flash. This was followed by a great red glare and high in the sky were three tremendous smoke rings. The highest was many hundreds of feet high. They swirled and twisted as though being agitated by a great force. The glare

*The Alamogordo press release, the first of many that would be necessary, was prepared with the help of the *New York Times*'s science reporter Bill Laurence, to give it, in Groves' words, "a more objective touch." The Office of Censorship saw to it that no news of the explosion made it into any Eastern newspaper, except a few lines in one Washington paper. On the Pacific Coast, however, it got picked up by radio and got a lot of play.

lasted about three minutes and then everything was dark again, with the dawn breaking in the east."

Whether Groves liked it or not—he reportedly cautioned security that he wanted the test results kept under wraps—the bomb was big news at Los Alamos. It was all anyone could talk about, especially the wives. After so many months of grim silence, they now wanted to know everything. At last, they could ask questions and expect answers. Even the most timid suddenly became vocal and were talking to everyone, and pestering the scientists and officers who had remained behind, to fill them in. But they had to be careful. Censorship was still very much in force, as Dorothy learned the hard way when G-2 showed up on her doorstep first thing that Monday morning and inquired about a suspicious call tipping her off about the time of the Trinity test. "We've been monitoring your telephone," the agent informed her, "and something was said about 'seeing something.' " Well acquainted with the ways of army intelligence, Dorothy quickly babbled something about her poodle "usually having the manners to sleep while she was sleeping," but explained that on this particular morning he had wanted out so badly that she was forced to rise at 5:30 and so just happened to see "the lights." The agents were clearly not satisfied with her explanation, but had too much to do to bother with her that day. Despite their best efforts, it was impossible for security to clamp down on the eruption of curiosity and chatter. It was as if a dam had broken. All Monday afternoon, friends converged on balconies to trade information and whisper snatches of what they had managed to pry out of the men returning from the viewing sites.

Everyone waited on tenterhooks for the buses to return from Trinity. By early evening, the first scientists from base camp began staggering back to the Hill looking filthy and tired but exuberant. Many of them were too exhausted to talk and shook off the barrage of questions and headed straight for a shower and bed. But some found sleep impossible and came streaming into the dining room in Fuller Lodge, flashing wide grins and greeting everyone by raising their fingers in the "V for Victory" sign. It was clear from their faces that they had experienced something unprecedented and profound and they were tense with excitement. Others were still badly shaken by the terrifying spectacle they had witnessed and needed to talk.

One by one their stories came tumbling out. They told of seeing the searing bright light and the churning mushroom cloud that had risen to 10,000 feet before fading and, for what seemed like a long time afterward, hearing the endless rolling thunder from the blast echoing in the distant hills. They talked of how, afterward, Fermi and a small crew of scientists had ventured into ground zero in two lead-lined Sherman tanks and discovered the sand had been turned to glass, as green and glossy as jade, by the heat of the bomb. They found a 1,200-foot crater the bomb had dug in the desert floor and another hollow, where the tower had stood, that measured 130 feet in diameter and 6 feet deep in the center. The tower itself had vanished, the steel vaporized and remnants of its charred skeleton—the twisted reinforced girders from the stump of the tower—were scattered on the ground.

They were still in awe of the energy yield of the weapon. Immediately after the detonation, Fermi had scrambled to his feet and performed a simple test to determine the explosion's force, dropping small pieces of torn paper, which instead of falling straight down were carried by the force of the shock wave. By measuring the distance they were displaced, he was able to make a crude estimate of the strength of the blast wave, and therefore the amount of energy released. Fermi had arrived at the figure of 10,000 tons. But by the end of the day, after all the data registered by their many instruments had been collected, it was evident that the explosion was far more massive: the energy released was equivalent to 20,000 tons of TNT. Rabi, who had guessed 18,000 tons, came the closest and was immensely proud of the fact that he had won the betting pool's $102 pot.

The post-test hysteria was catching, and the whole town went wild. "The place was a madhouse," said Anne Wilson. "All the scientists, their SEDs and assistants, and even the GIs who were there—they were so pleased with themselves. They had worked so hard, and they had done it. There was lots of rejoicing. Feynman got his bongo drums out and led a snake dance through the whole Tech Area." People sat on the hoods of jeeps and led noisy parades down the main streets.

"Everybody had parties, and we all ran around," recalled Feynman. "But one man I remember, Bob Wilson, was just sitting there moping." When he asked him what was wrong, Wilson replied, "It's a terrible thing that we made." Feynman, who was still walking on air, protested, "But you started it. You got us into it." At the time, Wilson's behavior struck him as very odd, but what he failed to realize in his boyish enthusiasm, Feynman wrote later, was that they had all been so caught up in their calculations and experiments that they had lost sight of what the project was really about: "You see what happened to me—what happened to the rest of us—is we *started* for a good reason, then you're working very hard to accomplish something and it's a pleasure, it's excitement. And you stop thinking, you know; you just *stop*."

Emotions on the mesa were running very high. "They were all just ecstatic," said Emily Morrison. "It was strange, but thrilling in a way. They were all so relieved and happy. But it didn't last." Reports came in that the moving cloud had deposited dust and radioactive materials as far as 120 miles away and that some parts of the desert were still too "hot" for humans to enter. The gadget, it seemed, had lethal coattails. Some stray cattle had been seen in the vicinity of the test area with large gray radiation burns on their brown coats, signaling their death sentence.

At the end of the week, Bill Penney, the British expert on blast effects, gave a seminar translating the weapon's yield into the brutal statistics of mass destruction, detailing exactly what it would mean in terms of the number of buildings destroyed and bodies incinerated. The meeting had a sobering effect on the physicists, many of whom had put the bomb's murderous power out of their minds while they had concentrated on the task at hand. Now there was no avoiding the awful reality. "The next day they had to go back to work to get the bomb ready to drop on Japan," said Emily Morrison, whose husband, Phil, was one of the small crew of men packing their footlockers and preparing to leave for the Pacific war zone. "There was one British physicist who realized how terrible it was going to be and went home. Bob Wilson was pretty upset, though he didn't leave. But I'm pretty sure he thought about it."

By Friday night, when the Oppenheimers threw a party for the senior Tech Area staff, as well as some of the army officials who had presided over the test, more than a few of the physicists had begun to have second thoughts. Overwhelming success has its own hangover, and after the post-test exaltation on the mesa had died down, remorse had begun to set in. There was plenty of drinking and dancing that night, but as Eleanor Jette recalled, a certain grimness showed through the surface gaiety. "Cyril [Smith] and Joe Kennedy stood talking together most of the evening," she recalled. "Neither man looked as though he'd ever smile again."

Oppie's mood had also turned somber. While some physicists thought they recognized a new swagger in their director's stride after his triumph at Trinity, Anne Wilson never detected any cockiness in his demeanor. "I saw him for so long every day—*every day*—afterwards and I never saw any of that arrogance or conceit," she said. "If anything, he was slightly depressed thinking about what was going to happen. It was, Oh God, what have we done! All this work, and people are going to die in the thousands."

Three days after the detonation, on July 19, Oppenheimer sent a telegram to Groves in Washington that read in part: "Should like to be quite sure that the cost of going through with our present program is understood by you." The bulk of the cable covered technical issues, which could have conceivably provided grounds to delay the assault on Japan until an improved bomb design could be implemented, allowing additional bombs to be built. But by that time Groves had become convinced that two bombs would probably get the job done. He would brook no delay. The Trinity test had not allayed all doubts about the bomb, he wrote in his memoir, "It proved merely that one implosion-type, plutonium bomb had worked; it did not prove that another would or that a uranium bomb of the gun type would." They were confident enough of the original gun-type bomb that no one had argued against using it in combat without first completing a test. "In any case," wrote Groves, "we simply had to take the chance." His reply ruled out any revision of the planned bombing schedule, which could now be fixed around the first of August:

> Factors beyond our control prevent us from considering any decision, other than to proceed according to the existing schedule for the time being.

By the time the gadget was detonated at Trinity, detailed arrangements were already in place for the use of a subsequent bomb as soon as enough material had been produced, and a carefully planned sequence of events began to unfold in quick succession. Just hours after Trinity, the cruiser USS *Indianapolis* sailed out of San Francisco Bay with Little Boy, the uranium bomb assembly, en route to Tinian, near Guam in the western Pacific and less than 1,500 air miles from Tokyo. A second shipment, containing the last necessary piece of uranium, soon followed by air. Fat Man would be ready around the first week in August and, once the Japanese submarine threat became obvious, would travel by plane. A second plutonium weapon would be ready two weeks after that, with more in production, as needs dictated.

In the hours immediately after the test explosion, Groves worked out with Oppenheimer what he would report by cable to Secretary of War Stimson, who was with the president at Potsdam. At 7:30 A.M. Washington time, Groves phoned his secretary in the War Department and dictated, with "guarded brevity," the pertinent facts, making use of a special code sheet, the only other copy of which was in her possession. He left base camp as soon as it was established that radioactive fallout in the area did not pose a problem and no one would have to be evacuated. Groves headed straight back to D.C. on the afternoon of July 16, accompanied by Bush, Conant, Lawrence, and Tolman, whom he observed were "still upset by what they had seen and could talk of little else." As for himself, Groves noted, "my thoughts were now completely wrapped up with the preparations for the coming climax in Japan."

Two days later, Groves followed up his first cable with an expanded account of the Trinity test that was not, as he advised Stimson, "a concise, formal military report, but an attempt to recite what I would have told you if you had been here on my return from New Mexico." Groves' report was sent by courier and reached Potsdam on July 21. Stimson read it to President Truman and Secretary of State James Byrnes that afternoon and later brought it to a meeting with Churchill and Lord Cherwell. It was "an immensely powerful document," Stimson noted in his diary that day. "It gave a pretty full and eloquent report of the tremendous success of the test and revealed far greater destructive power than we expected in S-1." His diary records Trinity's immediate impact on history:

> Churchill read Groves' report in full. He told me that he had noticed at the meeting of the Three yesterday that Truman was much fortified by something that had happened, that he had stood up to the Russians in a most emphatic and decisive manner, telling them as to certain demands that they could not have and that the United States was entirely against them. He said, "Now I know what happened to Truman yesterday. I couldn't understand it. When he got to the meeting after having read this report, he was a changed man. He told the Russians just where they got on and off and generally bossed the whole meeting." Churchill said he now understood how this pepping up had taken place and he felt the same way.

Assured two more bombs would be ready by the end of the month, Churchill and Truman finalized their plans for action against Japan. On July 24, they approved the November 1 deadline for the invasion of Kyushu. Before the session was over, Truman, according to the Interim Committee's recommendation, was supposed to inform Stalin of their atomic progress. But the new president was in no mood to tip his hand to the Russians and instead made only a passing comment to the effect that the United States possessed "a new weapon of unusual destructive force." Exactly what Stalin made of the cryptic comment was unclear, as his reply was elaborately casual. All he said, according to Truman, was that "he was glad to hear it and hoped we would make 'good use of it against the Japanese.' " It is unlikely, however, that he was surprised, given the Soviets access to atomic secrets through Klaus Fuchs and other sources.

The following day, Churchill and Truman gave Stimson and Marshall approval to move ahead with the operational orders to use the first atomic bomb as soon after August 3 as weather permitted. The Target Committee, together with Stimson's military staff, had drawn up a final list of suggested target cities. Stimson had struck Kyoto from the list, even though it was deemed a strategic military target, because it had been an ancient Japanese capital and was considered a sacred shrine. He had approved four other targets, including Hiroshima, Kokura, Niigata, and Nagasaki. The order had been given to spare these cities from the continuing saturation bombing so the effects of the atomic bomb could be clearly seen. As Stimson would later argue in an open letter to *Harper's Magazine,* entitled "The Decision to Use the Atomic Bomb," all four were active working parts of the Japanese war effort. In particular, "Hiroshima was the headquarters of the Japanese Army defending southern Japan and was a major military storage and assembly point," he wrote. They would strike where it hurt the Japanese military the most.

On July 26, Churchill and Truman issued the Potsdam Declaration. At Los Alamos, people listened to the ominous terms of the ultimatum being broadcast on the radio. It called for the Japanese government to proclaim "unconditional surrender" of all their armed forces and to provide proper and adequate assurances of their good faith, and it warned that the alternative was "prompt and utter destruction." Two days later, the premier of Japan, Baron Kantaro Suzuki, rejected the Potsdam Declaration and called it "unworthy of public notice."

After that, Dorothy left her kitchen radio on all the time. She did not know when or where it was going to happen, but like everyone at Los Alamos, she sensed *something* was going to happen soon. Security remained very tight, and MPs patrolled outside the office. G-2 returned several times to question her about the suspicious phone call on the day of the test shot, but she stuck to her story. Incredibly, life returned to normal. The *Daily Bulletin* was full of the usual scolding notices—unless customers return their empty bottles immediately, there will be an inadequate supply of Coca-Cola for sale; unless crutches and canes are returned to the hospital ASAP, a shortage will ensue—and banal news about Scout hikes, picnics, and Saturday night dances. There was no mention of Trinity, the bomb, or the impending assault. Life took on a slightly unreal quality. "We just followed the same routine, taking care of the children and playing cards with our friends," said Marge Schreiber.

The war might continue for many months, or it might end in an hour. One friend who could not stand the interminable anxiety complained that the problem with life on the Hill was that "it was so daily," recalled Shreiber: "I always thought that pretty much summed it up."

They were all supposed to go on about their business while they waited to see if the Japanese would surrender, knowing all the while that the deadliest weapon in the world's history was waiting in the wings on some small Pacific island, only striking distance away. Of course, it was forbidden to talk about the bomb, but people did anyway. A contingent of sixty to seventy people from Los Alamos had been assigned the task of assembling the bomb components and preparing Little Boy for airlift. They had been sent to Tinian, a B-29 base in the Mariana Islands that was the staging area for the bomb operation. The post grapevine, efficient as ever, buzzed with the names of the men who had left the mesa: Deke Parsons, Luis Alvarez, Phil Morrison, Harold Agnew, Bob Serber, Bill Penney, and Norman Ramsey, among many others.

Work in the laboratory had slowed considerably, and those left behind had little to do but document what they had done and debate the pros and cons of using the weapon. Wouldn't it be better if the Japanese were invited to watch a demonstration on an uninhabited island? But what display could convince such an ancient, honor-bound people to agree to the Allied demand for unconditional surrender? On the other hand, wasn't it better to destroy one Japanese city if it served to save the lives of many more Americans, as well as Japanese, who would die in the invasion slaughter, which, in the absence of the atom bomb, was the only way to terminate the war?

Others, like Kistiakowsky, felt the bomb was "no worse than the fire raids." The Japanese had lost a hundred thousand civilians in just one night of the war, during the firebombing of Tokyo on March 9, 1945. More than a million people had been wounded in that demonstration of might, and yet the Japanese, with their ancient tradition of honor that made unconditional surrender unthinkable, had fought on. How many lives would have to be sacrificed in battle before the Japanese military leaders, who had attacked and sunk the U.S. Pacific Fleet at Pearl Harbor, were humbled? As proof of their defiance, on July 30, a Japanese submarine sank the *Indianapolis* on her way to the Philippines. It was the single greatest loss in the history of the U.S. Navy, but at the time, few people knew how close to disaster the country had really come—only four days earlier, the cruiser had delivered the uranium bomb to Tinian. The scientists argued the issue endlessly, with some insisting that they should take a moral stand like Szilard and Franck, while others took the view it was out of their hands, and that "the cobbler should stick to his last." Dorothy heard enough of these discussions to know they could get pretty lively.

For Oppenheimer, who was intimately involved with the final preparations for the bomb delivery, it was not a happy time. In the days after the Trinity test, the momentum propelling the combat use of the bomb accelerated. At the same time, Oppenheimer was not immune to the inner turmoil afflicting many of the mesa scientists. He, too, dreaded the fearsome demonstration of force to come. But at no time did he, or any of the president's chief advisors—including Stimson, General Marshall, Groves, Bush, and Conant—recommend calling a halt to the operation. "After Trinity, there was no slowing down," said Anne Wilson. "If anything, it speeded up. They were selecting targets for bombing. It was a very busy time. We even had some of the pilots come up to the Hill, maps were brought in, and there were discussions of the whole business." One day, as they walked together to the Tech Area office, Wilson noticed that Oppie appeared "very distressed." When she asked him about it, he just shook his head and said, "I just keep thinking about all those poor little people."

On the morning of Monday, August 6, Dorothy received word "to bring a radio along to work." She asked no questions, but brought her small radio from home, put it on her desk, and kept it turned on very low. "It was a typical New Mexico day of summer," she recalled. "Sunshine of golden brilliance, sky bluer than paint. . . ." At 11:00 A.M., there was a news flash from the White House, and Harry Truman's voice came over the radio:

> Sixteen hours ago an American airplane dropped one bomb on Hiroshima, an important Japanese Army base. That bomb had more power than 20,000 tons of T.N.T. It had more than two thousand times the blast power of the British "Grand Slam" which is the largest bomb ever yet used in the history of warfare. . . .

> It is an atomic bomb. It is a harnessing of the basic power of the universe. The force from which the sun draws its power has been loosed against those who brought war to the Far East.

The president explained that the Germans had been working feverishly on ways to add atomic energy to their other "engines of war," with which they "hoped to enslave the world." By pooling their knowledge and resources, the greatest American and British scientific brains were able to beat the Germans in "the race of discovery." Employment in the bomb project, at several secret plants, during peak construction, numbered 125,000, and more than 65,000 were presently employed at the plants. His reference to "the many who had been working for two and a half years," while "few knew what they have been producing," was not lost on Dorothy. "We have spent two billion dollars on the greatest scientific gamble in history—and won," he said. "The battle of the laboratories held fateful risks for us as well as the battles of the air, land and sea, and we have now won the battle of the laboratories as we have won the other battles."

There was more. If the Japanese did not accept the terms of the Potsdam ultimatum, the president promised, "They may expect a rain of ruin from the air, the like of which has never been seen on this earth." Dorothy sat and listened to his speech, scarcely breathing. She listened to all the radio broadcasts that followed, the announcers so excited they could hardly control their voices. Colonel Paul Tibbets, the boyish-looking air force officer she had allowed up to the Hill, had piloted the *Enola Gay* over Hiroshima, and the bomb had been dropped at 7:15 P.M. on August 5. While she had been eating dinner, the port city had been wiped out, and more than 100,000 Japanese were thought to have been instantly killed. It had been morning in Hiroshima, and they had awoken to death and destruction on a scale no one had ever seen before. Four square miles of the city had been disemboweled. A news bulletin interrupted to report that when Tibbets landed, he was awarded the Distinguished Service Cross for his daring strike on Hiroshima, for "carrying for the first time a bomb totally new to modern warfare." He had safely turned his plane away from a "monstrous" column of smoke that had erupted from the ground while aerial shocks, like bursts of flak, rocked his plane. . . . Dorothy listened to report after report until she could not stand to listen anymore.

At noon, she made a beeline back to the house, where she knew Kevin would be returning from his summer job for lunch. They both worked only five minutes from home and usually ate together. She sat down and, struggling to appear calmer than she felt, faced him across the dining room table. She had never told him anything more than that the laboratory was connected to the war effort and he could never mention its existence or his visits there. "Keep your mouth shut," she had told him a hundred times. Now at fourteen, he was tall and gangly, a boy in a man's body, but old enough to be told the truth. The words came out in a rush: "Kevvy, the President of the United States announced this morning to the whole world that the bomb that was dropped on Hiroshima was manufactured in the hills up in northern New Mexico."

His eyes quickly met hers. For a moment, neither of them said anything. Then she spoke again, adding quietly, "That's our bomb."

A long silence followed between them. Then, as she recorded in her memoir, she was gripped by panic. No sooner had the words left her lips than she wanted to call them back, as "the realization of what she had said, she, conditioned for more than two years not to so much as whisper the name of Los Alamos." Her hand shot across the table and caught his wrist. This time when she spoke, her voice was full of alarm: "*But don't tell anyone!*"

It did not take long for Dorothy to realize that the secret was out. Her phone was ringing off the hook with friends and family demanding to know if this was the same "government project" she had been working for all along. Had she been up to the Los Alamos laboratory? Had she known from the beginning that they were building a bomb? Had she witnessed the Trinity test? How could she not have told anyone? So many people called with so many questions that she felt overwhelmed by it all.

The president's announcement had lifted the veil on their mysterious encampment, and the army propaganda office was busy working overtime distributing a general history of the project, complete with quaint details of life behind barbed wire, to the press. The late editions of Monday's *New Mexican* ran front-page stories under the banner headlines:

"LOS ALAMOS SECRET DISCLOSED BY TRUMAN; ATOMIC BOMBS [*sic*] DROP ON JAPAN." The first line in the local story about the laboratory was: "Santa Fe learned officially today of a city of 6,000 in its own front yard." Dorothy read with fascination as the classified bomb project that had been her life for the past twenty-seven months was demystified in a series of cold facts and statistics. Most of the reporting about the bomb's development and test detonation was new to her, as was the great potential promised by the "Atomic Age" and the "tool to end wars." Groves was heralded as the head of the atomic bomb project, and Oppenheimer credited with "achieving the implementation of atomic power for military purposes."

It was a very strange and unsettling feeling to see the name "J. Robert Oppenheimer" staring out at her in inky black type in the newspaper. His name, his very presence in Santa Fe, had been her most closely guarded secret for more than two years. And now the whole world knew. After being turned inward for so long, and trained to keep every detail of her work and association with the laboratory locked up inside her, it was stunning to have everything suddenly exposed to the light of day. It was as if the heavy door of secrecy that had shut behind her in the spring of 1943 had been thrown open, and she found herself surprised and confused to be the center of so much flattering attention. It had all worked out better than she had dared to believe when, those many months ago, she had signed on to work for Oppenheimer in hope and ignorance. She took great pride in the scientists' achievement and, to a lesser degree, her own small part in the project's success. They had brought the war to an early end. It would all be over in a matter of days, as much as six months to a year earlier than expected, and they had saved the lives of innumerable American and British soldiers. "She was thrilled and proud about the whole thing," said Kevin. "The local papers were full it. Just to think of all the people that might have been killed by the Japanese."

Up on the Hill, one big party was in progress. The town was delirious, and there was dancing in the streets. There were even bigger, noisier parades than after Trinity, and children ran in and out of the apartment complexes beating pots and pans and applauding their suddenly famous fathers. Parsons, Alvarez, Agnew—the whole bomb crew—were on the way home and being hailed as conquering heroes. There was a big assembly that night at Fuller Lodge, and when Oppenheimer strode in, they all rose and applauded. As he stepped up to the podium, he clasped his hands above his head in triumph. But even in the midst of the euphoria, many of the scientists experienced a rising anguish, and the bilious aftertaste stayed with them. Frank Oppenheimer had been standing just outside his brother's office when the Hiroshima announcement came over the loudspeaker. "The first reaction was thank God it wasn't a dud," he recalled. "But before the whole sentence of the broadcast was finished one suddenly got this horror of all the people who had been killed. I don't know why up to then we—I—hadn't really thought of all those flattened people." He had no doubt his brother's reaction was the same. "The image of all those people was really pretty awful."

Otto Frisch was working in the Tech Area when he heard the sound of yelling and running feet. "Somebody opened my door and shouted, 'Hiroshima has been destroyed.' " Years later, he still vividly remembered the moment and "the feeling of unease, indeed nausea" when he saw how many of his friends were "rushing to the telephone to book tables at the La Fonda hotel in Santa Fe, in order to celebrate."

But Little Boy did not end the war. The great fireball that incinerated the city of Hiroshima had not brought a quick capitulation. The U.S. Army prepared for a second raid, and millions of leaflets were dropped over Japan warning them what lay in store and urging them to petition their Emperor to sue for peace. Two days later Russia declared war on Japan. Robert Patterson, under secretary of war, sent a telegram "To the Men and Women of the Manhattan District," lauding them for their help in developing "the most devastating military weapon that any country has ever been able to turn against its enemy." It went on to remind them "to keep the secrets you have kept so well. The need for security and continued effort is fully as great now as it ever was."

On Thursday, August 9, the White House announced that another bomb, Fat Man, had been dropped on Nagasaki. To maximize the shock effect, it was delivered as soon after the first as possible. But the weather was not favorable over Japan, and the second mission did not go as smoothly. For many hours there was no word of the fate of the plane, the *Bock's Car,* and its crew. Although it was a major seaport that contained several large industrial plants, Nagasaki had not been the intended target. The primary target was the large arsenal at Kokura on the northern coast of Kyushu. But because of poor visibility, Major Charles Sweeney, the pilot of *Bock's Car,* was forced to keep circling, burning up fuel reserves and attracting the attention of Japanese fighter planes. Finally, he made a desperate run for Nagasaki. At the last minute a hole opened in the cloud cover, and he unloaded Fat Man, missing his target by two miles. It landed in a valley, in the middle of an industrial area, and a fiery yellow ball erupted, followed by a massive black mushroom cloud that billowed up to 20,000 feet.

Later that same day, President Truman sent a cable to the Hill with his thanks from "a grateful nation." At Los Alamos, there was nothing they could do but pray the Japanese military leaders would come to their senses and realize the utter hopelessness of their position. The scientists were in for yet another agonizing wait. They had lost control of their creation, and as the president had foretold, the force of the sun was being "loosed" on the world. "One of the things we learned was that once a machinery like Los Alamos is started, it takes on a life of its own," said Rose Bethe. "We could no more have stopped it than we could have stopped a boulder from rolling downhill."

Two bombs proved sufficient. Fat Man, as predicted, was more powerful than Little Boy, with a yield of roughly 22 kilotons, yet did less damage, landing wide of the target and obliterating approximately a third of the city and half as many civilians. But it was enough to shock the Japanese into surrender. Twelve hours later, Nagasaki was still engulfed in flames. At 7:00 A.M. August 10, Tokyo time, Japan sent a cable to its minister in Switzerland agreeing to the Potsdam terms, but insisting on one condition—that Emperor Hirohito be allowed to retain his sovereignty. While Suzuki awaited an official response, a rogue faction of the military, unable to accept the Emperor's shameful capitulation, attempted a coup, and it was quashed. After a day of confusion and wild speculation, during which the president negotiated the terms with the British, Chinese, and Russians, a compromise was struck: the Supreme Allied Military Commander, in all probability General Douglas MacArthur, would rule Japan through the authority of the Emperor. At noon on August 15, the Emperor ended his imperial silence and gave an unprecedented radio broadcast informing the Japanese public that he had no choice but to "order the acceptance of the provisions of the Joint Declaration of the Powers," and blaming "a new and most cruel bomb."

Los Alamos residents reacted with a mixture of relief and sorrow, delight and profound distress. Their emotions were in such a tumultuous state it was hard to know how to feel or act. Only the GIs seemed to celebrate with total abandon, piling en masse onto army trucks and jeeps and riding around cheering wildly. George Kistiakowsky drunkenly saluted the long-awaited end of the war by firing off twenty-one boxes of Composition B in an empty field. Other celebrations followed, but there were fewer of them, and they were more subdued. There was something unseemly, almost "ghoulish," as Frisch put it, "about celebrating the sudden death of so many people, even if they were 'enemies.' "

The conviction that Japan had been very far from surrender had maintained the emotional cohesion of the laboratory staff up to that point and had enabled most of them to rationalize the bombing of Hiroshima. But that same belief did not apply to Nagasaki. None of them could escape the thought that Nagasaki had served to prove beyond a question of a doubt that the new plutonium weapon was superior, automatically rendering the first bomb obsolete and putting the world on notice. It had been a political and strategic exercise of hideous proportions. "Few of us could see any moral reason for dropping a second bomb," admitted Frisch. "Most of us thought the Japanese would have surrendered in a few days anyhow."

Oppenheimer attended the round of victory dinners, but recalled that as he left one gathering, he found one of his best group leaders vomiting in the bushes and told himself, "The reaction has begun." Bob Wilson, who looked much younger than his thirty years, was, by his own admission, "ill, just sick"—he was so overwhelmed by what had happened. At the start of the project, he had been caught up in the patriotic fever of the war effort and had thrown himself into the weapons work, but now he felt only revulsion at his part in the holocaust. His wife, Jane, returned from a trip to San Francisco to find him "very depressed." V-J Day, when it finally came, could not have differed more from V-E Day, when they had hosted a victory bash that ended with his merrily "throwing garbage cans around." This time, she said, "We didn't have a party."

They had known a second strike was a possibility, but it had still caught many of them by surprise. "It took me back. It just seemed like too much," said Shirley Barnett. "The reasons for using the first bomb were valid. I didn't have any doubts about it. But I did feel bad about Nagasaki. The biggest sadness of my life, and that of many others, was the dropping of the second bomb."

For many, the excitement that the war was over was supplanted by horror at what they had wrought. Oppenheimer organized a party of scientists to go to Japan and study the devastation caused by the bomb's fireball—the gutted rail stations, bridges, and buildings, the burnt and blackened radius of the heat storm, and the radioactivity. It was a macabre task, but Serber, Bill Penney, and the two post doctors, Henry Barnett and Jim Nolan, among a handful of others, agreed to go and document the carnage. They started in Nagasaki, where Serber recorded his first impression: "Everything flattened and burnt over in the residential and business area." He remembered following the line of charred telephone poles out beyond two miles from ground zero. They picked among the wreckage, measuring the pressure it took to squash a five-gallon gas can and, based on how things were blown out, calculating the wind velocity behind the blast front, and measuring the flash burns on the walls of the Hiroshima post office building to estimate the size of the fireball. "The ruins were hard enough to endure, but the really harrowing experience was a visit to the Nagasaki hospital with Henry," Serber recalled. "This was five weeks after the bombing and the patients were mostly suffering from flash burn or radiation sickness."

As the horrifying statistics began to sink in—at Hiroshima alone, 100,000 were dead, 130,000 wounded, and as many as 8,000 missing—along with more information about the lingering effects of radiation, the soul-searching began.* This was not the dawn of a better world. The magnitude of the tragedy that had befallen Hiroshima and Nagasaki was beyond their comprehension. They thought of themselves as physicists, quiet

*The death toll steadily went up as more information was available and as fatalities due to injuries and radiation exposure accumulated: the total number of deaths was closer to 140,000, with the five-year total estimated to reach 200,000.

and reflective intellectuals, and when they read the bombings described in a single issue of *The New York Times* as "mass murder," "sheer terrorism," and a "stain upon our national life," they felt like pariahs. The Vatican registered its disapproval of the new weapon and harkened back to Leonardo da Vinci, who had suppressed his invention of the original submarine because of the harm it might cause. As Laura Fermi observed, in some ways the Los Alamos scientists had not foreseen the full consequences of their work. "Helped by the physical separation of Los Alamos from the world, they worked in certain isolation," she wrote. "Perhaps they were not emotionally prepared for the absence of a time interval between scientific completion and the actual use of their discovery. I don't believe they had visualized a destruction whose equivalent in tons of TNT they had calculated with utmost accuracy." Wives watched in bewilderment as their dedicated husbands suddenly became riddled with guilt and remorse. Where they had once focused all their attention on their research, they now talked of nothing but the bomb. To Laura Fermi, it seemed that "they assumed for themselves the responsibility for Hiroshima and Nagasaki, for the evils that atomic power might cause anywhere, at any time."

Not all the physicists felt penitent. Alvarez, who had flown on the Hiroshima mission, returned from Tinian flush with victory only to find his colleagues mired in gloom. "Many of my friends felt responsible for killing Japanese civilians, and it upset them terribly," he recalled. "I could muster very little sympathy for their point of view; few of them had any direct experience with war or the people who had to fight it." He seriously doubted any technical demonstration would have convinced the Japanese high command the war was lost. Right to the end, the Japanese were trying to negotiate for the continued authority of the Emperor and, short of that, were determined to fight to the finish for the honor of their homeland. By that point in time, most Americans were hell-bent on unconditional surrender and considered Hirohito "a master war criminal who deserved to be hanged." In such a bloodthirsty atmosphere, Alvarez noted, "Qualifying our demand for unconditional surrender was politically impossible." Alvarez shared his fellow scientists' regret over the loss of life on both sides, but neither their moral queasiness nor their second thoughts at the dropping of the two bombs that had brought the conflict to a swift conclusion.

Among the smoldering ruins of Hiroshima, "almost directly under the point of explosion," he noted with grim satisfaction, was the Mitsubishi factory that "had made the torpedoes that devastated Pearl Harbor."

Conant remained committed to the idea that the surprise attack was necessary not only to shorten the war but because it was the only way to alert the world to the need for international control of such an indiscriminate and barbaric weapon. This belief, which became his postwar credo, formed the underlying argument of the influential open letter, "The Decision to Use the Atomic Bomb," which was orchestrated by Conant, drafted with his supervision by McGeorge Bundy, and which ultimately appeared on the cover of *Harper's* magazine in February 1947 under the name of the venerable former secretary of war Henry Stimson. Conceived by Conant as an authoritative defense of the bombing on military grounds—and justification of the human cost in the face of growing public doubts stirred, in part, by John Hersey's moving report on the victims of the bomb in the *New Yorker*—the Stimson article strove to define the lesson of Hiroshima for history. Solemnly maintaining that the use of the bomb was the only way to end the war and save "over a million" American lives, the article concludes, "No man, in our position, and subject to our responsibilities, holding in his hands a weapon of such possibility for accomplishing this purpose and saving those lives, could have failed to use it and afterwards looked his countrymen in the face."

Dorothy, like almost every mother of a son, had wanted them to do whatever was necessary to bring American boys safely home. She had too many friends and relatives with sons, husbands, and brothers overseas to think otherwise. "They wouldn't have hit Nagasaki unless the Japanese ruler had been so stubborn about it," she said later. "They hadn't wanted to do it." For her part, she felt no guilt, only immense gratitude that the war was finally over.

There were signs that the years of waging war had worked a change on Oppenheimer. After the surrender, he seemed overcome by weariness. In mid-August, he gave an interview in which he admitted to a local reporter that he felt "greatly relieved" after the Trinity test assured that the laboratory had accomplished its mission. "It has been an extremely tight and difficult program," he said, "and I was aware of the many possibilities that one of the integral parts would not work out." He made no apologies for his creation, saying, "We were at war and it was necessary and right for us to make bombs." He even sounded a hopeful note, explaining that the peacetime possibilities of atomic power could be of lasting benefit. "If our discovery is wisely used politically, it may help to reduce the chances of future war. This is a matter for the statesmen—the statesmen supported by the peoples of the world."

Oppenheimer was no longer the same quiet, gentle physicist she had first met in the lobby of La Fonda, who talked to her of poetry, the mysteries of life, and unlocking the secrets of nature. He was now the "Father of the Atomic Bomb," the most brilliant of the Manhattan Project's brilliant men. He had become the leader his country needed him to be; he had made history and was a national hero. His life had changed completely and irrevocably. These were dangerous times, and he was being drawn onto the world's stage. She knew he longed to get away to his ranch in the upper Pecos Valley for a few days of rest and solitary reflection. But he was expected back in Washington, where he would present the Scientific Panel's report on postwar planning to the secretary of war. "He had done his job," she said. "They had done what they had started out to do. But he smoked constantly, constantly, constantly."

On his return from Washington, Oppenheimer headed straight for Perro Caliente, and his first holiday in almost three years. Taking stock over the next few days, he took the time to answer the mountain of congratulatory letters, including one from his old teacher Herbert Smith. In his fond reply, Oppenheimer gave voice to his own doubts about the days ahead. "It seemed appropriate, & very sweet, that your good note should reach me on the Pecos," he wrote, "like so many of the beautiful things of which I first learned from you, the love of it grows with the years."

> Your words were good to have. You will believe that this undertaking has not been without its misgivings; they are heavy on us today, when the future, which has so many elements of high promise, is yet only a stone's throw from despair. Thus the good which this work has perhaps contributed to make in the ending of the war looms very large to us, because it is there for sure.

NINETEEN

By Our Works We Are Committed

AFTER HIROSHIMA, the army relaxed the strict security governing life on the post. Once freed from General Groves' repressive rule, the scientists and their wives descended en masse on the tiny town, crowding the locale cafés, the Paris Theater, and Pasatiempo, the popular dance hall on Agua Fria Street. They were finally at liberty to come and go from their barbed-wire compound, and a recent edition of the *Daily Bulletin* had instructed them that they were even permitted to say they lived at Los Alamos. When they went down to Santa Fe, they found themselves the object of intense curiosity and were treated like local celebrities. People pointed and stared at the scientists as they stood in line at the telegraph office waiting send off dozens of messages to friends and family back home informing them at last of their whereabouts and what they had been doing for the past two years.

Dorothy's office was mobbed from morning to night with Santa Feans eager to know more about the "hidden city" that had sprung up outside of town, somewhere beyond the sleepy villages and pine-stippled ridges. How close could they get to the reservation by car? How much could be seen from the fences? Could she give them day passes to get by the guard station? Could she, by any chance, arrange a tour? At the end of one particularly hectic day, Mrs. Martha Field, the owner of the building at 109 East Palace, which Oppenheimer had rented under an assumed name, came roaring in the back door full of righteous indignation. "So *that's* what you've been doing," she said, confronting Dorothy and demanding a full explanation of what had been going on for two years on under her own roof.

There was such a tremendous demand for information and access to the site that Dorothy and her small staff were deluged. Los Alamos had been the best-kept secret of the war, and it was immediately apparent that the project leaders had woefully underestimated the impact its sudden revelation would have on the public. In the days afterward, the army hastened to get its press operation up to speed, and on August 12, 1945, *The Smyth Report,* a comprehensive overview of the Manhattan Project, which had been written in advance by Henry DeWolf Smyth, chairman of the Princeton physics department, was released to the press and sold in book form. The first copies sold out in Santa Fe in one day. But *The Smyth Report,* far from satisfying the desire for information, only seemed to pique people's curiosity further. The number of inquiries increased exponentially, and the chaos at 109 in the weeks after the war ended was almost enough to make Dorothy miss the days when G-2 chased nosey strangers down the street.

Phyllis Fisher realized their cocoonlike existence was over when she answered a knock on the door, only to find a traveling salesman smiling back at her. She almost fainted from surprise and then, out of habit, so mercilessly interrogated the poor man about how he had managed to get a temporary pass to the site that she never found out what he had for sale. The sudden change in their status from top secret to world famous was jarring for everyone. People walked around in a state of heightened excitement bordering on hysteria, and the post hospital was overrun with scientists requesting headache cures, sleeping pills, and palliatives for nausea and other stomach ailments. In the end, more medication had to be ordered from Bruns Army Hospital in Santa Fe. It turned out that they had grown used to their odd wartime regimen—to armed guards, security passes, censorship, and the predictable shortages of gas, milk, and eggs. Peace, with all its opportunities and uncertainties, was not without its traumas. Anne Wilson had become so accustomed to their insular little world, populated by the most intelligent and diverting individuals she had ever met, that she was initially reluctant to leave, like a bird that has become too attached to its cage. "I found it nerve-wracking," she said. "Up there, there was safety within the fence."

Many others, however, could not wait to kick up their heels. Oppenheimer's secretary, Shirley Barnett, had never thought life on the post had many charms, and it did not take much inducement to convince her it was time to clear out. When she discovered a navy plane was flying to California over Labor Day weekend, she decided a little out-of-town trip would be just the way to celebrate their new liberty. She was in charge of scheduling the flight, so she and Charlotte Serber hitched a ride out to the West Coast. They spent a few days in Pasadena and then went to Los Angeles, where they wound up at a house party at Ira Gershwin's, at which they heard the famous radio comedian Abe Burrows improvise songs at the piano, including one about the bomb. "We had a wonderful time," recalled Barnett, "but when we came back, I got a blistering note from Parsons, who was in charge of the plane, saying he had been apprised of the fact that Charlotte and I had gone to Pasadena." If it had not been for the fact that both their husbands were in Japan, they might not have gotten off with just a slap on the wrist. It later turned out the FBI had tracked their misuse of navy aircraft, and the whole escapade was written up in Oppenheimer's FBI file, according to Bob Serber, "as evidence of the kind of untrustworthy people he hung out with."

In September, the British embassy sent hearty congratulations and money for a grand celebration in honor of "the Birth of the Atomic Era." Rudolf Peierls had been awarded the Order of the British Empire for his contribution to the project, and Oppenheimer was to receive America's highest civilian award, the Medal of Merit, from Truman, so this promised to be the party to end all parties. All the members of the British mission, many of whom were set to leave in a handful of days, pooled their ration points and threw themselves into planning skits spoofing security and the Trinity test, and writing hilarious toasts to guarantee that it would be a memorable evening. Engraved invitations were sent out—a marked improvement over the usual mimeographed sheet—and Fuller Lodge was decorated for the gala occasion with the flags of the two Allied nations. The guests assembled at eight, resplendent in their best approximations of black tie and ball gowns—a few even managed white tie and tails—and bellowed their respective anthems, "God Save the King" followed by "The Star Spangled Banner." "The women worked all day preparing the feast of pork pie and peach truffle," recalled Dorothy. "The climax was reached when everyone rose to their feet, brandishing their paper cups, and drank the King's health with sparkling Burgundy." After the entertainment was over, there was dancing. As she looked up at the "high table," where the Peierlses and Oppenheimers were seated in good British boarding-school fashion, and at all the familiar smiling faces around the room, she could not help feeling sad that they would all soon be going their separate ways.

The exodus, which had begun in the heat of August, had reached truly startling proportions. The main road from Los Alamos to Santa Fe was bumper to bumper with the line of cars and trailers of departing workers. They looked like refugees fleeing the parched, drought-menaced plateau, abandoning the mountain as the Indians had centuries before them. The scores of builders, machinists, engineers, electricians, and other specialists were finally returning to their homes and families. The graduate students and SEDs were also going, eager to get back to school before the start of the winter term. Everyone was packing up and preparing to return to the real world. For the first time in more than two years, they had time to think about the future and make plans. They wanted to put their bomb work behind them and pick up the pieces of their old lives. People talked of university jobs, research fellowships, and possible careers in industry. Most of the senior scientists were on leave and would be returning to their former teaching positions, or had plenty of lucrative new offers, and were helping younger colleagues find places. The grapevine was full of who was going where, when, and with whom.

The academic recruiting had begun even before the war was over, with representatives from the major universities competing for the top brains at the Manhattan Project, eager to secure their postwar prestige and scientific advances for their institutions. Arthur Compton had already begun talking to Robert Hutchins, the forward-looking president of the University of Chicago, about forming a research institute after the war "to preserve the spirit of Los Alamos." The basic idea was to keep together some of the laboratory physicists, chemists, engineers, and metallurgists who had enjoyed working together, and form a cooperative endeavor led by Fermi that would allow them to take their developments beyond the narrow and pragmatic purpose of the wartime project.

By mid-July, it was decided a meeting was necessary between Fermi, and a few of his interested Los Alamos colleagues, and three principals of the university, including Harold Urey. Since, unbeknownst to the university officials, the Trinity test had been completed only days earlier, and no civilians were being allowed up to the classified site, Dorothy gave Fermi permission to use her home for their private conference. No one was to know of the Saturday meeting, and they did not want to be spotted together in town. Dorothy made arrangements for the Chicago party to be picked up at Lamy and taken directly to her home, and then she made herself scarce. Fermi, Sam Allison, and Cyril Smith drove straight from the Hill to her house for the all-day meeting, during which the plans for the three Institutes for Basic Research took shape. "The six men met in Santa Fe, on the terrace of Dorothy McKibbin's home, on top of a hill with a view of the golden vastness of the desert between the town and the distant mountains," recalled Laura Fermi. "Over a lunch of sandwiches packed at Fuller Lodge on the mesa the policies for the future institute were discussed."

Some of the Los Alamos physicists were adamant they would never again do defense work and were determined to lock themselves in their proverbial ivory towers. Others felt reconciled to the idea that atomic energy was now a reality: its ramifications had to be explored, along with the many new promising areas of research that could benefit society. They argued that it was impossible to stop progress and that the evil lay in the hearts of men and not in scientific discovery. They felt Hiroshima and Nagasaki were proof that the atomic bomb was so destructive that they could now annihilate all mankind, and that nuclear war was therefore no longer a rational means by which countries could settle their differences. After the Potsdam conference, the president had given a radio address in which he had said that what we had done to Japan was "only a small fraction of what would happen to the world in a third world war." Truman had said the United Nations would work to see that it could never happen again, that "there shall be no next war." Most of the Los Alamos scientists dearly wanted to believe this was true, and the president's speech gave them cause for hope.

There was a growing consensus that the atomic era could not be one of hegemony in the field of atomic weapons, but must be one of international cooperation and the sharing of knowledge for peaceful purposes. "Hallways, offices, and living rooms teemed with talk," wrote Alice Smith. "War was now so terrible that effective steps to control it would have to be taken. He [Oppenheimer] recognized the flowering of seeds that he and Niels Bohr had planted even though he had felt duty-bound to discourage organized discussion until the laboratory's mission was accomplished. Out of his hearing, much of the talk was prefaced by, 'Oppie says. . . .' "

The dangers inherent in the new atomic era had come home forcefully to them all after one of the young project scientists was involved in a Tech Area accident and exposed to a lethal dose of radiation. On the night of August 21, Harry Daghlian had been working alone on assembling a mass of uranium 235 when a large twelve-pound chunk of heavy metal slipped out of his fingers and fell on the nearly completed assembly. Although he quickly swept it aside, he saw the blue glow of ionized air as his assembly momentarily went critical. Daghlian called for help, but grew sick even as the ambulance rushed him to the post hospital. There was nothing Louis Hempelmann could do to help him, and Daghlian lingered in excruciating pain for weeks. His hands were burned raw, and became blistered and gangrenous. It took almost a month for him to die. On September 15, with his blood count way down, his body gave out.

The twenty-six-year-old physicist was Los Alamos's first casualty, but he served as a terrible reminder of the thousands of Japanese who had survived the blast but were exposed to a massive dose of radiation and were sentenced to the same slow, horrible death. There was little comfort to be had in the knowledge that the Los Alamos accident had been avoidable. Daghlian had broken two strict Omega site rules: first, by working alone; and second, by failing to use a safe method of assembly so that if any material were dropped it would not cause a chain reaction. For security reasons, his condition was hushed up, though the senior staff knew what had happened and friends went to see him in the hospital. Louis Slotin, his team leader, sat with him every day until he died.*

Guilt, exhaustion, and a restless desire to move on made Oppenheimer impatient to leave the mesa. He announced his intention of returning to academic life as soon as possible, casting the future of the Los Alamos Laboratory in doubt. His decision left many of his loyal followers feeling bereft. "Everyone was very pleased with Oppenheimer and the job he did," said Harold Agnew, who years later would helm the laboratory. "Los Alamos was very loyal to him." They were hurt by the rumors that he felt Los Alamos had outlived its usefulness and should be shut down. Those who wanted to stay on wondered who could possibly fill his shoes and redefine the laboratory's postwar mission.

But Oppenheimer could not be swayed. He was physically and spiritually depleted. "There was not much left in me at the moment," the forty-one-year-old physicist explained later. His recent trip to Washington had left him feeling depressed, not only about the international situation and what he had learned from Stimson about Potsdam's failure to enlist the Russians in cooperation in establishing postwar arms control but also about the president's injunction against any discussion or disclosures about the atomic bomb. How would they be able to warn the world of the dangers of this new weapon, and work toward responsible international control, if the scientists were all silenced?

Oppenheimer had also quarreled with his old friend Lawrence about returning to Berkeley, perhaps sensing that the campus was not big enough for two of the project's brightest stars, and was distractedly entertaining offers from Conant, who was trying to lure him to Harvard, and from his old colleagues at Caltech. If there was any real chance to make a positive contribution to future atomic national policy, he wanted to help,

*In a bizarre twist, Slotin was involved in a similar accident on the same day one year later. In October 1946, he was showing a group of physicists a plutonium assembly at Los Alamos and was using a screwdriver to lower one of the hemispheres of beryllium on the core when the tool slipped, the pieces came into contact, and the assembly went critical. Five others in the room were irradiated, but only Slotin, who was closest, died.

he wrote Lawrence. Summarizing his report to Stimson, he added, "All of us would earnestly do whatever was in the national interest, no matter how desperate or disagreeable; but we felt reluctant to promise that much real good could come of continuing the atomic bomb work just like the poison gasses after the last war."

Oppenheimer wanted out of the Manhattan Project and the bomb-making business, even though it was apparent even then that Lawrence, not to mention others, would have "very strong, very negative reactions" to his misgivings. It was not the first time he and Lawrence had disagreed about Oppie's politics and championing of causes, and it would not be the last. Uncannily foreshadowing the breach that would soon end their friendship, and ultimately divide the field of physics for decades to come, Oppie wrote that Lawrence would do well to remember "how much more of an underdogger I have always been than you. That is a part of me that is unlikely to change, for I am not ashamed of it; it is responsible for such differences as we have had in the past, I think; I should have thought in the long years it would not be new to you."

It had been several years since the old Oppenheimer, the Berkeley campus idealist, had been heard from. But now that he had discharged his duty at Los Alamos, he was returning to a more familiar role. The younger laboratory scientists, deeply concerned about how the U.S. government would use its terrible new power, had banded together to form the Association of Los Alamos Scientists (ALAS), and had approached Oppenheimer about helping them bring their views to the press.* Its purpose, set forth in a newsletter, was to "urge and in every way sponsor the initiation of international discussion leading to a world authority in which would be vested the control of nuclear energy."

In the aftermath of Hiroshima, the newspapers had been filled with sermons railing against the bomb, and resolutions calling for abolishing all kinds of atomic energy, which had come to be seen only as a destructive power. All kinds of ill-conceived schemes were being put forward. Faced with this alarming national trend, ALAS felt an obligation to edu-

*ALAS later merged with similar groups to become the Federation of American Scientists.

cate both the policy makers in government and the American public about the bomb so that they could make informed decisions about its future use. They were optimistic they could work with the United Nations to establish a system of international control that would avoid an atomic arms race. ALAS members began holding meetings, drafting statements of principle, publishing articles, and giving speeches in Albuquerque and Santa Fe. With his tenure as head of the weapons laboratory almost over, Oppenheimer felt a little freer to express his own views and agreed to take up their cause. As soon as he formally resigned that fall, he joined the ranks of ALAS.

In the meantime, Oppenheimer forwarded the ALAS statement of purpose to George Harrison, Stimson's aide, and attached a polite cover letter stating that he had been informed that "the views expressed in the statement are held very nearly unanimously; that of all the civilian scientists (something over three hundred) who could be reached, only three felt they could not sign the statement." He added, "[Although I had] no part in the organization of the group or preparation of the statement, you will probably recognize that the views presented are in closest harmony with those I have discussed with the Interim Committee."

It was hardly the ringing endorsement his Los Alamos colleagues might have hoped for, but Oppenheimer was still convinced that the political and military leaders were intelligent and reasonable people, and that he could be effective in his role as policy advisor and help them come to responsible decisions. So he did not object when the ALAS statement, which was duly shown to the Cabinet, was immediately classified and its contents sealed. Much to the Los Alamos scientists' dismay, the public was not to know of their dissent until later. As Alice Smith observed of Oppenheimer's first forays into diplomacy, he was trying to establish himself as "an inside scientist" and work within established channels. He had developed great confidence in Stimson and his War Department staff, wrote Smith, and "did not realize, with Stimson on the point of retirement, how rapidly their influence was being supplanted by that of advocates of the cold war posture."

Not all the physicists at Los Alamos felt that Oppenheimer represented their views. A number of the émigré scientists, including Fermi and Teller, were far from convinced that public understanding and an international arms control authority would solve the world's problems. There was always the danger another tyrant would come along who would not care how much damage he inflicted in order to impose his will. They took exception to Oppenheimer's philosophizing and, in particular, to a number of almost religious quotes that had been picked up by the media and widely circulated and were becoming a kind of mournful motif for the last months of the project. Oppenheimer was always referring to the Bhagavad Gita, and after Hiroshima, importance was invested in his allusions to the Hindu sacred text. When some years later, during an address at MIT, he said, "Physicists have known sin," many of the project scientists were upset and strenuously objected that he had no right to speak for them, especially as they viewed themselves as having done an honorable service for their country. Charles Critchfield, who had been recruited to work at Los Alamos in late 1942 by both Oppenheimer and Teller, shrugged off the Sanskrit quotations as just another example of Oppie's penchant for grandstanding. "He had a great sense of the dramatic," he recalled. "He loved to make statements that were completely absurd without any warning if they were supposed to be funny or whether they meant something that you could take seriously."

This desire to provoke, which Critchfield noted "might be called a weakness," had often gotten Oppenheimer into trouble, but after the war it "proved his undoing." His displays of arrogance—or snobbishness, or narcissism—earned him a reputation for being difficult in Washington, and a number of influential administration insiders, including Lewis Strauss and Alfred Lee Loomis, formed a lasting dislike and distrust of him. Critchfield cited the story of Oppenheimer's first meeting with Harry Truman, when he walked into the oval office and declared theatrically, "Mr. President, I have blood on my hands." Put off, Truman offered him a handkerchief and asked him if he wanted "to wipe them." Afterward, the president reportedly told his aides to put a lid on the Los Alamos director. "Damn it, he hasn't half as much blood on his hands as I have," Truman snapped. "You just don't go around bellyaching about it."

This same tendency particularly worked against him with Teller, who never had any appreciation for Oppie's poetic license, particularly as he had often been the object of his slings and arrows, and grew increasingly impatient with what he saw as the director's consuming need for redemption. In his memoir, Teller recalled that the day after peace was established, Oppenheimer came into his office at Los Alamos and told him that "with the war over, there is no reason to continue work on the hydrogen bomb." Teller was stricken. Work on the hydrogen bomb had begun in earnest only two months earlier, following the Trinity test, when Oppenheimer had appointed Bethe and Fermi to head up the fusion bomb program. "His statement was unexpected," wrote Teller. "It was also final. There was no way I could argue; no way I could change Oppenheimer's mind."

> Beginning with his strange comment after the Trinity test (a quotation from Bhagavad Gita, "I am become the Destroyer of Worlds"), Oppenheimer had seemed to lose his sense of balance, his perspective. After seeing the pictures from Hiroshima, he appeared determined that Los Alamos, the unique and outstanding laboratory he had created, should vanish. When asked about the future, he responded, "Give it back to the Indians."

Teller was greatly disturbed by the growing sentiment among Oppenheimer's division leaders and senior staff to turn against military work. "The emotion," Teller observed, "seemed especially strong among those who had been most enthusiastic about using the bomb before the actual bombing." His bitter childhood memories of Hungary—and the violent antisemitism that led to the deaths of thousands of Jews—had made him less sanguine about the prospects for peace, and he wanted to see the atomic bomb program continue so that the United States could maintain its technological superiority in nuclear weaponry. After having been forced by Oppenheimer to take a backseat on the wartime project, he was more determined than ever to realize his ambition to develop the Super. But given the present mood at the laboratory, he knew there would be little support for a program to develop fission bombs, let alone a fusion weapon, which would be a hundred times as powerful. An interim successor for Oppenheimer as director had been named, but Teller had little faith in the mild-mannered Norris Bradbury, a Stanford University physicist who had joined the project midway. Bradbury wanted to turn Los Alamos into a peacetime research facility to study the military applications of nuclear energy and had asked Teller to remain on as head of the Theoretical Division. Teller chose to go to Chicago with his friends instead. "Bradbury," he observed, "maintained a cautious approach, then and throughout his career as director."

Teller's ambition to do work on the hydrogen bomb was well known on the mesa, and he was regarded, as much by himself as by everyone else, as virtually the sole proponent and defender of this work. "They were all against it, everybody except Teller," said Dorothy, recalling the strong feelings on the mesa at the time. "They thought it was shocking, excessive and unnecessary, and we could destroy the world easily enough with our little atomic bomb." Dorothy knew that Oppenheimer thought it was an unconscionable weapon. They had all had enough death as it was. "But Teller just thought for the sake of science we ought to do it . . . he kept hammering, hammering, hammering until he finally got it," she said. "We hated the idea of it."

At the farewell party for Deke Parsons, who had just been promoted, Oppenheimer greeted Teller jovially and said, "Now that you've decided to go to Chicago, don't you feel better?" Teller did not feel better, and he told Oppenheimer so, adding that he thought their "war time work was only a beginning." Oppenheimer brushed off the statement. "We have done a wonderful job here," he replied. "It will be many years before anyone can improve on our work in any way." Years later, Teller remained bitter about Oppenheimer's decision to send him packing. "He wanted to stop me," Teller said in an interview "He was very kind about it, but he was giving me a command. He wanted me to go back to Chicago and do physics." Teller went grudgingly, "but I was already wondering about his attitude." No two men could have been more at odds about the future development of nuclear weapons and which path the country should follow. Given their past experience, Oppenheimer should have known that this would not be the end of it. Teller would not be so easily dismissed.

On October 16, on a flawless New Mexico afternoon, the laboratory was awarded the Army-Navy "E" (for excellence) Award. A grandstand had been erected in front of Fuller Lodge for the occasion, and draped in red, white, and blue bunting. Beyond, on the horizon, were the Sangre de Cristo Mountains, cloaked in fall shades of lavender and blue and as serenely beautiful as ever. There was no brass band, but a military guard marched in and put on a good show, making for quite a change from the laughable SED parades. An enormous crowd had gathered, including nearly all the remaining scientists, who now numbered about a thousand, as well as hundreds of WACs and GIs, clusters of Indian housemaids in their colorful shawls, and many of the Spanish laborers who had worked at the laboratory and had come out of respect to "Señor Oppenheimer." They were an eclectic lot, as usual, and it was amazing to think that they had ever managed to come together and work effectively as a team. Dorothy, who had come up to the Hill for the ceremony, was amused by the improbable sight of a radio crew struggling to set up their equipment to broadcast the speeches to the world at large. Much had changed in their secret city in the two months since the war ended.

As she joined the milling throng, shaking the hands of people she had hired over the past two years and hugging departing friends, Dorothy glimpsed Oppenheimer, a lean figure off by himself in the distance, dressed in an ill-fitting suit, his porkpie hat hiding the crew cut which had begun to show the first hint of gray. "Robert was pacing along," she recalled, "he was within himself, and I knew that because when he's within himself he's not conscious of anyone else." She went over to him and said, "Hello." When he looked up, there was "a rather glazed look in his eyes," and she realized he was still composing the speech he was due to give in a few moments' time. For some reason, that image of him became fixed in her mind. It was the way she always liked to remember him—wandering the mesa, lost in thought. That was "the best portrait of him," she said years later, "and it was one of the best speeches that has ever been done."

Dorothy sat in one of the rows of folding chairs in the dusty field and watched Oppenheimer on the stage, shaking Groves' hand and accepting the Certificate of Appreciation from the secretary of war. Next to the squat figure of the general, stuffed into his dress uniform, and the row of well-fed dignitaries, Oppie stood apart, looking strikingly tall and thin, like the member of a separate, attenuated race. He had kept his hat on, and the wide brim cast a shadow across his face as he stood to speak. He was the one they had all come to see, and a hush fell over the crowd as he briefly and very eloquently summed up what they all felt in their hearts, but had not been able to express. After thanking the men and women of Los Alamos for their work, Oppenheimer told them, "It is our hope that in years to come we may look at this scroll, and all that it signifies, with pride."

> Today that pride must be tempered with concern. If atomic bombs are to be added as new weapons to the arsenals of the warring world, or to the arsenals of nations preparing for war, then the time will come when mankind will curse the names of Los Alamos and Hiroshima.
>
> The peoples of this world must unite or they will perish. This war that has ravaged so much of the earth has written these words. The atomic bomb has spelled them out for all men to understand. Other men have spoken them, in other times, of other wars, of other weapons. They have not prevailed. They are misled by a false sense of human history who hold that they will not prevail today. It is not for us to believe that. By our works we are committed to a world united, before this common peril, in law, and in humanity.

Among the various awards handed out, every laboratory employee received a tiny sterling silver pin, no bigger than a dime, bearing a large "A" and in smaller letters, the word "BOMB." It was a strangely diminutive token to commemorate such a monumental undertaking, but they probably had their penny-pinching general to thank for that. According to the papers, Groves was already busy defending their $2 billion budget and charges that their man-made town was so expensive that the handful of aspens planted by the Army Corps of Engineers were referred to as "twenty-four-carat trees." But that did not matter. While Groves may have been head of the Manhattan District, and the University of California their actual employer, according to their paychecks, Dorothy, like most people on the mesa, had always thought of herself as working for

Oppie. He had been their undisputed leader, loved or despised, and was already well on his way to becoming a legendary figure.

Now that Oppenheimer was leaving, and had handed over the directorship to Norris Bradbury, the project was officially over. The laboratory might continue to exist, but it would never be the same. Only those who had been there from the very beginning, who had answered his urgent call for help, who had been sworn to silence and had labored long hours under conditions of extreme secrecy, could know what an adventure it had been. "It was sort of like falling in love," Dorothy told an interviewer years later. "You carry on and everything, and then you're aware of this fact that's hit you." Smiling, she shook her head, as if trying, and failing, to describe an unknown phenomenon, the Oppenheimer mystique. "I don't mean to compare this project—the atomic bomb—to love," she added. "I'm just trying to tell you . . ."

On another occasion, she put it all down to Oppenheimer's peculiarly vivid blue eyes, a visionary's eyes, the exact color of the gentians that carpeted the valleys, and the matchless New Mexico sky. "He hypnotized me with those blue eyes," she said.

He mesmerized them all. Dorothy's allegiance to Oppenheimer was deeply personal to the end, but it was not unique. In many ways, it was emblematic of his leadership style and the profound mark he made on everyone who came to the Hill. He inspired love, loyalty, hard work, and dedication. He seemed to expect no less, and he reciprocated with his warmth and solicitude, and by living up to his own high standards and never dictating what should be done. "He brought out the best in all of us," said Hans Bethe, when he spoke at Oppenheimer's memorial service in 1967. "Los Alamos might have succeeded without him, but certainly only with much greater strain, less enthusiasm, and less speed. As it was, it was an unforgettable experience for all the members of the laboratory. There were other wartime laboratories . . . but I have never observed in any of these other groups quite the spirit of belonging together, quite the urge to reminisce about the days of the laboratory, quite the feeling that this was really the great time of their lives."

TWENTY

Elysian Dreamer

With the project winding down, the pressure at 109 East Palace eased. Dorothy, like everyone else, was preoccupied with tidying up loose ends. The laboratory, like the country, would have to find a new way to go forward. A number of physicists who had fallen in love with the setting, and still savored the technical challenges, wanted to stay on. It looked as if the government might allow Los Alamos, which the papers had dubbed "Uncle Sam's town," to continue as a weapons research center for at least a few years, producing more bombs for the country's stockpile and working on necessary improvements.

Most of the younger scientists were dead set against this and insisted the laboratory should be closed down. A group of them had even taken a bus to Washington to protest the May-Johnson bill, hastily introduced before Congress on October 3, 1945, which left open the possibility for military control over all atomic energy work. The problem was that the vaguely worded bill, which was being rushed through hearings, did not go far enough in promising that nuclear research would be wrested from the military and put in the hands of a civilian commission. Complicating matters, Oppenheimer, along with the other members of the scientific panel, had sided with Groves and endorsed the legislation. Older, wiser, and weary, Oppie thought it would make for a smoother transition to leave the bomb initially in military control. He maintained that this was only an interim measure, and he believed the military authority would eventually yield to a world body. He was already drafting an amendment he hoped would explicate this and placate his critics.

At Los Alamos, politics had replaced physics as the topic of conversation, and as the arguments intensified, both sides became more entrenched. Morale was at a lower point that at any time during the war. The scientists seemed to have lost their sense of purpose and direction. The dissent that fomented on the mesa reflected the growing public tension and confusion about the country's atomic policy in the months after the war. In the meantime, with Los Alamos's future far from certain, Dorothy had been asked by Norris Bradbury to remain on as manager of the Santa Fe office. She agreed, girding herself for the troubled times ahead.

Almost immediately after the Army-Navy E Award ceremony, Oppenheimer went back to Washington. Dorothy knew it was the first of many good-byes leading up to his final departure for Pasadena later that fall. It was a time of many tearful farewells, but there was none she dreaded more. But at the same time, she could see how badly he wanted to get away. His beloved Sangre de Cristo Mountains were now "under the shadow of the mushroom cloud," and the mountains and wooded trails would never again promise the same sweet escape and tranquility. There was nothing to hold him there anymore. He had made a tremendous sacrifice for his country, he had paid for it in pounds of flesh, and the time had come for him to move on to new challenges. War was a time when personal desires were of necessity subjugated to a singleness of purpose, and she had satisfied herself with being his most loyal and devoted lieutenant, his confidante, and his friend. She had done so readily, and without regret.

Oppenheimer had told her that after much indecision, he had finally made up his mind to rejoin the faculty at Caltech. It had not been an easy decision. He had turned down an invitation to go to Harvard, sending Conant a heartfelt note saying that his "one regret" was that he would not have him as a boss. "I would like to go back to California for the rest of my days," he wrote. "I have a sense of belonging there which I will probably not get over." He had still been undecided about returning to Berkeley, however, writing to his old professor Raymond Birge, "how hard it would be" not to return, but at the same time confessing real doubts about the kind of welcome he could expect. He had clashed with officers of the university when he was running Los Alamos and was wary that the high-profile role he sought as presidential advisor on atomic policy might result in further conflict. He added, "As you can see, I am worried about the wild oats of all kinds which I have sown in the past; nor am I quite willing in the future to be part of any institution which has any essential distrust or essential lack of confidence in me."

Although Oppenheimer was anxious to join the arms control issue along political lines, realizing that the opportunity to use his influence might not come along again, some of his devoted staff members worried about how he would fare on the treacherous path from Los Alamos to Washington. His new secretary, Anne Wilson, who had been accompanying him to the capital, where he was testifying at Senate and House committee hearings, confided her doubts to her predecessor, Priscilla Greene Duffield, commenting that the theoretical physicist might not be cut out for the rough and tumble world of politics. "She said, 'He'd better be careful. He is going to get into terrible trouble,' " recalled Priscilla. "I think she was referring to the fact that he spoke out immediately about what he was feeling, and that he perhaps had the wrong [mentors], the people he was admiring in Washington initially were the wrong people. She really was concerned about him and obviously knew what she was talking about."

Wilson remembered her sense of foreboding that autumn. Oppenheimer's new celebrity and talent for clear exposition made him an obvious choice for the administration, and he would not be able to resist the lure of the corridors of power. "He was riding high and enjoying it," she said. "And he was very intense about trying to make something out of all this that was not all bad." But Oppenheimer had more enemies than he knew, and she wondered who in Washington would watch his back. Wilson was particularly sensitive on this score as she had been adopted by the Tellers upon arriving at Los Alamos and had observed Edward's deep antagonism toward Oppenheimer, even though she could see that the obstinate Hungarian had brought many of his problems on himself. Oppie had a gift for drawing people to him, like moths to a flame, and they either basked in his glow or got burnt. "The woods were always thick with people who had nasty things to say about Robert," she said, noting that everyone knew that in some sense what he had achieved at Los Alamos went beyond any Nobel Prize. Jealousy was already afoot on the mesa: "There were always people who were vying for his attention, and those who felt snubbed by him, or felt hurt because they thought Robert didn't love them anymore."

Oppenheimer's verbal knife play, which he had used over the years to dazzle or wound, had given more offense than he realized, and after the war it became more of a problem. His reputation as a great humanist, and his new seerlike role in world affairs, gave even his offhand comments an edge. Even some of his old friends, who were familiar with his sharp tongue and his habit of poking holes in people, including those he liked and admired, were finding it hard to overlook some of his behavior and had cooled toward him. "He was so arrogant after the bomb—his triumph," said Emily Morrison. "This confirmed to many people, particularly those who never liked him, their worst suspicions about what he was really like. For his students, of course, he was still God, and they went on worshipping him." But Morrison recalled that when Oppenheimer started casually dropping the names of four-star generals in conversation, and took to calling General Marshall "George," even her husband, a longtime acolyte, decided to stop idolizing him. His fame as "Father of the Atomic Bomb" was having an intoxicating effect on him. "Oppie changed," she said. "He was interested in being a great man and dealing with other great men." Phil Morrison did not dispute his wife's appraisal. "After the war, Oppenheimer thought he was powerful," he said. "But he was not as powerful as he thought."

Many of the project scientists were appalled by Oppenheimer's support of the May-Johnson bill—which Harold Urey labeled "either a Communist bill or a Nazi bill, which ever you think is worse"—and felt betrayed by him. A few outraged young physicists argued that their former leader must have been duped and was unwittingly being used as a pawn by the War Department. How could the wizardlike Oppenheimer, who had achieved folk hero status as the philosopher-king of Los Alamos, acquiesce to the political elite? Leave science to the scientists, he had lectured them, and politics to the politicians. But for all his speeches, the only cause he appeared to be championing was his own. His name was all over the newspapers, and he talked sagely of the bomb peril, and constructive versus deconstructive uses of atomic power, but in the eyes of many physicists, he was achieving a dubious kind of notoriety. While putting himself forward as the nation's atomic expert, he was fanning public fears about the deadly new threat facing the world and adding to the growing state of congressional alarm.

When a senator asked him if it was true that one raid on a U.S. city could kill 40 million Americans, Oppenheimer said: "I am afraid it is." He was quoted in *Time* as suggesting that in the long run, the bomb would actually weaken, rather than strengthen, the U.S. military and international position because "atomic weapons ten or twenty years from now will be very cheap," so America—and presumably Russia—would be able to afford to accumulate stockpiles of bombs in the near future. There was no limitation on man's ability to destroy his fellow man, he warned, and future bombs would be "terribly more terrible."

Watching Oppenheimer's perplexing performance on Capitol Hill, in which he alternated between taking credit for the bomb's creation and calling it "an evil thing," as he told the National Academy of Sciences, even his loyalists at Los Alamos felt he had done a poor job of presenting to the public what he proposed to do about the weapon that had only recently decimated Hiroshima. Not surprisingly, neither the May-Johnson bill, which Oppie campaigned for, nor any of his amendments came to fruition. A few months later, a rival bill, introduced by Senator Brian McMahon of Connecticut and supported by the dissenting scientists, became law. Although Conant, who with Vannevar Bush helped draft the May-Johnson bill, maintained that it was never meant to turn atomic energy affairs over to the military, the bill did provide that one of the positions on the committee be filled by a military officer. Everyone assumed—however incorrectly—that the seat would automatically go to Groves, whom they suspected of trying to extend his wartime authority over atomic energy. As Conant noted, "The scientists could contemplate such a possibility only with extreme horror." Oppenheimer had not proved an effective advocate for the cause of international control of nuclear weapons, and in part because of his close association with Groves, his reputation in the scientific community was somewhat tarnished.

The tempting possibilities of being involved in planning the future management and development of atomic energy had not materialized. Feeling demoralized, Oppenheimer returned to Los Alamos in early November and gave a farewell address to the members of ALAS. A crowd of more than five hundred packed the largest of the two theaters to hear him speak. He began, as he usually did when speaking in public, in a quiet voice and slowly warmed to his theme of the enormous impact of the atom bomb, which "arrived in the world with such a shattering reality and suddenness." Oppenheimer may not have succeeded as a statesman, but he still greatly impressed his fellow scientists as one of the most far-sighted thinkers, in a league only with Bohr. Even many years later, those who were present recall it as one of the most deeply affecting speeches they had ever heard.

Oppenheimer resumed his professorship at Caltech. He took up his old research in cosmic rays, but found it hard to shut the door on the exciting world of international postwar policy, and all too soon was drawn back into the fray. Looking back on that time, he admitted that teaching had lost its luster compared with the chance to play a pivotal role in the mounting conflict between the United States and Russia. "I was asked over and over both by the Executive and Congress for advice on atomic energy," he said, sounding the part of the reluctant bride. "I had a feeling of deep responsibility, interest and concern."

His opportunity came in January 1946, when the United Nations General Assembly convened in London and Russia voted along with the rest of the world to exchange atomic information and to establish the United Nations Atomic Energy Commission (UNAEC). Secretary of State James Byrnes wanted a committee to formulate American policy. Byrnes asked Undersecretary Dean Acheson to head it and appointed Bush, Conant, Groves, and John J. McCloy as members. Acheson persuaded David Lilienthal, then head of the Tennessee Valley Authority, and a superb administrator, to chair an advisory panel on all the facts bearing on international arms control. Oppenheimer's name was immediately put forward as a logical choice.

Herbert S. Marks, a brilliant legal aide to Acheson, was to assist the panel, and he introduced Oppenheimer to Lilienthal. Marks, who would soon marry Oppie's pretty young secretary, Anne Wilson, arranged for Lilienthal and the physicist to meet at the Shoreham Hotel in Washington, D.C. The erudite Los Alamos director immediately struck Lilienthal as "an extraordinary personage," as he noted in his journal: "He [Oppie] walked back and forth making funny 'hugh' sounds between sentences and phrases . . . a mannerism quite strange, very strange. . . . I left liking him, greatly impressed with his flash of mind, but rather disturbed by the flow of words." After another meeting the next day, Lilienthal was so impressed with Oppenheimer he later gushed: "He is worth living a lifetime just to know mankind has been able to produce such a being. We may have to wait another hundred years for the second one to come off the line."

Oppenheimer was recruited to join what became known as the famous Lilienthal board of consultants. On January 28, 1947, the group went to work. Oppenheimer acted as the advisory panel's tutor and began what amounted to an intensive course in nuclear physics. By March 7, after long and hard labor, they had completed what came to be called the Acheson-Lilienthal report, though it was largely authored by Oppenheimer. Oppie rejected earlier proposals suggested after Hiroshima that all nations outlaw the bomb and create international inspectors to ensure no nation was manufacturing one. Instead, he argued that the only workable system of safeguards required an international agency with authority over all atomic energy work, including the construction of reactors, separation plants, and laboratories, as well as the raw materials needed to make bombs. In typical Oppie fashion, he swayed everyone to his way of thinking, and the committee advanced his proposals with almost complete unanimity. After thorough discussion, and the incorporation of a number of changes, Acheson endorsed the report as "a foundation on which to build," and the plan was adopted as official U.S. policy.

Groves, who had objected to the scientific panel in the first place, knew exactly what he was up against. He had utilized Oppenheimer's persuasive powers to his advantage in the past, and now found himself in the unfortunate position of arguing against him on the issue of international control. "Everyone genuflected," he complained later. "Lilienthal got so bad he would consult Oppie on what tie to wear in the morning." Groves regarded the plan as overly optimistic and unworkable. It recommended "step-by-step co-operation with the other powers rather than for the first steps to be taken by the United States," Groves noted, adding, "This was not our usual diplomatic approach."

Even though the response to the Acheson-Lilienthal report was overwhelmingly favorable, with Alfred Friendly of *The Washington Post* writing that the statesmanlike report offered hope for lifting the "Great Fear" that had descended over the world since Hiroshima, it came in for some sharp criticism. Conservatives denounced the plan as a transparent scheme to hand over atomic secrets to the Russians. The columnist Dorothy Thompson dismissed Oppenheimer's vision of an international authority as an "Elysian daydream" and suggested he was better suited to those mythical fields than the real-world business of making new law.

Much to the Acheson-Lilienthal group's dismay, President Truman appointed the feisty seventy-five-year-old financier Bernard Baruch to present the plan to the United Nations Atomic Energy Commission. A respected, conservative elder statesman, Baruch had little or no knowledge of the technical complexities of the subject. He immediately proceeded to introduce changes that distorted Oppenheimer's whole plan, minimizing the joint task of developing atomic power for peaceful uses and emphasizing the punishment of violators of the control treaty. Oppenheimer, who advocated a conciliatory approach to Russia, was so distressed by the choice of the hard-liner Baruch that he refused to act as his chief scientific advisor. "That was the day I gave up hope," he said later, "but that was not the time for me to say so publicly."

Making matters worse, just two weeks after Baruch unveiled the American plan for international control on June 14, dramatically announcing to the delegates of the United Nations Atomic Energy Commission that they were gathered there "to make a choice between the quick and the dead," the United States went ahead with the first of two atomic bomb tests at Bikini Atoll, in the Pacific. Oppenheimer had objected as soon as the tests were announced and wrote Truman a strongly worded memo questioning the wisdom of conducting the tests while simultaneously trying to seek military control. He argued that the trials were both counterproductive and a waste of money. The Bikini tests would cost $100 million, and for less than a million the scientists could simulate the blast's effect on a warship. But after Hiroshima, the navy had convinced the administration that it was vital that they learn exactly what a nuclear weapon could do to their fleet, and the president had agreed. So on July 1, 1946, eleven and a half months after Trinity, America dropped an atomic bomb over a group of warships anchored in the lagoon of Bikini Atoll. Somewhat disappointingly, the first test bomb—which produced the classic purple mushroom cloud form—succeeded only in sinking a destroyer and two transport vessels. A beneath-the-surface test followed on July 25, producing a huge column of water that was reportedly far more impressive to the official observers, including military experts, congressmen, scientists, foreign dignitaries, and an entire shipload of journalists.

However ill timed the Bikini tests were, Oppenheimer regarded Baruch as an even greater threat to the success of the negotiations. Baruch's dogmatic approach, and his insistence upon the principle that violators of the treaty should have no power of veto to protect themselves from punishment, stalemated the talks. He insisted all decisions had to be carried by a majority vote, or the organization would become "no more than a debating society." The Russians may never have accepted the agreement, but Baruch's heavy-handed tactics gave them the excuse they needed to say that the Americans were negotiating in bad faith. They dug in their heels, refusing to consider the removal of the veto—and privately counted the months until they could duplicate what the Manhattan Project had produced. The Soviets, thanks to Klaus Fuchs' funneling of Los Alamos's atomic secrets, were confident the American monopoly on nuclear weapons was only temporary. Baruch did not succeed in forcing a decision. When the delegates of the United Nations Atomic Energy Commission finally voted on the plan on the last day of December, the Soviet Union abstained.

Oppenheimer sat on the sidelines at the first meeting of the UN Atomic Energy Commission, held in the gymnasium of Hunter College, and watched the proceedings with a sinking heart. He knew the Russians would reject the plan as presented and saw his last slim hope of warding off a nuclear arms race disappear. America would keep building more and better bombs, assuring itself of its military superiority in the transition period, until a treaty could be ratified. It would continue to hold tight to its atomic secrets in the vain belief that they actually were secrets and that other countries would not soon catch up. Disarmament looked to be a hopeless cause. Oppenheimer's dream of controlling his creation, and using it to build a lasting world peace, was all but dead.

Tired and disillusioned, he returned to his ranch in the Pecos for solace. During his weeklong holiday, Dorothy joined him and Kitty for dinner one night at La Fonda. At one point, Oppie had to leave the table to take an important call from Washington. When he returned, he had lost his appetite. The caller had just informed him of the Soviets' latest obdurate stance at the United Nations. "It's finished," he said with disgust, pushing his plate away. "Russia wants world conquest."

Over the previous year, Dorothy and the remaining members of the laboratory staff had followed Oppenheimer's adventures in Washington with both admiration and trepidation. They had applauded his efforts to control weapons of mass destruction and prevent future wars, quoted from his speeches and interviews with pride, and blamed his failures on the political climate. Many of the old guard, like Dorothy, had come to regard Oppenheimer as the patron saint of the mesa, spreading the word that physicists should leave behind the "devil's work of making armaments" and get back to their "real work—the sober, consecrated task of penetrating the unknown." But while he was off doing God's work, Los Alamos had gone into a steady decline, and it was hard for some not to feel that their former leader had abandoned them. As 1945 came to a close with no clear resolution, and no congressional vote on atomic energy legislation, people began to question the laboratory's reason for being. The division leaders had all left for greener pastures, to the promise of better salaries and housing, leaving only a few experienced hands to man an operation that was seriously shorthanded. Army interference had increased, and there was widespread discontent.

By early 1946, Los Alamos was on the brink of collapse. The infrastructure of the hastily constructed town was failing. The generators were on their last legs, and the power outages became more frequent than when the laboratory was in full gear. The water shortage, a chronic problem in the war years, had worsened. The winter had been particularly cold, and the plastic pipeline carrying the water from Guaje Canyon to the post, which for reasons of economy had been laid on the surface of the ground, had frozen. The army instituted its usual conservation orders, and for weeks the use of all toilets and bathing facilities was suspended. At the height of the emergency, trucks had to haul untreated water up the mountain from the Rio Grande, and the army doled it out in pails. Doc Barnett, who had been dispensing typhoid booster shots at the post hospital, finally phoned Washington and warned that he would not be held responsible for the public health consequences if something was not done at once.

Before it was over, the crisis caused both the army and Los Alamos authorities considerable embarrassment, and wrecked a long-planned celebration formally introducing the secret community to Santa Fe society. Dorothy, working with a Santa Fe citizens' committee, had helped arrange the special evening at the Museum of Anthropology, opposite her house. There was a program of speeches featuring Phil Morrison and Victor Weisskopf, to be followed by a question-and-answer period and an elegant dinner in one of the museum rooms. Afterward, the scientists were besieged with invitations to come to the homes of the city's cultivated benefactors, who were fascinated by the atomic scientists who had been kept under wraps for so long. As usual, Dorothy acted as go-between, sending word that two dozen scientists and their wives had been invited to a dinner, and posting sign-up sheets on the bulletin board of the ALAS office.

To reciprocate, Rose Bethe and Jean Bacher persuaded security to allow them to throw a party and treat a group of Santa Fe's leading citizens to a tour of their mystery town. "When permission was given, Dorothy McKibbin helped us make out the list of guests and phoned them from her office at 109 East Palace," recalled Bernice Brode. Dorothy briefed the visitors on security and the pass system, and arranged transportation. Then the water shortage became acute, and with the taps dry and toilets out of order, the party had to be canceled. Dorothy, who had to be the bearer of the bad news, reported back that the good citizens of Santa Fe were crushed. Feeling thoroughly chastened, the Hill wives decided to make the best of a bad situation and go ahead with the planned festivities. Their visitors would have to take them as they were, unbathed and unshaven. Dorothy called everyone back to say the party was on again, and the Santa Feans started arriving at the East Gate for their promised tour and cocktails at the Tellers'. Another of the project stars, Johnny von Neumann, was on hand to charm the old ladies.

As it turned out, Los Alamos made quite an impression. "All the guests had been so appalled at our town that they were in a state of shock," recalled Brode. One woman told her, "I can't get over it. Such nice people living in such a place all this time. Incredible, my dear, unbelievable, we had no idea." But Dorothy made everyone on the Hill feel much better when she phoned a couple of days later with an update. "The city of Santa Fe was now divided into two parts," she reported, "those who had seen and those who had not seen."

To remedy the situation, Groves ordered the construction of wells, pipelines, and pumping stations to bring water to a new one-million-gallon steel storage tank. He also got approval for the construction of three hundred units of commercial-grade housing, so the laboratory scientists could move out of their hovels. This was the first step in what would be a long process to turn the hodgepodge military installation into a civilized town with sidewalks, paved roads, planned business centers, and utilities. His move precipitated by several months an executive order, issued on the last day of 1946, ensuring that the Los Alamos site would become permanent. It was decided the routine military work of producing bombs and preparing them for Bikini bomb tests, which had drained the lab's resources for the past year, would no longer be handled by the laboratory. Los Alamos would be devoted to research on atomic bombs, as well as the prospects for more powerful weaponry, including Teller's Super. For reasons of morale, there would also be studies on applications of atomic energy for peaceful uses. The lab would still be doing classified work, but security would not be as stringent as during the war.

On July 12, 1947, Los Alamos was transferred to civilian control under the auspices of the AEC. With its future no longer in doubt, and the living conditions becoming more attractive, new scientists began coming to the laboratory, and high-level military leaders and VIPs again became frequent visitors. Teller returned on a part-time basis to continue his work on the hydrogen bomb, and familiar faces like Bethe, Fermi, and Lawrence often came through to see how he was progressing. Dorothy continued welcoming newcomers to the Hill, screening employees, and running the one-of-a-kind information bureau she had pioneered during the early days of the project. Over time, her duties increased as she continued to help service the needs of the expanding, clamoring city on the Hill. After the war, in an effort to improve the quality of life for those remaining on the mesa, she established a shopping service that tracked down items not stocked in the army PX. As more new personnel started arriving, she also started an official hostess program to provide information and assistance to project employees. Even when the construction workers' ugly old temporary quarters were razed, and new upscale bedroom communities began to grow up around Los Alamos, housing remained a problem, and Dorothy continued to help people find adequate accommodations.

Tourists continued to poke their heads into her office at 109 and pester her with questions about Los Alamos. She blamed Hollywood for fanning the flames of notoriety. The 1946 March of Time newsreel about the project, *Atomic Power*, had brought droves of people to her door, but it was nothing compared to the numbers who came after Metro-Goldwyn-Mayer's A-bomb blockbuster *The Beginning or the End*, which premiered to much fanfare in Washington, D.C., in February 1947. Hume Cronyn starred as Oppenheimer and Brian Donlevy as General Groves, and publicity posters of the duo were plastered all over town. Dorothy thought Cronyn did not begin to capture Oppie's good looks or charisma, and she agreed with the *Times'* critic who found the film to be imbecilic. She was pleased as punch when the picture flopped.

As the pace slowed, there was more time to socialize with the project scientists who had become her close friends. Dorothy would invite them to her home for dinner parties, and they finally had a chance to meet her friends from Santa Fe, including the painter Cady Wells, the writer Witter Bynner, and other local luminaries. For the wartime scientists, who had spent two years living next to the centuries-old town without knowing a soul other than Dorothy, it was a wonderful opportunity to finally meet members of the artistic community they had heard and read so much about. Up till then, they had had to satisfy themselves with pouring over guidebooks and taking walking tours of the town, peering through the gates of the old pink adobe houses, and at the lush walled gardens tucked inside, and wondering about the poets or painters who lived there.

As one of Santa Fe's most prominent citizens, Dorothy was now much in demand, and she became a leading figure in the city's cultural affairs, sitting on the boards of several museums and the opera, and through her the separate communities became better acquainted and formed enduring ties. The "Atomic Lady," as she was affectionately known, was a public relations force, and she did much to secure the laboratory's good reputation at a time when many communities were unhappy about having bomb factories next door. With her encouragement, the Santa Fe Committee for Atomic Energy was formed, with Witter Bynner as its president. He even dedicated a series of poems in praise of the new atomic age.

Dorothy's wedding chapel, which had served thirteen couples during the war years when there was no place else for them to go, continued to do a flourishing business. Many of the young project scientists chose to return to Santa Fe to get married at her beautiful adobe farmhouse, where they had such happy memories. One of these couples, Becky and Ben Diven, exchanged vows in her garden in 1951. "Dorothy was in many ways the heart of the project," said Becky Diven, and "we all considered both Oppie and Dorothy to be part of our most intimate family." Many others echoed the sentiment. "We were all so young in the beginning, and the town was so new, we sort of grew up together," said Marge Bradner, who was secretly married there in 1943. In that intense twenty-seven month period, 173 babies were born on the Hill, and Dorothy helped look after them all. "They were like our parents," agreed Marguerite Schreiber, "Dorothy and Oppie. They got us through it."

Dorothy remained close to Oppenheimer, and they kept up a warm, steady correspondence. For the first few years after the war, he was a regular visitor. Whenever he was in town, he would come to the house and they would fix their ritual dinner of steak and asparagus and sip the strong martinis, expertly prepared by him, on the patio. He and the children would often stay the night with her on their way to and from their ranch in the Pecos, and she kept a storeroom full of their sleeping bags and other camping supplies. Other times, Dorothy and Kevin would go up and spend the day with them at Perro Caliente and go on hikes and picnics as in the old days. She had promised to keep an eye on the ranch for Oppie and often drove up with mutual friends and reported back to him about broken windows and water pipes. As the years went by, she saw less and less of him. In the fall of 1947, Oppenheimer assumed the prestigious post of director of the Institute for Advanced Study at Princeton. He bought a large, stately white house, called Olden Manor, and added a greenhouse so that Kitty, who studied mycology, could finally grow the plants she had tried in vain to coax out of the arid soil at Los Alamos. They spent their last extended summer at Perro Caliente in 1950. Beginning the following year, the Oppenheimers started vacationing in the Caribbean. The island of St. John, where they built a house, became their preferred holiday retreat.

Dorothy's feelings for Oppenheimer were not diminished. "In a word, she loved him," said Kevin. "He was the finest man she would meet, and she devoted herself to him. He took my father's place, no question about it. He filled a void in her life. Beyond the devotion there was a genuine understanding and acceptance of the man, his family, and their many problems—the whole complicated equation. Kitty knew this, and came to rely on it. Over the years the two women became closer, establishing an entente only they understood. Perhaps because they shared a love for the same man, they learned to share his triumphs and failures, happiness and heartache.

Kitty continued to be in a bad way, and Oppenheimer blamed himself for her problems. Dorothy would hear both sides of the sad story, first from one and then from the other. Kitty's nerves were not good, and it did not help that she was drinking heavily. (Less charitable members of the Princeton faculty had renamed the director's house "Bourbon Manor.") Kitty had also begun to suffer from the recurrent intestinal ailments—"la grippe," as Dorothy called it in consoling letters—that would often leave her bedridden and unable to travel. On those occasions, Dorothy would accompany Oppie on trips, as she did when she went with him to the inaugural "Conference on the Foundations of Quantum Mechanics," held on Shelter Island, New York, in June 1947. All during those first years after the war, when Oppenheimer was commuting to Washington, and Kitty was alone and depressed, she took to calling Dorothy. Kevin remembered seeing his mother taking the phone into another room and disappearing for hours while she patiently listened to Kitty pour out her troubles. "It might have been her husband's difficulties or her own demons," he said. "Either way, my mother would try to talk her down."

As the 1940s came to a close, Oppenheimer was the most admired and certainly best known of the World War II physicists, his gaunt visage with the haunted eyes almost as recognizable as Einstein's, who was one of his colleagues at Princeton. Oppenheimer had become a highly influential figure in Washington and a much-sought-after confidential advisor in international relations, which, as he had accurately foreseen, was now dominated by nuclear policy. He had been appointed to a six-year term on the AEC's General Advisory Committee (GAC), along with eight other distinguished scientists including Conant, Fermi, and Rabi. At their first meeting, they unanimously elected him chairman. In their first year, Oppenheimer's committee worked with Groves to set a new, productive course for the Manhattan Project's moribund laboratories and, most significantly, recommended that Los Alamos develop a wide range of nuclear weapons.

As director of the Institute for Advanced Study, Oppenheimer was regarded as the intellectual guru to an amazing roster of talent that ran the gamut from Nobel Prize–winning physicists Niels Bohr and Paul Dirac to the poet T. S. Eliot and the historian Arnold Toynbee. As *Life* magazine wrote in a reverent cover story in October 1949, "Unlike many scientists, Oppenheimer has a Da Vincian range of interests and knowledge, encompassing the arts and humanities, the social sciences, current affairs and oriental philosophy. He is a linguist who finds himself at home in half a dozen languages including Sanskrit. His own rhetoric, both written and ad lib, is rich and exquisite. Over and above his scholarly achievements he has revealed himself as a graceful executive and diplomat, as astute and imaginative in the performance of his public role as leader of the nation's scientists."

In the article, one of his institute colleagues pointed out that Oppenheimer encompassed "two antithetic but complementary natures—the man of science and the man of affairs, creator and administrator, introvert and exponent." Three years later, those contradictory sides of his personality would be subjected to merciless national scrutiny.

TWENTY-ONE

Scorpions in a Bottle

THE YEAR 1950 began a dangerous decade in American politics. Senator Joseph McCarthy's rabid Cold War anti-communism and campaign of vicious denunciation was spreading fear and uncertainty throughout the country. The sensational allegations drummed up by the House Un-American Activities Committee (HUAC) hearings in the fall of 1947, which pressured motion picture studio heads to "name names" of prominent Communists in the film business and resulted in the blacklisting of hundreds of writers, actors, and directors, had already cut a wide swath through Hollywood. The following August, Washington was rocked by Whittaker Chambers' accusation during a HUAC hearing that Alger Hiss, a senior State Department aide, was a member of a Communist cell scheming to infiltrate the American government. Hiss's guilty verdict on January 21, 1950, followed less than two weeks later by the shocking news from London that Klaus Fuchs, one of the Los Alamos atomic physicists, had confessed to being a Soviet spy, threw the government into a state of virtual panic.

Exploiting the turmoil, the junior senator from Wisconsin decided he could make headlines by exposing the Communist infiltration of the State Department and the army. On February 12, 1950, after a speech to the Ohio County Women's Republican Club in Wheeling, West Virginia, McCarthy tauntingly waved in front of reporters a sheet of paper that he claimed contained "a list of 205" names of known Communists who were "working and shaping policy in the State Department." He made the same charge again and again in the days to come, changing the number of "bad risks" to 57 and then 81, as the press clamored for details of McCarthy's witch hunt.

Given the virulent times, it was a wonder Oppenheimer was able to go on operating in Washington as well as he did, advising both the State and Defense Departments, and preaching his increasingly unrealistic and unpopular doctrine of international control. He had more reason than most to be apprehensive, if not downright paranoid, given the complex and strained personal and professional entanglements that littered his past. It was the heyday of wild rumors, informers, and smear campaigns, and Oppenheimer was beginning to realize, possibly for the first time, that he was terribly vulnerable. Despite the fact that he knew he had many enemies, he made no attempt to temper his rhetoric. But the more he bucked the tide, and defied public and political opinion by holding fast to Bohr's humanist message to stop building bombs for the sake of the future, the more he drew fire. As McCarthy's fanaticism swept the halls of government and hardened the lines of debate, Oppenheimer's usual ability to charm and cajole deserted him, and he began to lash out at critics he saw as stupid or ignorant with his acid tongue, humiliating antagonists in ways they never forgave or forgot. In private, he admitted to close friends that the ruthless "Red scare" tactics and political thuggery of the new Cold War era frightened him, and he had already begun looking nervously over his shoulder to see if McCarthy's bloodhounds were on his trail.

Oppenheimer had already had a narrow escape after his appointment to the AEC in 1946, when FBI director J. Edgar Hoover, who was now in the possession of all the Manhattan Project security files, personally sent a voluminous file of "derogatory information" to Lilienthal, charging that the nation's foremost physicist might be a Communist and could represent a dangerous security risk. Lilienthal did not believe Oppenheimer was a security risk, but after reading the dossier, he and his four AEC commissioners recognized they had a political disaster on their hands if any of it were leaked to the press. Lilienthal quickly called Conant and Bush for character references, and the two senior scientists confirmed that there was no doubt as to Oppenheimer's loyalty and patriotism. Conant put his opinion in a letter to Lilienthal later that month:

> I can say without hesitation that there can be absolutely no question of Dr. Oppenheimer's loyalty. Furthermore, I can state categorically that, in my opinion, his attitude about the future course of the United States government in matters of high policy is in accordance with the soundest American tradition. . . . Any rumor that Dr. Oppenheimer is a Communist or toward Russia is an absurdity. As I wrote above, I base this statement on what I consider intimate knowledge of the workings of his mind.

Conant and Bush both put their reputations on the line for Oppenheimer and quietly obtained similar endorsements from Groves and Secretary of War Robert Patterson. The AEC chairman, on the basis on these assurances, agreed that there appeared to be "no immediate hazard" requiring an official inquiry and, on August 6, 1947, formally ruled that Oppenheimer's security clearance could continue. But even as Oppenheimer squeaked by, his brother, Frank, was collared. A month before the AEC decided to overlook Oppenheimer's past, the *Washington Times-Herald* ran a front-page story on July 12 under the headline U.S. ATOM SCIENTIST'S BROTHER EXPOSED AS COMMUNIST WHO WORKED ON A-BOMB. Frank denied the story, but he stood publicly accused of being a former party member, and HUAC now had him in its sights.

A year later, the raging controversy over Hiss's alleged espionage sent nervous tremors through government as people waited to see who would be brought down in HUAC's next purge. As part of its widening investigation, HUAC, and its chairman, J. Thomas Parnell, were now on a mission to discover the "Alger Hiss of science." They began looking into rehashed charges and old Manhattan Project security memos concerning leftist physicists on Berkeley's campus who might have spilled atomic secrets and posed a security risk during the war. No one with any Communist ties was safe, and many of Oppenheimer's old friends and campus colleagues fell under suspicion. Robert Serber was fingered in July 1948 and subjected to an AEC Personnel Security Board hearing after an investigation into his "character, associations and loyalty" had raised doubts about the continuance of his security clearance at the Rad Lab. After accusing both his brother-in-law and father-in-law of being Communists, and attempting to grill him on all his subversive Berkeley associates, including Oppenheimer and Chevalier, the AEC board excused him. Months later Oppenheimer told Serber he had seen the final report and assured him that he "passed with glowing colors."

Picking through old intelligence files, HUAC subpoenaed a group that had been identified as Berkeley campus radicals, including Frank Oppenheimer, Joe Weinberg, David Bohm, Bernard Peters, and Rossi Lomanitz. The latter four names were part of that same element that Oppenheimer, at both Lawrence's and Lansdale's urging, had distanced himself from after being named director of Los Alamos. Mistakenly believing his record would show that, Oppenheimer appeared before HUAC on June 7, 1949, and confirmed much of what was in the files, acknowledging that he had once characterized Peters as "quite Red" and a "dangerous man" and that Haakon Chevalier had told him about Eltenton. He also confirmed the substance of his interviews with Berkeley security officers. When asked about his brother, Oppenheimer replied firmly that Frank was not currently a party member, but then turned his blue eyes on the chairman and, with a look of exquisite suffering, appealed to him to stop his line of questioning. "I will answer, if asked," he said, "but I beg you not to ask me these questions." It was vintage Oppenheimer, and the embarrassed committee counsel quickly withdrew the question.

The committee was so under his spell that at the end of his testimony, they all rose to come forward and shake his hand. One of the committee members, a young Richard Nixon, gushed with admiration: "Before we adjourn, I would just like to say—and I am sure this is the sense of all who are here—I have noted for some time the work done by Dr. Oppenheimer and I think we all have been tremendously impressed with him and are mighty happy we have him in the position he has in our program."

Among the many former colleagues who crowded into the HUAC hearing that spring morning was Dorothy, who was in town visiting friends. She had seen the Oppenheimers only a few days earlier and had been aghast to hear that he had been called before the committee. She decided that Monday morning to cancel her plans and attend the hearing in a show of support. In her memoir she recorded her horror at the circuslike proceedings in the Senate caucus room:

> On a chair facing the six inquisitors ensconced on their dais sat Robert, sprayed by strong lights, a lone figure. I slipped into a seat midway in the hall and purely by chance found myself beside Anne Wilson Marks. I whispered to her, "How can he see his notes with lights in his eyes?" Anne's answer was, "He has no notes." I was shocked.

Dorothy could not stand to watch the witch hunt for long. She left at the lunch break, slipping out as unobtrusively as she had entered. She did not wait to speak to Robert, but dashed off a line telling him she had been there. "Robert," she added grimly, "that was an experience I won't forget."

Two days later, it was Frank's turn, and he did not fare as well. The committee had been sitting on his damning dossier for some time, and since they could not get Oppenheimer, they went after his younger brother instead. Frank was forced to publicly admit that he had lied about his prewar party membership: "My wife and I," he testified in the grim caucus room, "joined the Communist Party in 1937, seeking an answer to the problems of unemployment and want in the wealthiest and most productive country in the world. We did not find in the Communist Party the vehicle through which to accomplish the progressive changes we were interested in so we left it about three and a half years later and never rejoined." The next day, the University of Minnesota announced it had accepted Frank's resignation. Ruined, unable to continue his research in cosmic rays, and an implicit threat to his brother, he fled with his family to his ranch in a remote part of southwest Colorado. It would be ten years before he was able to return to his work as a scientist.

While Oppenheimer had not been ensnared along with Frank, he could not avoid being dragged back before another tribunal a week later, on June 13. Senator Bourke Hickenlooper, chairman of the Joint Committee on Atomic Energy, feeling empowered by the public hue and cry about security, decided to take a page out of HUAC's hearings and embarked on his own campaign to discredit the fledgling AEC. He was put up to this by Lewis Strauss, a self-made financier and conservative AEC commissioner, who had been growing increasingly vexed with Oppenheimer. Strauss was a trustee of the Institute for Advanced Study at Princeton and, despite being acquainted with the contents of his FBI file, had offered Oppenheimer the plum job in 1947. But now he found himself at cross-purposes with his appointee on the subject of research isotopes. The AEC had been steadily sending these overseas, filling more than two thousand requests in the spirit of Oppenheimer's international cooperation. Most of these were small samples of nonsecret radioisotopes to be used in pure physics research, but Strauss objected that foreign countries could conceivably apply them to atomic energy for industrial purposes or radioactive warfare and, worse yet, build reactors to build bombs. He had been roundly outvoted on the issue two years earlier, and he now saw a way of discrediting Oppenheimer's committee and getting his way. Strauss had discovered that a millicurie of iron 59 had been sent to Norway's Defense Research Establishment for physics experiments, and further, he had information that one of the Norwegians on the research team was a suspected Communist.

Senator Hickenlooper, who was apparently under the misapprehension that all this meant fissionable material was being shipped overseas, called Oppenheimer to the stand and tried to get him to admit to the Joint Committee that the shipment of the millicurie of iron was proof of gross mismanagement and a violation of the Atomic Energy Act. Oppenheimer almost laughed him out of the room. "No one can force me to say you cannot use those isotopes for atomic energy," he began incredulously. "You can use a shovel for atomic energy. In fact you do. You can use a bottle of beer for atomic energy. In fact you do. But to get some perspective, the fact is that during the war and after the war these materials have played no significant part, and in my knowledge no part at all."

Oppenheimer was putting on a good show, and there was a titter of amusement in the hearing room. But Strauss, who had come by expecting to gloat, was red-faced with anger, his jaw muscles working furiously. For all his accomplishments, Strauss had never finished high school. He still seethed with insecurity about his lack of formal education, and here was the great Robert Oppenheimer making a public fool of him. Oppenheimer kept up his clever performance, ridiculing several more of Hickenlooper's blundering questions, leaving the gallery in stitches, and turning the hearings on their head. When it came to the last, routine query about endowing security personnel with more authority, Oppenheimer responded with one word, "Morbid," as though the entire boorish line of questioning had rendered him almost speechless. Oppenheimer spotted Joseph Volpe, one of the AEC's lawyers. "How did I do?" he asked. "Too well, Robert, too well," Volpe told him.

If Oppenheimer had wanted to eviscerate his inquisitors, he succeeded admirably, but his victory came at a price. The next day, HUAC, with its power to command headlines, exposed Frank. His face and name were splashed across the nation's morning newspapers, accompanied by the sordid details of his life as a "closet Communist." Over the next few weeks, HUAC also leaked bits of Oppenheimer's secret testimony to the press, revealing that he had acknowledged that a number of prewar friends and associates were radicals or Communists. When the story broke in the Rochester, New York, *Times-Union* that Oppenheimer had given evidence against his talented former student Bernard Peters in the HUAC hearing, it turned people's stomachs. It may have been that Oppenheimer felt put upon and harassed, but not only did he not lift a finger to help Peters, he appeared to have betrayed him in an effort to save himself. Many of the old Berkeley crowd, who remembered the old days, found his behavior especially galling. After Bethe and Weisskopf advised him of the intense negative reaction among his peers, Oppenheimer wrote a letter to the paper praising Peters' "high ethical standards" despite his left-wing activities. But the damage was done. Oppenheimer still did not seem to grasp that he was bound by the same laws as everyone else. As he had committed more than a few indiscretions of his own, some considered it only just that he suffer the same fate as those he had so easily condemned.

Oppenheimer made himself even more of a target by his outspoken opposition to the hydrogen bomb. He had always been skeptical about the feasibility of the thermonuclear bomb, and believed the technical challenges and prohibitive costs involved in the bomb program made it inadvisable.

Throughout the fall of 1949, after Russia exploded its first atomic bomb at the end of August and nervous government officials seized on Edward Teller and Ernest Lawrence's proposal to develop hydrogen weapons, Oppenheimer argued America should not deliberately step up the arms race. When Teller had first called him in a state of extreme excitement after news broke of the Soviet test, he had responded brusquely, "Keep your shirt on." Oppenheimer did not agree that this was a reason to rush the hydrogen bomb into being and engage in the all-out production of dueling doomsday machines. On October 29, 1949, he led the General Advisory Committee's strong recommendation against the crash development of the Super, supported by all seven members present that morning, especially Conant, who memorably declared the thermonuclear weapon would be built "over my dead body." Oppenheimer commended the group for firmly opposing the weapons development and boasted of their "surprising unanimity,"* a claim his critics would later disparage.

During the three months of fierce debate, Oppenheimer's statements against the hydrogen bomb program once again brought him into direct conflict with Teller, whose hatred of his old Los Alamos boss had become almost pathological. Teller accused Conant of being "unduly influenced by Oppenheimer"—though in fact the opposite may have been true—and referred to the conspiracy of Los Alamos loyalists as "Oppie's machine." This time, however, Teller sensed he had an advantage. Playing on fears of an imminent Cold War showdown, he galvanized support for the Super and won powerful allies in ardent conservatives like Lewis Strauss and Senator Brian McMahon, chairman of the Joint Committee on Atomic Energy, as well as a militarist wing of the air force, which desperately wanted to protect the millions of dollars promised for the Strategic Air Command. With his advisors deadlocked, President Truman stalled. Then word came of Fuchs' confession, his funneling of atomic secrets to the Russians while he was at Los Alamos, and his presence at high-level meetings in which the fusion weapon had been discussed. There was now no longer any doubt

*Two members—Fermi and Rabi—agreed, but made their renunciation of the hydrogen bomb conditional on the Soviets' agreement to do the same. Glenn Seaborg, who did not attend, sent a letter that indicated that he was undecided.

about whether the Soviet Union was in a race for nuclear supremacy. On January 31, 1950, Truman decided that, given the Soviet threat, America had "no choice" and ordered the AEC to pursue "all forms of atomic weapons, including the so-called hydrogen or superbomb."

In the wake of the president's decision, Oppenheimer delivered a widely quoted speech before the Council on Foreign Relations warning of a tense, protracted Cold War ahead. "During this period the atomic clock ticks faster and faster," he lectured. "We may anticipate a state of affairs in which two Great Powers will each be in a position to put an end to the civilization and life of the other, though not without risking its own. We may be likened to two scorpions in a bottle, each capable of killing the other, but only at risk of his own life."

In the speech, Oppenheimer went out of his way to attack the air force for being more interested in protecting its Strategic Air Command that in any attempt "to protect the country." Angered by his continued resistance, Oppenheimer's emboldened critics exacted revenge for his many scornful slights, and articles began to appear describing his persistent campaign to reverse U.S. military strategy, endowing his views with the patina of disloyalty. He was personally blamed for allowing the communists to get ahead. The climate in Washington was so poisonous that on February 14, Conant wrote Oppenheimer, "I hope you are standing up under the strain of these trying times as well as usual," and enclosed a copy of a letter he had received from a reporter disclosing that a Republican senator was spreading the story that the GAC opposed the H-bomb on "moral grounds."

Oppenheimer received a letter of encouragement from Dorothy, who had read the text of his speech in the papers. She wanted him to know that it had the ring of conviction and had reassured her that "there is light ahead."

> There stand you, the beautiful Robert, the open mind, and your thoughts and suggestions which would quiet the din and still all hatred. Stand Robert, with the clarity and courage the world aches for. You speak, with the power of poetry and music.

Truman's announcement to go forward with the hydrogen bomb marked a turning point in Robert Oppenheimer's life. His stature and influence, after such an obvious rebuff, was sharply curtailed. Both he and Conant considered resigning from the GAC. According to Conant, one consideration that stopped them was not wanting "to do anything that would seem to indicate we were not good soldiers and did not want to carry out orders of the President."

It could not have been a complete coincidence that Paul and Sylvia Crouch, former officers of the American Communist Party, chose that moment to inform on Oppenheimer, acting as paid informants for the Justice Department. In May 1950, Sylvia Crouch told the California Committee on Un-American Activities that in the summer of 1941, Oppenheimer had hosted a "session of a top drawer Communist group known as a special section, a group so important that its make-up was kept secret from ordinary Communists." She claimed that it had taken place at his house in Berkeley and that she and her husband had been there. The story made headlines in newspapers across the country, and reporters began hounding Oppenheimer at his home in Princeton. Then the FBI paid him a call, returning a second time that same week. Oppenheimer denied the Crouches' story, and told the FBI that if he ever attended such a gathering—and he did recollect going to one where William Schneiderman, a leading Communist functionary, was present—it would probably have been at the home of his friend Haakon Chevalier. He was quoted in the FBI report as saying he was "greatly concerned with the allegations against him due to their possible effect on his reputation."

At the urging of his lawyer, Joe Volpe, Oppenheimer began collecting evidence to disprove the charges, and in the spring of 1952 he wrote Dorothy and asked if she would meet with Volpe and help with the process. Not put off by the taint of scandal, she readily agreed. The difficulty they faced was in trying to reconstruct a series of events that had taken place over a decade earlier, with very little to go on but his own faulty memory. Fortunately, it was a hard summer to forget, a "bad-luck summer," as Robert Serber later called it.*

*Neither Dorothy nor Oppie make any mention of the Serbers visit that summer. It may have been that because of Charlotte's own Communist ties—both her brother and father were identified by the FBI as party members—Oppenheimer decided to omit any mention of them.

Oppenheimer recalled that both he and his wife were unwell that spring of 1941. He was tired and overworked and their son, Peter, was born in May, and Kitty, who was a long time recovering her strength, was still feeling weak in July. It was thus extremely unlikely that they had thrown any large parties at their home. At some point, they went to their ranch in the Pecos to recuperate, and Oppenheimer was able to give Dorothy a number of leads to work with in establishing their movements. He recalled that Hans and Rose Bethe had visited, and they had stopped by to see his old friend Katherine Page at her guest ranch. Around that same time, he was mucking about in the corral when a horse kicked him in the knee, causing a painful enough injury that he was worried the knee might be broken and sought medical attention. Not long afterward Kitty, while driving her Cadillac convertible from Santa Fe to the ranch, had a car accident. She slammed on the brakes to avoid hitting a truck that had stopped short in front of her on a sharp curve and collided with the truck, hurting her leg and causing considerable damage to the front of the car.

Oppenheimer's assistant, Priscilla Greene Duffield, also did what she could to pin down where he was from late July to early August 1941 and remembered Dorothy's intrepid detective work on his behalf. "I was impressed by how close they were, and how much he relied on her," said Greene. "When he got involved in all the security stuff, she started helping him. She went all around Santa Fe asking questions, and getting dates of prescriptions, and even tracked one down at the Capital Pharmacy to prove he had been in New Mexico at a certain point."

Dorothy's weeks of legwork paid off. She was able to establish that the Oppenheimers, accompanied by Frank and Jackie, arrived at Katherine Page's ranch on the evening of Friday, July 11, and departed the following Tuesday, when they went to Perro Caliente. The local grocery store still had records going back that far, and they showed the Oppenheimers had made purchases on five separate occasions between July 12 and 29. In addition to the pharmacy receipt, she also dug up a receipt for the X-rays that were taken of Oppenheimer's bruised knee at St. Vincent Hospital on July 25. It turned out that both the Bethes clearly remembered the incident with Oppie's knee, and Rose had even taken a photograph of the "kicking horse," which she had marked and dated. It was submitted as evidence, along with another receipt Dorothy dug up from the garage in Pecos that repaired Kitty's car. Kevin recalled the stacks of yellow notepads that his mother filled with dates, times, and places. "She really threw herself into the detective work," he recalled. "She would have done anything for him."

Dorothy and Oppie were in contact more often during this period. Following the outbreak of the Korean War in June 1950, Kevin had enlisted in the army, and after he was stationed overseas in the winter of 1952, and Robert called often to see how she was doing. It had only been seven years since the "war to end all wars" had concluded, and Dorothy had never imagined she would live to see another one. Now she had sent her only son off to battle. His news from the front filled her with "pangs of dreadful horror," she wrote Oppie, and she could already see how the experience had "aged and saddened and matured him." Oppenheimer knew how frightened she was for Kevin, and how much she missed him, and did his best to reassure her. At the time, Kitty was not getting along with their own son, Peter, who was on the brink of adolescence, and because he thought it might do everyone some good, Oppie packed Peter off to Santa Fe to be with Dorothy. "I think she was very worried about me, and he was sort of a substitute for me while I was gone, which was fine," said Kevin. "She took Peter under her wing a little bit and transferred a lot of mothering to him."

In March 1953, Oppenheimer sent a heartfelt note to Dorothy thanking her for all her assistance and unfailing loyalty. The California committee had assembled quite a dossier on him, but no action had been taken. On a more personal note, he added that he had heard that Kevin had returned safely from Korea and wanted her to know that he was celebrating with her. He wrote he was "thinking, even though I find it hard to say, of all the nobility and simple courage which you have brought to the last years":

> This is the hour to put on paper, however inadequately, a word of the profound gratitude that Kitty and I have for all that you did last summer on our behalf. You can well think that this has played a decisive part in the course of events: and I must comfort

myself with the reflection that anything done so superbly and well must also have brought you a little pleasure. . . .

Our love, Dorothy, I hope we shall see one another very soon.

But Oppenheimer's troubles were far from over. By the end of 1953, McCarthy's reign of terror had already helped defeat the Democratic presidential candidate, Adlai Stevenson, in the 1952 election, and the newly elected Eisenhower did not have the stomach to take on the powerful demagogue. "I just will not—I *refuse*—to get down in the gutter with that guy," he repeatedly told his aides. With American boys now fighting Communists in Korea, however, McCarthy's reckless allegations that the country was at risk because of disloyal government officials had taken on new force. Anti-Communist hysteria was at its height. After a lurid trial, played out before a full gallery of reporters and propagandists, the prosecution, aided by Roy Cohn, soon to be McCarthy's chief counsel, sent Julius and Ethel Rosenberg to the electric chair on June 19, 1953. McCarthy, now in his second term, had become chairman of the Senate Committee on Government, and turned its Subcommittee on Investigations into an instrument of persecution, holding hearings, issuing subpoenas, and hunting down and destroying anyone accused of being a Red, a Communist, an enemy of the state, a fellow traveler, a traitor, or a spy. By the spring of 1954, he had succeeded in embroiling the Department of the Army in an embarrassing inquiry into who was responsible for "coddling Communists" in the rank and file. The case quickly blew up into a major clash between the senator and the administration, with the Republican Senate demanding an investigation of the feud between the army and McCarthy. The atmosphere was one of such unbelievable tension that only those who lived through it can attest to its madness.

By then, Conant had long since concluded that Washington had become a "lunatic asylum." When Eisenhower offered him the job of U.S. High Commissioner to West Germany in 1953, he leapt at the chance to get out of town. The bitter battle over the hydrogen bomb had divided the scientific community and made enemies of old friends, and a disgusted Conant, who had suffered the humiliation of having to withdraw his nomination as the next president of the National Academy of Sciences because of a well-orchestrated right-wing vendetta, washed his hands of the American scientific establishment for good. The prestigious post should have been the crowning achievement of Conant's scientific career, but instead his enemies used it as an opportunity for vengeance and payback. As George Kistiakowsky later described the double-cross in a letter to Conant's wife, Patty, a West Coast faction of Manhattan Project chemists had "ganged-up on him behind his back," and he never saw the knife coming:

> Probably the most painful incident in Jim's life as science leader occurred without warning to me and without my being able to take any steps to prevent it, an event which I see as a tragedy to American science as well as a disappointment I know to Jim. I refer to Jim's withdrawal from nomination as the next president of the National Academy of Sciences when suddenly being confronted by a small but secretly well organized group of little men who resented Jim's wartime leadership. The rest of us were unaware of what was being organized and thus were unable to demonstrate to Jim in good time the strong support which in fact would have been his.

With Conant gone, Oppenheimer had lost his staunchest ally and was in trouble. It turned out the man who had led the revolt against Conant was Wendell M. Latimer, dean of chemistry at Berkeley, who not only loathed Oppenheimer, but had helped organize the Berkeley contingent—Alvarez, Lawrence, Teller—who had lobbied for the hydrogen bomb. It was a "dirty deal," as Conant had put it, and there seemed to be nothing the ascending faction of scientists, who had favored the H-bomb, would not do to expel Oppenheimer and his "machine" from their positions of influence over U.S. atomic policy. In 1951, after the first successful test of the hydrogen bomb—a device based on a new design by Teller and Ulam—Oppenheimer finally gave up his opposition to further development. As even he admitted, their new approach was "technically so sweet that you could not argue about that." Anyway, it had become a hopeless cause. Lawrence, still smarting from Oppenheimer's decision after the war to choose Princeton over Berkeley, teamed up with Teller to start a second bomb laboratory, called Lawrence Livermore, to rival Los

Alamos. Like rejected suitors, the two physicists were determined to prevail over the man who had spurned them and their weapons. In the meantime, with the revelations about Fuchs' espionage spreading fear and innuendo throughout the entire nuclear establishment, and Oppenheimer's reputation wounded by reports that it had happened on his watch, McCarthy's bloodhounds were closing in.

In December 1953, Oppenheimer, who had just returned home from Europe after delivering the BBC's illustrious Reith Lectures, received a call from Lewis Strauss, now chairman of the AEC, requesting a meeting in his Washington office. When he arrived at Strauss' office on the afternoon of Monday, December 21, he was surprised to find AEC general manager Kenneth D. Nichols, one of the generals who had presided over the Trinity test with Groves, waiting for him. After a halfhearted attempt at polite conversation, the two men coolly informed him that the AEC had drafted a letter of charges stating the reason why Oppenheimer's security status was being questioned and that his clearance would be withdrawn in thirty days unless he requested a hearing.

Nichols and Strauss went on to provide Oppenheimer with a detailed summary of some two dozen charges or items of "derogatory information," among them an account of all his left-wing activities in the 1930s, a list of undesirable associates—including his former girlfriend, his wife, his brother, and his sister-in-law—and documentation of suspicious, contradictory, and misleading statements made by him to security in August 1943 regarding Communist Party members such as Giovanni Rossi Lomanitz and Haakon Chevalier, among others. Oppenheimer must have recognized that the final nail in his coffin was the last charge, which stated that in 1949 Oppenheimer had opposed the hydrogen bomb on "moral grounds," and that even after the GAC's vote against the crash program to develop the Super, he had continued to oppose it, was instrumental in persuading other outstanding scientists not to work on it, and had thereby "definitely slowed down its development." After Oppenheimer had reviewed Nichols' notification letter, Strauss laid out his rights under the AEC security procedures, and they discussed the relative merits of his resigning rather than facing a full-scale hearing.

Oppenheimer walked numbly out of Strauss' office. He had phoned his former assistant, Anne Wilson Marks, who lived close by on Olive Street in Georgetown, and she came directly and drove him to her house. "I knew something was wrong as soon as I heard his voice," she said. "He was very upset and shaken, but he wanted to fight it, too." Her husband, Herb Marks, a trusted friend and lawyer, met them at the house. They had a stiff drink and spent the next few hours huddled in the study as they discussed his predicament, and how best to mount a defense. After dinner, Marks' partner, Joe Volpe, joined them, as did Kitty, who had come from Princeton, and they talked into the night.

To everyone's surprise, Oppenheimer stood up abruptly at ten o'clock and announced he was going to bed. The next thing they heard was a resounding crash, and they rushed upstairs to find that he had passed out cold in the bathroom. "Apparently, Kitty was concerned that Robert get some rest and had given him some of her sleeping pills," said Wilson. "Well, the combination of the booze and pills did it." She called a doctor, who told her to make coffee. "He came over and started waking Robert up and kept him walking up and down for an hour. He was absolutely zonked." The next morning when she went downstairs, the newspapers had already caught wind of the security clearance story and were demanding to talk to Oppenheimer. Anne Wilson stood out on the landing in the pouring rain telling them over and over again that she had no idea where he was.

Oppenheimer spent the next day with his lawyers, debating whether he should try to clear his name or go quietly and lose his clearance, and with it his consultant's contract and considerable reputation. Volpe thought Oppenheimer should resign and avoid an embarrassing trial. But both Oppie and Marks felt that he could not take a route that, in Robert's words, would be conceding that he was "unworthy." Moreover, McCarthy was now a major power on Capitol Hill. He and Roy Cohn were holding hearings on "Communist influence" in the federal government and had already targeted the Voice of America, the Government Printing Office, and the Foreign Service. Because of the damage Fuchs' spying had done at Los Alamos, it would be easy for McCarthy to attack

Oppenheimer and cast him in a sinister light. Oppenheimer's lawyers advised him that if he did not submit to the AEC hearings, it was more than likely he would be called up before McCarthy's merciless Red-hunting subcommittee, which they all agreed was a far worse fate.

Strauss had given him one day to decide. On the morning of December 23, Oppenheimer sent a letter to Strauss saying he wanted to appear before the AEC's three-man Personnel Security Board. "I have thought most earnestly of the alternative suggested," he wrote. "Under the circumstances this course of action would mean that I accept and concur in the view that I am not fit to serve this government, that I have now served for some 12 years. This I cannot do." On Christmas Eve, two representatives from the AEC dropped by Olden Manor to present a letter that amounted to a preemptory strike by the prosecution, directing Oppenheimer to deliver any and all AEC documents in his possession.

Oppenheimer spent the holidays frantically conferring with lawyers and friends about how to deal with the security investigation. According to the hearing procedures, he had thirty days to decide on his course of action and frame a response. On March 4, 1954, he submitted a formal, forty-three-page reply to the charges. He concluded his deeply personal letter by saying that in the process of preparing it, he had reviewed two decades of his life. "I have recalled instances where I acted unwisely," he wrote. "What I have hoped was, not that I could wholly avoid error, but that I might learn from it. What I have learned has, I think, made me more fit to serve my country."

In the midst of all the preparations, Senator McCarthy suddenly caught wind of what was going on and weighed in on the closed-door proceedings in a television interview broadcast on April 7, setting the stage for what was to come. "If there were no Communists in our government why did we delay for eighteen months—delay our research on the hydrogen bomb?" McCarthy demanded. "Our nation may well die because of that eighteen months' deliberate delay. And I ask who caused it? Was it loyal Americans—or traitors in our government?"

The Personnel Security Board, chaired by Gordon Gray, former secretary of the army, commenced its closed-door sessions on April 12, 1954, ten days before Oppenheimer's fiftieth birthday. The hearing concluded on May 6, after three weeks of grueling testimony, and thirty-nine witnesses had been called, both for and against Oppenheimer. Diana Trilling, who wrote about the hearing in *Partisan Review,* described the kaleidoscope of characters from Oppenheimer's past who paraded into room 2022 of AEC Building T-3 to determine his fate:

> The conflict between the H-bomb proponents and the H-bomb opponents is undoubtedly the climax of dissension but we cannot fail to be aware of the multitude of alignments and groupings, commitments and cross-purposes, all of which are unmistakably bound up with Dr. Oppenheimer's predicament. The scientific "in" versus the scientific "outs," the friends of Dr. Oppenheimer versus the friends of Dr. Teller, the military versus the academics, the Air Force versus the Army versus the Navy, the Strategic Air Command versus the Air Defense Command, Los Alamos versus Livermore, MIT versus the University of Chicago, Dr. Oppenheimer the individual versus individuals of a dramatically different disposition of mind. . . .

The only person to be heard from who was devoid of professional interest in or conflict with Oppenheimer's career was his wife, who was not asked to testify to his character but to her Communist past. Oppenheimer's oldest friend and associate, Ernest Lawrence, too afraid of the controversial proceedings, and perhaps of what he might say, begged off, citing ill health. Conant, however, was determined to testify on his colleague's behalf, despite a chilling warning from Secretary of State John Foster Dulles that "factors unknown to [him]" made his public involvement "undesirable."

Herb Marks had felt strongly that Oppenheimer needed a "big name" to defend him, and he hired Lloyd K. Garrison, an eminent New York lawyer and the great-grandson of the abolitionist William Lloyd Garrison. But while Garrison was a good and principled white-shoe attorney, he was not a veteran of these stealthy security proceedings, as Marks was, and he failed to protect his client from the cutthroat maneuvering of the AEC attorney, Roger Robb.

Oppenheimer testified first and made a muddle of it. To settle the question of whether the former director of Los Alamos was a threat to the security of the United States, Robb zeroed in on Oppenheimer's veracity and conduct. The prime piece of evidence being used against him was the Chevalier incident, in which Oppie himself had admitted to Berkeley security officers that he had been approached about a way to transmit atomic secrets to the Soviets, that Chevalier had suggested a mutual acquaintance by the name of George Eltenton, that he then failed to report the incident immediately, and that when he did report it, he refused to identify his source, covering for his close friend Chevalier. Worse yet, Oppenheimer had lied about several of the details to the intelligence officer, Colonel Pash. Confronted by the bullying prosecutor and the convoluted record of his tape-recorded interviews with Lansdale and Pash, Oppenheimer stepped into trap after trap, and seemed to come unraveled on the stand. After Robb got him to admit that what Eltenton was doing amounted to treason and that he had "invented a cock-and-bull story," he asked Oppenheimer why he had lied. Oppenehimer answered, "Because I was an idiot."

At the conclusion of the first week of the hearing, which resembled a trial without the same promise of objectivity or fairness, Herb Marks was crushed. He had tried to warn Garrison that Robb would use the Chevalier incident to make Oppenheimer look bad and, in the insidious way these hearings were always conducted, win by planting the suggestion of, rather than proving, his disloyalty. Marks watched as Robb made a fool of Oppenheimer, pulling out every trick in the book to trip him up. He furiously scribbled notes warning Garrison to head off a dangerous line of questioning, which Garrison pocketed without reading. "Herb was just sick," said Anne Wilson. "He felt he could have saved Robert. He never got over it."

The reaction on Capitol Hill that weekend was guarded, as both the nation's politicians and press waited for the shroud of secrecy surrounding the extraordinary hearings to lift. It was still inconceivable to many that the man on "trial" was the same scientist to whom the country had turned a decade earlier to undertake what the *New York Times* called "the most awesome mission ever assigned to a private citizen—the mission to open the door to the atomic age." Most editorial writers adopted a wait-and-see attitude and indicated a hope that the Gray board would find a way to reaffirm Oppenheimer's fitness for public service. But with congressional elections already under way, the administration's attitude did not seem to be one of forbearance. One of the few administration officials to comment publicly was Secretary of Defense Charles E. Wilson, who defended the removal of Oppenheimer at a press conference and commented, "I have great sympathy for people who have made a mistake and have reformed, but we don't think we ought to reform them in the military establishment." McCarthy could not resist gloating: "It was long overdue," he told reporters. "I think it took considerable courage to suspend the so-called untouchable scientist—Oppenheimer. I give Strauss credit for that."

The first of the bomb scientists called to testify was Hans Bethe, who spoke glowingly of Oppenheimer's achievement in creating Los Alamos, concluding that he thought "nobody else could have done this than Dr. Oppenheimer." Under cross-examination, Robb ruthlessly moved to undercut Bethe's credibility, asking: "Which division was Klaus Fuchs in?"

Bethe answered, "He was in my division which was the Theoretical Division."

Robb, finished with the witness, said, "Thank you. That is all."

Few of the prominent scientists called testified more firmly on Oppenheimer's behalf than Conant, who argued that as they had worked closely together on most of the GAC's policies, and held the same beliefs, it was something of a paradox that Oppenheimer alone stood accused of disloyalty. Conant opened his testimony in "an aggressive mood," and his anger at the tenor of the hearings, and the tactics of the AEC attorney, was evident. When asked by Lloyd Garrison what he thought of the commission's letter containing the derogatory charges that initiated the hearings, the usually restrained Conant did not hold back: "Well, it seems to me that letter must have been carelessly drafted, if I may say so." He noted that the letter subscribed to:

> an impossible position to hold in this country; namely, that a person who expressed views about an important matter before him, as a member of the General Advisory Committee, could then be ineligible because of a security risk for subsequent work in connection with the government. I am sure that argument would not have been intended. If it did, it would apply to me because I op-

> posed it strongly, as strongly as anybody on that committee, that is, the development of the hydrogen bomb.

Robb, trying to weaken Conant's testimony, eventually got him to concede that as he had not heard all of Oppenheimer's testimony, "the board may be in a better position to judge." It may simply have been a figure of speech, but in agreeing with the prosecutor, Conant was, in effect, acknowledging that there might be things about Oppenheimer that he did not know and that only the board held all the cards.

Rabi proved a more wily witness, and try as he might, Robb could not get the same admission from him. Stating that the board "may be in possession of information which is not now available to you," Robb suggested Oppenheimer might have lied to intelligence officers, and therefore Rabi could not pass judgment on his character. Rabi held firm, underscoring the difficulty of judging a man's behavior back in 1943 and reaffirming the great store he set by Oppenheimer: "I am in possession of a long experience with this man, going back to 1929, which is 25 years, and there is a kind of seat of the pants feeling in which I myself lay great weight. In other words, I might even venture to differ from the judgment of the board without impugning their integrity at all. . . ."

When Robb tried to shake Rabi's confidence, telling him to confine his answer to the one incident in question, in which the board had damaging testimony from "Oppenheimer's own lips," Rabi was unmoved. "You have to take the whole story," he said finally. "That is what novels are about. There is a dramatic moment and the history of the man, what made him act, what he did, and what sort of person he was. That is what you are really doing here. You are writing a man's life."

Under Herb Marks' redirect, Rabi's voice grew querulous with fury as he argued that he thought Strauss' suspension of Oppenheimer's clearance was wrong and should not have been done: "He is a consultant, and if you don't want him to consult, you don't consult him, period," Rabi stated. "There is a real positive record. . . . We have the A bomb and a whole series of it . . . and what more do you want, mermaids? This is just a tremendous achievement. If the end of that road is this kind of hearing, which can't help but be humiliating, I thought [sic] that was a pretty bad show."

When Oppenheimer was recalled to the stand near the end of the hearing, Robb walked him through his "tissue of lies" and again got him to admit he had "embroidered" his story of the Eltenton incident when he reported it to security. Oppenheimer made one last attempt to explain his conduct: "I think I need to say that it was essential that I tell this story, that I should have told it at once and I should have told it completely accurately, but that it was matter of conflict for me." By the end, collapsing in weariness, he said, "I wish I could explain to you better why I falsified and fabricated."

The hearing reached its dramatic culmination when Teller lumbered into the gloomy room. Long and dark, it was set up with the tables in a U-shaped configuration, with the witness chair at the open end. In the back, just behind the witness chair, was a sofa where Oppenheimer sat, isolated and alone. Speaking slowly, with his heavy Hungarian accent, Teller first affirmed Oppenheimer's loyalty and then, in a series of remarks that were no doubt carefully choreographed in advance, vilified him with vague innuendo:

> In a great number of cases I have seen Dr. Oppenheimer act—I understand that Dr. Oppenheimer has acted—in a way that is for me extremely hard to understand. I thoroughly disagreed with him in numerous issues and his actions frankly appeared to me confused and complicated. To this extent I feel that I would like to see the vital interests of this country in hands which I understand better and therefore trust more. In this very limited sense I would like to express a feeling that I would feel personally more secure if public matters would rest in other hands.

As Oppenheimer's friends awaited the AEC's determination, anger over Robb's prosecutorial style, Teller's betrayal, and McCarthy's witch hunts, of which this was certainly a sordid last act, dominated their conversations. The senator from Wisconsin was even then getting his comeuppance at the hands of Joseph Welch in the ongoing Army-McCarthy hearing, and had been subjected to a verbal hiding by Edward R. Murrow on CBS's *See It Now,* but this was all coming too late to help Oppenheimer. He had been cornered and had had no choice but to fight and to hope that his many powerful friends and his service to his country would count to his credit. The Gray board had clearly been stacked against Oppenheimer, but in the end, he was his own worst enemy. As *Time* magazine concluded, "In the list of witnesses against J. Robert Oppenheimer, the most effective was J. Robert Oppenheimer himself."

Few had thought he would win, including Oppenheimer himself, but the announcement was no less of a shock for being expected. On May 28, 1954, the Gray board voted two to one to strip Oppenheimer of his security clearance. As soon as it was released, the Gray board's decision was dissected, analyzed, and discussed by scientists everywhere. The three-man panel agreed that out of the twenty-three charges concerning Oppenheimer's Communist ties, twenty-two were true. Only two judges believed he shared any guilt in the Chevalier case. The panel also found he had been insufficiently enthusiastic about the hydrogen bomb and that, whatever his motivation, "the security interests of the United States were affected." Two of the judges, Gordon Gray and Thomas Morgan, in a lengthy decision that ran 25,000 words, voted to deny Oppenheimer clearance. Ward Evans, the only scientist on the panel, voted to reinstate him, stating his minority opinion in eight succinct paragraphs. Writing of the shift that Oppenheimer's attitude toward the hydrogen bomb underwent from 1944 to 1945, when he changed his position from one of favor to one of opposition, Evans wrote: "After 1945 he did not favor it for some years perhaps on moral, political or technical grounds. Only time will prove whether he was wrong on the moral and political grounds."

On June 29, the AEC commissioners, by a vote of four to one, confirmed the panel's conclusion that Oppenheimer was a security risk. The only bone they threw Oppenheimer was that although they agreed that he was, as Strauss had phrased it, unfit to serve his country, they found no evidence that he was "disloyal to the United States." Once again, the only scientist on the AEC, Henry DeWolf Smyth, dissented, arguing that there was "no indication in the entire record that Dr. Oppenheimer has ever divulged any secret information."

At Los Alamos, Oppenheimer's faithful were devastated. "Our reaction was real fury," recalled Rose Bethe, "at the waste of people, the terrible waste. Oppie had had a hand in his own undoing. He had been very arbitrary in his treatment of people. But we didn't know at the time that it was mostly Strauss' doing." The trial, in their view, had been a dirty business. There had been a series of revelations in the press about the inequities of the process, from delays in providing his lawyers with documents to General Nicholas' secret excoriating letter to the AEC prior to their verdict. "Even before the full story was known, the attempt to discredit Oppenheimer was vigorously denounced, not only by his admirers, but by people who had never liked him," wrote Alice Smith. "He was seen by some observers as a martyr to McCarthyism, by others as a partner to a kind of Faustian bargain."

Dorothy did not attend the hearing, but Rabi had related the proceedings in great detail, and he told her that Teller had given a brilliant performance—he had given them everything they needed to hang Oppie. Afterward, when Teller ran into Oppie outside the AEC building, Teller had told him, "Good luck!" Oppenheimer told Rabi that he just looked at Teller and said, "I don't know what you mean after what you have said." To Dorothy, Teller's conduct, as relayed by Rabi, was almost indecent. In July, she wrote Oppenheimer a long, impassioned letter, making it clear that her commitment to him, and acceptance of him, was unaltered: "Have not read all the 'testimony' which Rabi gave me because it makes me so sad," she began, adding in a chastising tone that if he had told her about the hearing in advance, she "would have come on to testify" if he had let her:

> If you had had half a dozen women testify on your behalf I think the result would have been different! We would have injected a touch of humanity into the hearings. Have spoken of gratitude. Has this nation no gratitude for the man who saved it? Could they judge you by performance? What's wrong with judging you by performance? The miracle of personal achievement which you accomplished has never been equaled in the history of the world.
>
> Oh Robert, Robert, we who know you grieve. The world is sick. And you are well and strong. And some day not too distant you will lead us again.

TWENTY-TWO

Fallout

AFTER THE HEARINGS, Oppenheimer went back to the Institute for Advanced Study, but he was not the same man. He had built his entire postwar career around his powerful advisory positions in Washington, and now had he been cast out like a thief in the night. "The trouble with Oppenheimer is that he loves a woman who doesn't love him—the United States Government," Einstein observed of his Princeton colleague, with whom he was never particularly close.

Oppenheimer had been humbled, but not destroyed. He was widely regarded as a martyr, and as usual he played his new part well, with an air of saintly resignation and a touch of wry humor. He continued to write and teach in the years that followed, and remained a dignified, if poignant, figure. But to old friends he seemed smaller somehow, shrunken, and profoundly sad. The hearings "hurt his spirit," said Dorothy. "It absolutely hurt him. He was wounded." She knew there had been mistakes on his part, but nothing that warranted his being found unworthy. Looking back on what happened, it was hard for her to escape the conclusion that Oppenheimer, a man who spoke "with the power of poetry and music," was never meant for the cold halls of government. He was a scientist and an intellectual, not a statesman. Only in Washington could a man of his mind and spirit, imagination and feeling, fail to be understood and appreciated. "When he wrote that he had been 'naïve in politics,' well, he *was* naïve in politics," she said. "He wasn't a sharp politician, or his career would have been quite different."

Rabi, like many of the scientists who had rallied to Oppenheimer's side, felt "a bit angry" that he had "let it happen," that he had allowed himself be taken for such a ride by the prosecution. Despite his quickness of mind, amazing ability to synthesize ideas, and enormous gifts and capacities of every kind, Oppenheimer appeared strangely defenseless on the stand. Rabi came to agree with Oppenheimer's own analysis of the problem when he observed after the trial that he had "very little sense of self." Oppenheimer was a man of many roles, and many faces, and there were too many facets to his complex personality for him to present a coherent picture of himself in the adversarial setting of the hearing room. "Oppenheimer was a man who was put together of many bright shining splinters," said Rabi. For all his brilliance, he had a flaw, and once his opponents found it, they kept hammering away it. In the end, he shattered.

In the spring of 1963, nine years after the Gray board debacle, President John F. Kennedy decided to try to undo the great wrong that had been done to one of the country's foremost scientists. The Cold War hysteria was in decline, Kennedy's close advisors—McGeorge Bundy, Dean Rusk, and Arthur Schlesinger, Jr.—admired Oppenheimer, and his lectures and writing had begun to attract considerable attention and praise again. Glenn Seaborg, chairman of the AEC, announced they would be awarding Oppenheimer the Enrico Fermi Award for his outstanding contributions to theoretical physics, to the development of the atomic bomb, and to the peaceful applications of atomic energy. In addition to the citation and medal, he would receive a check for $50,000.

On November 22, it was announced that Kennedy wanted to present the award personally in a ceremony scheduled for December 2, the anniversary of Fermi's first self-sustaining chain reaction. When Oppenheimer called Dorothy to tell her the news, and asked her to be with him at the White House event in his honor, she was overjoyed. Like most of the old mesa scientists, she felt an honor such as this was long overdue. The evil wind that had blown in with the 1950s had finally begun to shift, and she thought she could foresee a time in the near future when the commission's 1954 verdict would be overturned and Oppenheimer's good name would be restored. She immediately sent him a congratulatory telegram: WONDERFUL NEWS ABOUT FERMI AWARD HAPPINESS INEXPRESSIBLE.

On the afternoon of November 22, Kennedy was assassinated in Dallas, and the country was plunged into mourning. All state occasions were canceled or indefinitely postponed. Given that the announcement of Oppenheimer's award had met with strong opposition from the right wing, which still demonized him, there was some doubt as to whether it would take place at all. When President Lyndon Johnson decided at the last minute to go ahead with the ceremony as planned, Dorothy was still too devastated by Kennedy's death to face a trip to the capital. On the day of the ceremony, she sat down and wrote Oppenheimer that she knew she "should be there" and was "truly heartbroken."

On December 2, Oppenheimer, Kitty, and a throng of dignitaries and men of great accomplishment in science and government packed the Cabinet Room. Oppenheimer, slightly stooped but elegantly trim as always, his dark crew cut now a thin cap of silver, stepped forward to receive the plaque from Johnson. For several moments, he stood regarding it in silence before speaking. Addressing the crowd in his soft, intimate voice, he said, "In his later years, Jefferson often wrote of the 'brotherly spirit of science which unites into a family all its votaries.' . . . We have not, I know, always given evidence of that brotherly spirit of science. This is in part because we are engaged in this great enterprise of our time, testing whether men can . . . live without war as the great arbiter of history. . . ." Then, turning to Johnson, he added with a hint of a smile, "I think it is just possible, Mr. President, that it has taken some charity and some courage for you to make this award today."

The only thing that marred the occasion was that as soon as people crowded around Oppenheimer to congratulate him, Teller darted forward and grabbed his hand. The moment was caught by all the photographers, and it was Teller's round, grinning face beside Oppenheimer's spectral one that made the front page of the papers, and was the first thing Dorothy saw when she got up the next morning. No one had invited Teller, nor had anyone wanted him there. Teller was present because he had won the award a year earlier and all the previous honorees had been invited. Dorothy, appalled that he had dared to show his face, told Kitty the man had no shame. "He couldn't be out of the picture for anything," she said with disgust. "I told Kitty, if I'd have been there I would have dropped a brick on his head." Kitty, who had been beside herself at the thought of being in the same room with Teller, said she would like to have done just that. Like Dorothy, many of the scientists who attended the award ceremony that day never forgave Teller, and several, including Rabi, refused to shake his hand.

In an interview in June 2003, shortly before his death at the age of ninety-four, Teller said that the hurt and anger of his rift with Oppenheimer over the hydrogen bomb were still painfully fresh. After calling it "the worst experience of my life," he became extremely agitated and broke down mid-sentence, saying roughly, "I can't talk about, I can't talk about." In his memoir, published in 2001, he admitted to being stung by Oppenheimer's "hostile" reaction to his 1949 phone call after hearing about the Soviet atomic bomb, adding that what little contact they had afterward was "superficial and reflected no encouragement of interest in his part." Teller's deep resentment of Oppenheimer, compounded by the accumulated slights and runs-ins, does more to explain his testimony than the reason he provides in his book, where he argues that his initial "shock" over the accusations that the Los Alamos director had been asked to pass information to the Soviets, had waited months to report this, and had then lied about it, left him "unsure" that he could "trust [his] convictions about Oppenheimer." No one who knew Teller ever believed him to be the least bit "unsure" about his opinion of Oppenheimer, and as a consequence, a large part of the scientific community ostracized him.

Teller never understood how he came to be as much of a casualty of the hearings as Oppenheimer himself, and never got over the cruel "exile" he was subjected to for years afterward. In late June 1954, about a week after the transcripts of the hearings were published by the Government Printing Office, Teller had gone to Los Alamos to attend a meeting. When he walked into Fuller Lodge, old friends turned their back on him or coldly attacked him for his testimony. Teller found it staggering that so many of his fellow physicists regarded him, and not Oppenheimer, as the true villain. His wife, Mici, was so upset by the rough treatment that she spent the rest of the day in their room and became ill. Dorothy reported on the visit to Oppenheimer with grim satisfaction. "Edward and his family arrived Sunday, and his reception has been and is being chill," she wrote. "I do not want to see him, and if I have to, I am sure my wide mouth will open. I have taken his photograph down from these walls and placed it in the closet with dirty rags, and empty bottles, until I can return it to the Hill." Teller did not set foot in Los Alamos again for nine years.

Years later, when asked to compare Oppenheimer and Teller during an interview with an Albuquerque television station, Dorothy was unequivocal as usual. "Why, that's a charming question," she said, her eyes taking on a mischievous gleam. "You can't compare their character any more than you can compare an orchid to a dandelion. . . . An orchid is more finely designed, and built, and delicate, and subtle, and aromatic. And a dandelion is something you kick up with the heel of your shoe if it's going to take over your grass."

In May 1963, Oppenheimer returned to the Hill for the last time to give a memorial address for Niels Bohr. The ordeal of the past decade had aged him well beyond his years. He was frail and worn out. The violent, rasping cough that had persisted on and off since his bout with tuberculosis had worsened, and his doctors advised against making the long trip to New Mexico. Oppenheimer went anyway, explaining that it was something he had to do. He stayed with Dorothy, and as they drove up the broad, flat state highway to Los Alamos, they laughed about the days when the road had been formidable, and Miss Warner's their cherished sanctuary. The little teahouse was long gone, and Dorothy had written to tell him of Edith's death in the spring of 1951. Miss Warner had been forced to relocate when the new bridge was completed a year after the war ended, and many of the old laboratory staff had gone down to Otowi to help make the adobe bricks for her new house. Afterward, Tilano had complained that he had been forced to redo most of them because "physicists did not know how to make mud bricks."

Dorothy, distressed by Oppie's fragile state, tried to keep him amused with such tales as they approached the mesa, which was now home to the country's leading nuclear research laboratory and a sprawling suburban community with a population of 15,000, more than a third of them children. The old East Gate was now a relic, the Guard Tower preserved as a reminder of dangerous times, and the new front entrance resembled nothing more than a turnpike toll barrier, as passes had not been required since the AEC opened the site to the public in 1957. Standing apart from the modern stores, restaurants, and office buildings, Fuller Lodge remained as one of the few symbols of preatomic times, before Oppenheimer and his physicists brought momentous changes to the mountain. That night, addressing the huge crowd that jammed the Civic Auditorium, Norris Bradbury introduced Oppenheimer as "Mr. Los Alamos." He went on to credit him with building the bomb laboratory through "sheer force of personality and character" until he realized his words were being drowned out by the growing swell of applause. It grew louder and louder, spreading across the hall in waves, until a moment later the crowd were on their feet giving a standing ovation to their battered hero.

A month later, on June 28, 1963, Dorothy closed the wrought-iron gate in front of 109 East Palace for the last time. At the age of sixty-five, after suffering a mild heart attack, she decided it was finally time to relinquish her post. The closing of the office coincided with Dorothy's twentieth anniversary as the "front man" for Los Alamos and as the beloved link between the laboratory and the outside world. Unbeknownst to her, the gate to 109 had been replaced with a duplicate, and the original taken to the Hill for safekeeping, later to become part of a permanent museum exhibit telling her story and of the legend that was wartime Los Alamos.

In a brief ceremony attended by the remaining members of Oppenheimer's old staff, Norris Bradbury explained that it was only fitting that the office be closed by the same woman who had first opened it in March 1943, for to all those displaced scientists who had passed through it on their way to the secret city on the Hill, Dorothy McKibbin and 109 East Palace were synonymous. He then took wire cutters and removed the sign identifying the office—the old wooden board had long ago been discarded—to reveal a bronze plaque commemorating the historic building:

109 EAST PALACE
1943 SANTA FE OFFICE 1963
LOS ALAMOS SCIENTIFIC
LABORATORY
UNIVERSITY OF CALIFORNIA
ALL THE MEN AND WOMEN WHO
MADE THE FIRST ATOMIC BOMB
PASSED THROUGH THIS PORTAL TO
THEIR SECRET MISSION AT LOS ALAMOS
THEIR CREATION IN 27
MONTHS OF THE WEAPONS THAT ENDED
WORLD WAR II WAS ONE OF THE GREATEST
SCIENTIFIC ACHIEVEMENTS OF ALL TIME.

Not long afterward, Oppenheimer's health started to fail. The physicist, who was never without a cigarette or pipe, heard confirmation of what he had probably known for some time. "He called me up in the summer of '66 from Princeton and said he had cancer of the throat," recalled Dorothy. He told her doctors were using cobalt to treat the tumor, and they discussed it in clinical terms. Later, when he underwent radiation, he told her about the electrons from the betatron. It had always been his way to talk like that, without too much fuss or show of sentimentality. Oppenheimer had lost all his fight years ago, and he faded fast. He died on February 18, 1967. He was sixty-two. More than six hundred people attended his memorial services at Princeton and listened to the eulogies by Hans Bethe, Henry DeWolf Smyth, and George Kennan. Afterward, Kitty took his ashes to St. John and scattered them over the ocean near their house.

The next few years visited a great deal of tragedy upon a family Dorothy thought had already suffered enough. Kitty rapidly deteriorated and lasted only another five years. The harrowing years of investigations, subpoenas, and hearings proved too much for her. She had suffered from painful stomach ailments for years, and had a diseased pancreas, which required medication. Her drinking had never abated, and after too many rounds she became difficult and erratic. "She was always a brittle person, and eventually she just broke," said Priscilla Greene Duffield, who remembered that Kitty, in a fit of depression, once asked Dorothy to buy her a gun. Kitty had never been without a man in her life, and it was somehow fitting that she ended up with Oppenheimer's most faithful student, Bob Serber, whose wife, Charlotte, had passed away. In the spring of 1972, Kitty bought a new sailboat, a fifty-two-foot ketch named *Moonraker,* and she and Serber planned an extended transpacific trip beginning in the Caribbean, passing through Panama, and sailing on to Japan via the Galapagos Islands and Tahiti. Their trip was cut short, however, when Kitty, the ship's captain and navigator, came down with a severe intestinal infection. On October 17, she entered the hospital in Cristobal, at the Atlantic end of the Panama Canal, and died ten days later of an embolism. Like Oppie, she was just sixty-two at the time of her death. In his memoir, Serber noted that the name of Kitty's boat had two meanings: the topmost sail on a full-rigged ship, or "someone touched with madness."

The Oppenheimers' twenty-seven-year-old daughter, Toni, took Kitty's ashes back to St. John and scattered them in the sea, in the same place they had bid good-bye to her father. The Oppenheimers' two children did not have an easy time coming of age in the midst of such turmoil, and it left indelible scars. Toni, the apple of her father's eye, grew up to be a beautiful but troubled girl. After her mother's death, she remained on St. John, left her husband, and filed for divorce. She later remarried, but it did not last. In January 1977, three months after the breakup of her second marriage, she committed suicide. Her body was found hanging by the neck in one of the bedrooms of the house her father had built on St. John, along with several notes. As there was no phone at Perro Caliente, Dorothy drove up to the ranch in a heavy snowstorm to break the news to Peter.

Peter was always Dorothy's favorite. She treated him as a second son and tried to lavish the love and attention on him she knew his distracted parents did not always have time to give him. Peter had doted on his father and at the height of the trial had angrily scrawled on his classroom blackboard:

> The American Government is unfair to Acuse [sic] Certain People that I know of being unfair to them. Since this is true, I think that Certain People, and may I say, only Certain People in the US Government, should go to HELL.
>
> Yours truly, Certain People

His problems were compounded when, as a moody teenager, he began having real conflict with his mother. Although he was aware of the problems, Oppenheimer sided with Kitty, causing a serious breach that never completely healed. In the spring of 1958, in retaliation for Peter's poor marks and failure to gain entrance to Princeton, Oppenheimer refused to allow him to accompany the family on a grand tour of Israel, Greece, and Belgium. Before that summer term was over, Peter had dropped out of the prestigious George School in Pennslyvania and escaped out west, staying at Frank's ranch in Colorado and visiting Dorothy in Santa Fe. In late July, Dorothy wrote to Oppie, knowing how much he regretted Kitty's failures as a mother and that he blamed himself for any pain they might have caused their son. "Yesterday into this office walked Frank and Pete," she wrote. "I could not take my eyes off Pete. He is truly beautiful. Made me think of paintings by Renaissance masters. The cut of his face, his color, his eyes, his lovely shy and grave manner. I thought of you and Kitty, and of what you have created."

Peter eventually settled in Santa Fe, and Dorothy helped find him carpentry work and recommended him to all her friends. When it came time for him to be married, it was only natural that he asked that it be at her adobe farmhouse, with the yellow roses blooming in the courtyard. Nothing could have pleased her more. When Peter's oldest daughter was born, he named her Dorothy, he said, in honor of the woman he admired for her "tremendous good cheer, courage and good will."

In the last years of her life, Dorothy set out to write a book. After several cataract operations, she was hampered by failing eyesight and knew she could not tackle the project on her own. She called on a close friend, Dorothy Hughes, who was a noted local author, and the two agreed to collaborate on the project. Dorothy had accumulated boxes of newspaper clippings, magazine articles, and photographs about the atomic bomb project, and over a period of two years she sorted through the material, reading aloud relevant bits, while Hughes taped her remarks. Dorothy had high hopes for the manuscript, entitled "Under a Piñon Tree: The Story of Los Alamos," but after close friends advised her to scrap the rather dry, impersonal compilation of clippings in favor of a more personal account, she abandoned the project. It was never published.

Dorothy, who had not expected to live to see thirty, died on December 17, 1985, five days after her eighty-eighth birthday. She was asleep when the end came. She had already slipped away, wandering into the hills with a man in a porkpie hat, shrouded in the alkaline mist that reduces all desert shapes to ghosts.

At her memorial service, her old friend Peggy Pond Church read a poem she dedicated to Dorothy.

She yields her hair to the wind;
she yields her face to the sun;
her love, like the evening star,
shines clear for everyone.
Love is a light for her.
Love is a warming fire.
The hungry, the sad, the cold
she heals of their long desire.
Men have gone down to death
wearing her love like a rose,
and the tears that her own heart sheds
only her own heart knows.

AUTHOR'S NOTE ON SOURCES

Dorothy McKibbin's recollections of Los Alamos were taken from the following sources:

MANUSCRIPTS AND LETTERS

Manuscript of a brief reminiscence by Dorothy McKibbin, "The Santa Fe Office: 109 East Palace Avenue," 1946. It was printed with permission by *Los Alamos Scientific Laboratory News* and reprinted in 1963. In 1946, the manuscript was reshaped and edited into chapter form by Jane Wilson and Charlotte Serber, and compiled along with eight other women's tales into a book. The book never found a publisher, and thirty years later, Jane Wilson gave it to the Los Alamos Historical Museum. In 1987, the Los Alamos Historical Society (LAHS) published the collection of short essays in *Standing By and Making Do: The Women of Wartime Los Alamos.* I chose to quote from the original manuscript, before Dorothy McKibbin's words were edited by numerous parties over the years.

Text of a speech written by Dorothy McKibbin, November 23, 1959; courtesy of LAHS.

Unpublished manuscript, Dorothy Scarritt McKibbin and Dorothy Bell Hughes, "Under a Piñon Tree: The Story of Los Alamos," Santa Fe, NM, circa 1993. Reprinted with permission from Kevin McKibbin.

Letters written by Dorothy McKibbin, as well as those written to her by J. Robert Oppenheimer; courtesy of the J. Robert Oppenheimer Papers, Library of Congress. Reprinted with permission from Kevin McKibbin.

The poem by Peggy Pond Church is courtesy of Peggy Pond Church Papers, Center for Southwest Research, University Libraries, University of New Mexico. Permission to reprint poem, courtesy of Kathleen D. Church.

PUBLISHED INTERVIEWS

Documentary: *The Woman Who Kept a Secret,* interview by Hal Rhodes; edited and produced by Dale Sonnenberg; Albuquerque, NM; KNME-TV, 1982.

Oral history interview: transcript of interview with Dorothy McKibbin, conducted by Los Alamos National Laboratory, January 13, 1982; Historical Perspectives Educational Films, Santa Fe, NM; courtesy of LAHS.

Oral history interview: videotape of interview with Dorothy McKibbin, no. 33, "World of Enrico Fermi"; courtesy of the Center for History of Physics, the American Institute of Physics, New York.

QUOTATIONS FROM NEWSPAPER ARTICLES AND MAGAZINES

"Baggage, Babies and the Atom Bomb: The Unique 20 years of Dorothy McKibbin." *Los Alamos Scientific Laboratory News,* June 28, 1963.

Corbett, Peggy. "AEC Office in SF Closes." *The New Mexican,* July 30, 1963.

———. "Oppie's Vitality Swayed Santa Fean." *The New Mexican,* June 26, 1960.

Hall, Rosanna. "Dorothy McKibbin Remembers Early Days in NM." *The New Mexican,* June 11, 1981.

McMaho, June. " '109' to Close: Dorothy McKibbin Retirement Told." *The Los Alamos Monitor,* June 27, 1963.

McNulty, William. "World's Most Famed Scientists, En Route to Los Alamos Project, Go Through Ancient City Office." *Santa Fe New Mexican,* May 10, 1946.

Pillsbury, Dorothy L. "The Adobe House That Helped Build the Atom Bomb." *New Mexico Magazine,* March 1962.

———. "Santa Fe Woman Serves in War and Peace." *Christian Science Monitor,* July 14, 1958.

Poore, Anne. "A Special Lady Creates Special Memories." *Los Alamos Monitor,* June 5, 1983.

"She Was Den Mother to Early Los Alamos." *Associated Press,* July 16, 1985.

"Smiling 'Front Man' for Atomic Bomb." *The New Mexican,* June 30, 1963.

Staley, Elizabeth. "Close to Oppenheimer: 'Gatekeeper' Kept Secret for 40 Years." *Albuquerque Journal,* April 9, 1982.

Ward, Eugene. "Los Alamos: Then and Now." *Albuquerque Journal,* August 14, 1977.

Grateful acknowledgment is made to the following:

Hans Bethe, Philip Morrison, and Edward Teller: transcripts of 1999 oral history interviews for the documentary *The Moment in Time,* Los Alamos National Laboratory; courtesy of Los Alamos Historical Society, P.O. Box 43, Los Alamos, NM.

Marge Bradner: written reminiscences, correspondence with Dorothy McKibbin, Bradner family papers, San Diego, California.

Vannevar Bush: transcripts of oral history interviews, courtesy of the Niels Bohr Library, the American Institute of Physics, College Park, Maryland.

Kathleen D. Church: poems of Peggy Pond Church, Albuquerqe, NM.

James B. Conant: papers and correspondence; courtesy of Harvard University Archives, Cambridge. Conant family papers, Hanover, NH.

Priscilla Greene Duffield: transcript of oral history interview, conducted by LANL; courtesy of LAHS, P.O. Box 43, Los Alamos, NM.

Ernest O. Lawrence: letters and papers; courtesy of Ernest Orlando Lawrence Papers, the Bancroft Library, University of California, Berkeley.

Dorothy Scarritt McKibbin: biographical material and papers; courtesy of the Sophia Smith Collection, Smith College, Northampton, MA.

Dorothy McKibbin, Robert Oppenheimer, Katherine Page, and Edith Warner: correspondence; courtesy of Papers of Robert Oppenheimer, Library of Congress, Washington, DC.

Also, for their invaluable assistance, the archives of the American Institute of Physics, Children of the Manhattan Project/Manhattan Project Heritage Preservation Association, Kansas City Public Library, the Museum of New Mexico, Los Alamos National Laboratory, and the Zimmerman Library at the University of New Mexico.

SELECTED BIBLIOGRAPHY

Alvarez, Luis W. *Alvarez: Adventures of a Physicist.* New York: Basic Books, 1987.

Bacher, Robert F. *Robert Oppenheimer, 1904–1907.* Los Alamos, NM: Los Alamos Historical Society, 1999.

Ballen, Samuel B. *Without Reservations.* Santa Fe, NM: Ocean Tree Books, 1997.

Barnett, Lincoln. "J. Robert Oppenheimer." *Life.* October 10, 1949.

Baxter, James Phinney, III. *Scientists Against Time.* Boston: Little, Brown, 1947.

Behind Tall Fences: Stories and Experiences About Los Alamos at Its Beginning. Los Alamos, NM: Los Alamos Historical Society, 1996.

Bethe, Hans. "J. Robert Oppenheimer: April 22, 1904–February 18, 1967." *Biographical Memoirs of Fellows of the Royal Society.* London: The Royal Society, 1997.

Brandt, Karen Nilsson, and Sharon Niederman. *Living Treasures: Celebration of the Human Spirit.* Santa Fe, NM: Western Edge Press, 1997.

Brode, Bernice. *Tales of Los Alamos: Life of the Mesa, 1943–1945.* Los Alamos, NM: Los Alamos Historical Society, 1997.

Burns, Patrick, ed. *In the Shadow of Los Alamos: Selected Writings of Edith Warner.* Albuquerque: University of New Mexico Press, 2001.

Bush, Vannevar. *Pieces of the Action.* New York: Morrow, 1970.

Childs, Herbert. *An American Genius: The Life of Ernest Orlando Lawrence.* New York: E. P. Dutton, 1968.

Church, Fermor S., and Peggy Pond Church. *When Los Alamos Was a Ranch School.* Los Alamos, NM: Los Alamos Historical Society, 1998.

Church, Peggy Pond. *The House at Otowi Bridge: The Story of Edith Warner at Los Alamos.* Albuquerque: Univerity of New Mexico Press, 1959.

———. *Accidental Magic.* Ed Kathleen D. Church. Albuquerque, NM: Wildflower Press, 2004.

Compton, Arthur. *Atomic Quest.* New York: Oxford University Press, 1956.

Conant, James B. *My Several Lives: Memoirs of a Social Inventor.* New York: Harper & Row, 1970.

Corbin, Alice. *Red Earth: Poems of New Mexico.* Santa Fe, NM: Museum of New Mexico Press, 2003.

Dallet, Joe. *Letters from Spain.* New York: Workers Library Publishers, 1938.

Davis, Nuel Pharr. *Lawrence and Oppenheimer.* New York: Simon & Schuster, 1968.

Dubos, Rene, and Jean Dubos. *The White Plague: Tuberculosis, Man and Society.* New Brunswick, NJ: Rutgers University Press, 1996.

Fermi, Laura. *Atoms in the Family.* Chicago: University of Chicago Press, 1954.

Fermi, Rachel, and Esther Samra. *Picturing the Bomb: Photographs from the Secret World of the Manhattan Project.* New York: Harry N. Abrams, 1995.

Feynman, Richard P., with Ralph Leighton. *"Surely You're Joking, Mr. Feynman!": Adventures of a Curious Character* and *"What Do You Care What Other People Think?": Further Adventures of a Curious Character.* New York: Quality Paperback Book Club, 1991.

Fisher, Phyllis K. *Los Alamos Experience.* Tokyo: Japan Publications, 1985.

Frisch, Otto. *What Little I Remember.* Cambridge, England: Cambridge University Press, 1979.

Giovanitti, Len, and Fred Freed. *The Decision to Drop the Bomb: A Political History.* New York: Coward-McCann, 1965.

Goodchild, Peter. *J. Robert Oppenheimer: Shatterer of Worlds.* London: British Broadcasting Corporation, 1980.

Greenberg, Daniel S. *The Politics of Pure Science.* New York; New American Library, 1967.

Groeff, Stephane. *Manhattan Project: The Untold Story of the Making of the Atomic Bomb.* Boston: Little, Brown, 1967.

Groves, Leslie M. *Now It Can Be Told.* New York: Da Capo Press, 1962.

Hales, Peter Bacon. *Atomic Spaces: Living on the Manhattan Project.* Chicago: University of Illinois Press, 1997.

Harris, Richard. *National Trust Guide—Santa Fe: America's Guide for Architecture and History Travelers.* New York: John Wiley, 1997.

Hershberg, James. *James B. Conant: Harvard to Hiroshima and the Making of the Nuclear Bomb.* New York: Knopf, 1993.

Hewlett, Richard G., and Oscar E. Anderson, Jr. *The New World: 1939–1946.*

Vol. 1 of *A History of the United States Atomic Energy Commission.* University Park: Pennsylvania State University Press, 1962.

Hillerman, Anne. " 'Exile' in New Mexico Gave Poet New Life." *Albuquerque Journal,* July 4, 2003.

Hilton, James. *Lost Horizon* (1933). New York: Pocket Books, 1960.

Hmura, Merideth A. *Mountain View Ranch: 1915–1945.* Lockport, Illinois: Leaning Pine Publishing Company, 1996.

"Home Has Genuine Santa Fe Pedigree: History, Hospitality." *Journal North,* May 26, 1987.

Howes, Ruth H., and Caroline L. Herzenberg. *Their Day in the Sun: Women of the Manhattan Project.* Philadelphia: Temple University Press, 1999.

Jette, Eleanor. *Inside Box 1663.* Los Alamos, NM: Los Alamos Historical Society, 1977.

Johnston, Lawrence. "The War Years." In *Discovering Alvarez: Selected Works of Luis Alvarez,* edited by Peter W. Trower. Chicago: University of Chicago Press, 1987.

Jungt, Robert. *Brighter Than a Thousand Suns: A Personal History of the Atomic Scientists.* Translated by James Cleugh. New York: Harcourt Brace Jovanovich, 1958.

Keegan, John. *The Second World War.* New York: Viking, 1989.

Kernan, Michael. "Learning to Live with the Bomb: A Town on the Mesa." *Washington Post,* Sept. 30, 1973.

Kevles, Daniel J. *The Physicists.* New York: Knopf, 1978.

Kunetka, James W. *City of Fire: Los Alamos and the Atomic Age, 1943–1945.* Albuquerque: University of New Mexico Press, 1944.

Lamont, Lansing. *Day of Trinity.* New York: Atheneum, 1965.

Lang, Daniel. *Early Tales of the Atomic Age.* New York: Doubleday, 1948.

Lanouette, William. *Genius in the Shadows: A Biography of Leo Szilard, the Man Behind the Bomb.* With Bela Szilard. New York: Charles Scribner's Sons, 1992.

Laurence, William L. *Dawn over Zero.* New York: Knopf, 1946.

———. *Men and Atoms: The Discovery, the Uses, and the Future of Atomic Energy.* New York: Simon & Schuster, 1959.

Libby, Leona Marshall. *The Uranium People.* New York: Charles Scribner's Sons, 1979.

Los Alamos National Laboratory. *Los Alamos: Beginning of an Era, 1943–1945.* Los Alamos, NM: Los Alamos Historical Society, 1986.

Luhan, Mabel Dodge. *Edge of Taos Desert: An Escape to Reality.* Albuquerque: University of New Mexico Press, 1937.

Lyon, Fern, and Jacob Evans, eds. *Los Alamos: The First Forty Years.* Los Alamos, NM: Los Alamos Historical Society, 1984.

Manley, Kathleen E. B. "Women of Los Alamos During World War II: Some of Their Views." *New Mexico Historical Review.* Vol. 65, No. 2. April 1980.

Martin, Craig. *Los Alamos Place Names.* Los Alamos, NM: Los Alamos Historical Society, 1998.

———. *Quads, Shoeboxes, and Sunken Living Rooms: A History of Los Alamos Housing.* Los Alamos, NM: Los Alamos Historical Society, 2000.

Mason, Katrina R. *Children of Los Alamos: An Oral History of the Town Where the Atomic Age Began.* New York: Twayne Publishers, 1995.

McMillan, Elsie Blumer. *The Atom and Eve.* New York: Vintage, 1995.

McPhee, John. "Balloons of War." *New Yorker,* Jan. 29, 1996.

Meketa, Jacqueline. "Ladies of the Hill." *New Mexico Magazine,* April 1975.

Metropolis, N., Donald M. Kerr, and Gian-Carlo Rota, eds. *New Directions in Physics.* New York: Harcourt Brace Jovanovich, 1987.

Michelmore, Peter. *The Swift Years: The Robert Oppenheimer Story.* New York: Dodd, Mead, 1969.

Mikeshi, Robert C. *Japan's World War II Balloon Bomb Attacks on North America.* Washington: Smithsonian Institution Press, 1997.

Nichols, K. D. *The Road to Trinity.* New York: William Morrow, 1987.

Norris, Robert S. *Racing for the Bomb: General Leslie R. Groves, the Manhattan Project's Indispensable Man.* South Royalton, VT: Steerforth Press, 2002.

Pais, Abraham. *A Tale of Two Continents: A Physicist's Life in a Turbulent World.* Princeton, NJ: Princeton University Press, 1997.

———. *Niels Bohr's Times: In Physics, Philosophy, and Polity.* Oxford, England: Clarendon Press, 1991.

Peierls, Rudolf E. *Atomic Histories.* Woodbury, NY: American Institute of Physics, 1997.

———. *Bird of Passage: Recollections of a Physicist.* Princeton, NJ: Princeton University Press, 1985.

Pfaff, Daniel W. *Joseph Pulitzer II and the Post Dispatch: A Newspaperman's Life.* University Park: Pennsylvania State University Press, 1991.

Pisano, Marina. "The Flying School Girl: Aviator Katherine Stimson." *San Antonio Express,* May 1, 2001.

Polenberg, Richard, ed. *In the Matter of J. Robert Oppenheimer: The Security Clearance Hearing.* Ithaca, NY: Cornell University Press, 2002.

Powers, Thomas. *Heisenberg's War: The Secret History of the German Bomb.* New York: Knopf, 1993.

Rhodes, Richard. *The Making of the Atomic Bomb.* New York: Simon & Schuster, 1986.

———, ed. *The Los Alamos Primer: The First Lectures on How to Build an Atomic Bomb.* Annotated by Robert Serber. Berkeley: University of California Press, 1992.

Rigden, John. *Rabi: Scientist and Citizen.* New York: Basic Books, 1987.

Roensch, Eleanor (Jerry) Stone. *Life Within Limits.* Los Alamos, NM: LAHS, 1993.

Rothman, Hal K. *On Rims and Ridges: The Los Alamos Area Since 1880.* Lincoln: University of Nebraska Press, 1992.

Rothman, Sheila M. *Living in the Shadow of Death: Tuberculosis and the Social Experience of Illness in American History.* Baltimore, MD: Johns Hopkins University Press, 1994.

Russell, Inez. "Dorothy McKibbin, the 'Gatekeeper,' Dies at Her Home." *The New Mexican,* 1986.

Segrè, Emilio. *A Mind Always in Motion: The Autobiography of Emilio Segrè.* Berkeley: University of California Press, 1993.

Serber, Robert, with Robert P. Crease. *Peace and War: Reminiscences of a Life on the Frontiers of Science.* New York: Columbia University Press, 1998.

Sherwin, Martin J. *A World Destroyed: Hiroshima and the Origins of the Arms Race.* New York: Vintage Books, 1973.

Smith, Alice Kimball, and Charles Weiner, eds. *Robert Oppenheimer: Letters and Recollections.* Cambridge, MA: Harvard University Press, 1980.

Smith, Alice Kimball. *A Peril and a Hope: The Scientists' Movements in America, 1945–1947.* Chicago: University of Chicago Press, 1965.

———. "Teaching at Los Alamos, 1943–45." *Los Alamos Historical Society News.* March 1986.

Sparks, Ralph C. *Twilight Time: A Soldier's Role in the Manhattan Project at Los Alamos.* Los Alamos, NM: LAHS, 2000.

Spivey, Richard L. *Maria.* Flagstaff, AZ: Northland Publishing, 1979.

Steeper, Nancy Cook. *Dorothy Scarritt McKibbin: Gatekeeper to Los Alamos.* Los Alamos, NM: LAHS, 2003.

Stern, Philip M., with Harold P. Green. *The Oppenheimer Case: Security on Trial.* New York: Harper & Row, 1969.

Stimson, Henry L. "The Decision to Use the Atomic Bomb." *Harper's Magazine.* Vol. 194, February 1947.

Stimson, Henry L., and McGeorge Bundy. *On Active Service in War and Peace.* New York: Harper & Brothers, 1947.

Szasz, Ferenc Morton. *The Day the Sun Rose Twice: The Story of the Trinity*

Site Nuclear Explosion, July 16, 1945. Albuquerque: University of New Mexico Press, 1984.

Teller, Edward, with Judith Shoolery. *Memoirs: A Twentieth-Century Journey in Science and Politics.* Cambridge, MA: Perseus Publishing, 2001.

"Terribly More Terrible." *Time,* October 29, 1945.

Trilling, Diana. "The Oppenheimer Case: A Reading of the Testimony." *Partisan Review* 21, 1954.

Truslow, Edith C. *Manhattan District History: Nonscientific Aspects of Los Alamos Project Y, 1942 Through 1946.* Los Alamos, NM: Los Alamos Historical Society, 1997.

Ulam, S.M. *Adventures of a Mathematician.* Berkeley: University of California Press, 1976.

United States Atomic Energy Commission. *In the Matter of J. Robert Oppenheimer: Transcript of Hearing Before Personnel Security Board, Washington D.C., April 12, 1954, Through May 6, 1954.* Washington, DC: GPO, 1954. Paperback edition, Cambridge, MA: MIT Press, 1971.

"The War Ends: Burst of Atomic Bomb Brings Swift Surrender of Japanese." *Life,* August 20, 1945.

Waters, Frank. *Masked Gods.* Athens, NM: Swallow Press, 1950.

Weideman, Paul. "John Gaw Meem, the Santa Fe Architect." *Santa Fe New Mexican,* Jan. 5, 2003.

Williams, Robert Chadwell. *Klaus Fuchs, Atom Spy.* Cambridge, MA: Harvard University Press, 1987.

Wilson, Jane S., ed. *All in Our Time: The Reminiscences of Twelve Nuclear Pioneers.* Chicago: Bulletin of the Atomic Scientists, 1975.

Wilson, Jane S., and Charlotte S. Serber, eds. *Standing By and Making Do: Women of Wartime Los Alamos.* Los Alamos, NM: Los Alamos Historical Society, 1997.

DOCUMENTARIES

ABC News Special. *The Century: The Race.* New York: ABC News productions, April 1999

Else, Jon. *The Day After Trinity: J. Robert Oppenheimer and the Atomic Bomb.* Santa Chatsworth, CA: MMII Image Entertainment, 1980.

Los Alamos Historical Society, ed. *Remembering Los Alamos, World War II.* Los Alamos, NM: LAHS, 1993.

Los Alamos National Laboratory. *The Moment in Time: The Town That Never Was.*

Murrow, Edward R., and Fred Friendly. *See It Now: Dr. Oppenheimer.* New York: CBS News Productions, April 4, 1955.

The Woman Who Kept a Secret. Interview conducted by Hal Rhodes. Produced and edited by Dale Sonnenberg. Albuquerque: KNME-TV, 1982.

Yesterday's Witness in America: "The Day the Sun Blowed Up." Narrated by James Cameron. Produced by Stephen Peet and edited by Mike Appelt. London: BBC News Productions, 1970.

INTERVIEWS

Beverly Agnew
Harold Agnew
Ethel Ballen
Samuel L. Ballen
Shirley Barnett
Hans Bethe
Rose Bethe
Hugh Bradner
Marjorie Bradner
Kathleen D. Church
Ellen P. Conant
Theodore R. Conant
Becky Bradford Diven
Ben Diven
Priscilla Greene Duffield
Bill E. Hudgins
Elaine Kistiakowsky
Betty Lilienthal
Anne Wilson Marks
Kevin McKibbin
Emily Morrison
Philip Morrison
Peter Oppenheimer
Ellen Bradbury Reid
James Scarritt
Marguerite Schreiber
Sharon Snyder
Ralph C. Sparks
Edward Teller
Françoise Ulam
Jane Wilson
Nancy Meem Wirth

ACKNOWLEDGMENTS

Mark Twain once wrote that "the only difference between fiction and nonfiction is that fiction should be completely believable." I am indebted to countless people for helping me to document the incredible story of Los Alamos and to recapture the unbelievable optimism, loyalty, and dedication of the men and women who put their lives on hold and followed Robert Oppenheimer to that isolated mountaintop in New Mexico. I am most grateful to Kevin McKibbin for sharing with me his memories of his mother and their adventures on the Hill, and for kindly granting me access to her unpublished biography, letters, and photographs. His guided tour of the plaza and 109 East Palace, and detailed descriptions of Bandelier Monument, Los Alamos, and old Santa Fe, evoked the magical colors, sounds, and smells of New Mexico. I would also like to thank Robert Oppenheimer's son, Peter, for taking the time to talk to me and help fill in some personal details of his father's life.

It is impossible to acknowledge adequately the help I received from Robert Oppenheimer's three wartime secretaries, who generously provided their personal stories, recollections, and photographs, and patiently fielded my repeated calls and hundreds of questions. Priscilla Greene Duffield, Shirley Barnett, and Anne Wilson Marks are among the most astute and witty observers of people and events it has ever been my privilege to encounter. They brought the contradictions and complex relationships of Los Alamos to life—the altitude and awful mud, the exhilarating company and insidious shadow of security, and the endlessly intriguing and seductive leader who made it all possible, Robert Oppenheimer. Without them this book would not have been possible.

I would like to express my gratitude to the following for giving generously of their time and, in many cases, their hospitality: Beverly and Harold Agnew, Ethel and Samuel Ballen, Rose and Hans Bethe, Marjorie and Hugh Bradner, Becky and Ben Diven, Bill E. Hudgins, Elaine Kistiakowsky, Betty Lilienthal,

Emily and Philip Morrison, James Scarritt, Marguerite Schreiber, Ralph C. Sparks, Françoise Ulam, and Jane Wilson. Only a few months before he passed away, the late Edward Teller, who was blind and bound to a wheelchair, was gracious enough to grant me an audience. Although we disagreed about many things, I will never forget his passion for science or his pain at the memory of his "falling out with Oppie."

For access to the Los Alamos photographs, records, and related material, and their assistance over many years, I would like to thank Roger Meade and Linda Sandoval at the Los Alamos National Laboratory archive and Rebecca Collinsworth at the Los Alamos Historical Museum.

In the writing of this book, I am greatly indebted to Kristine Dahl, my literary agent, who has been of inestimable help as a sounding board and critic from the beginning, and to Barbara Kantrowitz, my friend and fellow writer, for her insight into this world and comments on the work in progress. I also benefited enormously from my continued collaboration with Ruth Tenenbaum, whose research skills and ingenuity greatly facilitated the long and arduous process of combing through archives for relevant material, unearthing forgotten interviews and misplaced documents, and tracking down the most obscure facts.

Too many friends offered encouragement and assistance along the way to name here, but the following were especially helpful: Marilyn Berger, Don Hewitt, Toni Goodale, Peter Jennings, and Linda Sylvester. Cavelle Sukhai is the miracle in my family's life that makes everything work. Above all, I would like to thank Perri Peltz for her unstinting enthusiasm and support. She was my champion on the first book and on this one, and I can hardly express my gratitude.

I must single out Alice Mayhew, my extraordinary editor, who immediately understood and embraced the idea of my writing about wartime Los Alamos and offered invaluable suggestions and guidance. Her wisdom, eye for detail, and concern for accuracy are reflected in these pages. At Simon & Schuster, I must also acknowledge the care and attention of Roger Labrie.

I deeply appreciate the assistance of my parents, Ellen and Theodore Conant, who loaned family books and papers, provided information, and confirmed past conversations with my grandfather, James B. Conant, and his feelings about many of the people and incidents in this book.

And finally, for their love and unfailing support, I owe a very special thanks to my husband, Steve Kroft, and son, John. They were forgiving of my absentmindedness, impaired domestic skills, and prolonged monopolization of the dining room table. I adore them both, and am forever grateful.

INDEX

PHOTO CREDITS

AIP Emilio Segrè Visual Archives, Segrè Collection: 16

Photograph by Ken Bainbridge, courtesy AIP Emilio Segrè Visual Archives, Bainbridge Collection: 33

Photograph by John P. Miller, courtesy AIP Emilio Segrè Visual Archives, Segrè Collection: 32

Photograph by Emilio Segrè, AIP Emilio Segrè Visual Archives, Segrè Collection: 31

Courtesy Amon Carter Museum, Fort Worth, Texas: 3

Courtesy Laura Gilpin Collection, Amon Carter Museum, Fort Worth, Texas: 2

Shirley Barnett: 30

Children of the Manhattan Project (www.childrenofthemanhattanproject.org): 4, 18, 24

Patricia Duffield: 25, 26, 27, 28

Los Alamos National Laboratory Archives: 1, 5, 6, 7, 9, 10, 11, 12, 13, 14, 15, 17, 20, 21, 22, 23, 29, 34, 35, 36

Alexander Lowry, Los Alamos National Laboratory Archives: 38

Time-Life Pictures/Getty Images: 8, 37, 39

Courtesy of Hagley Museum and Library: 19

ABOUT THE AUTHOR

JENNET CONANT is the author of *The New York Times* bestseller *Tuxedo Park: A Wall Street Tycoon and the Secret Palace of Science That Changed the Course of World War II.* A former journalist, she has written profiles for *Vanity Fair, Esquire, GQ, Newsweek,* and *The New York Times.* She lives in New York City and Sag Harbor, New York.

ALSO AVAILABLE FROM

Jennet Conant

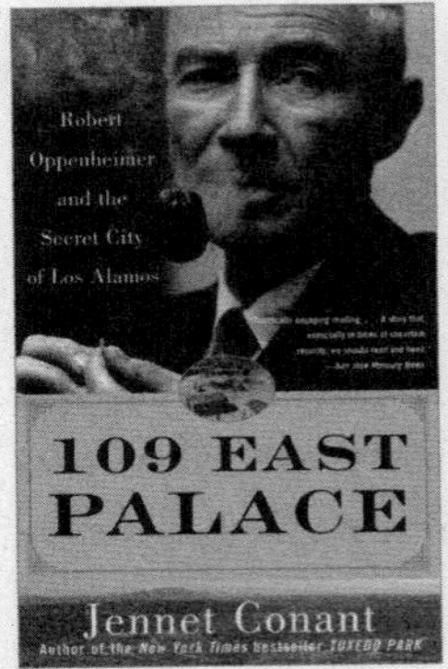

Pick up or download your copies today!